Challenges of Occupational Safety and Health

Editorial Board

About the Editor

Pradeep Chaturvedi, Chairman of the Institution of Engineers (India), Delhi State Centre, is a Mechanical Engineer and an energy and environment expert. He has been involved with sustainable development issues, mainly the energy policy and planning and implementation of environment friendly sustainable energy projects for over three decades in India and other Asian and Pacific countries. As a Member of the Perspective Planning Group of Planning Commission, preparing Development Vision 2020, he contributed 'India's Energy Vision 2020'. He has been actively involved with the formulation of Energy Conservation Act 2001 and the Electricity Bill 2001.

Mr. Chaturvedi has been involved with the Studies Programme of the World Energy Council since 1983. His understanding of energy issues of the South Asia Region, in the global context, has been widely appreciated.

Mr. Chaturvedi was deeply involved with the preparatory phase of the World Summit of Sustainable Development. He is also involved in the follow-up phase.

Mr. Chaturvedi is Vice-Chairman of the UNESCO sponsored World Renewable Energy Network; and Vice Chairman of the World Environment Foundation, both based in UK; and President of the Bio Energy Society of India. He is also a Visiting Faculty at IIT, Delhi.

Mr. Chaturvedi has authored six and edited 25 books on Energy Management and Sustainable Development. These books are widely referred to understand the situation in the country.

Challenges of Occupational Safety and Health

Thrust: Safety in Transportation

Proceedings of the Safety Convention on
Challenges of Occupational Safety and Health held at New Delhi

Edited by
Pradeep Chaturvedi

Published for

Safety and Quality Forum **Quality Council of India**

by

CONCEPT PUBLISHING COMPANY, NEW DELHI-110 059

ISBN 81-8069-284-1

First Published 2006

Published and Printed by
Ashok Kumar Mittal
Concept Publishing Company
A/15-16, Commercial Block, Mohan Garden
New Delhi-110059 (India)
Phones: 25351460, 25351794
Fax: 091-11-25357103
E-mail: publishing@conceptpub.com

Preface

Safety culture is basically a combination of a number of complex and hard to understand human aspect, yet these provide a clear direction that the positive safety culture is a visible demonstration of collective beliefs, attitudes and behavioural patterns, at an enterprise level towards safety issues at all functions and hierarchy of the organizations. In simple terms, the positive safety culture can be simply defined by the way we do things.

The safety culture improvement matrix typically comprises of the following three step-by-step elements that an organisation should follow in a progressive manner.

(i) Regulatory compliance driven safety management approach.
(ii) Self-regulation/process management driven based safety management approach.
(iii) Behavioural based safety management driven approach.

There is a reasonable degree of assumption here that organizations having adequate technological/process safety and regulatory compliance in place are the only best candidates for progression into behavioural-based safety management approach.

Governments and corporates are deeply concerned at improving occupational health and safety performance owing to a variety of internal and external drivers in terms of legislative pressures, employee expectations, customer demands etc. Also pro-active motivational shift should recognize that good occupational health and safety performance is a vital prerequisite to build a sustainable enterprise. However, at corporate level, in developing countries in particular, majority of the enterprises yet considerably struggle to figure out the specificities of the whole or integrated safety management approach to be persuaded as a long term and systematic strategy within the framework of overall organizational vision and strategy to support the building of proactive and employee owned positive safety culture.

Safety considerations are important for all activities of social and economic importance. Safety practitioners do not have line responsibilities to direct employees and supervisor to react safely, but they must generally rely on motivation to promote the safety concept in practice in the organisation. Motivation is the actual initiation of safety action by any organisation's personnel. Its purpose is to fulfil a programme objective. Planning and organizing are necessary to find out if the objectives are being reached. Motivation is how we impel others to carry out these objectives.

There are two primary approaches to psychologically motivate people to improve performance. Each approach includes ways of promoting and explaining personality changes and has proponents who adapt the techniques in various ways. The two approaches to motivating change are termed as behavioural and humanistic psychology. It is the viewpoint that each can be used to improve the overall safe performance of the entire organisation, from the safety practitioner's own performance to the performance of other employees.

Transportation sector is considered as one of the hazardous industry in terms of occupational safety and health. Rapid technological innovation has changed the transport sector over the last century, often enabling it to go faster than overall economic activity. The globalization of production and changes in management strategies have increased the demand for mobility. The sector is changing as a result of economic integration at regional and world levels and also because of more stringent environmental regulations and the drive for sustainable, seamless transport system. Further advancemans in information and communication technologies, expanded satellite systems and e-commerce are expected to increase integration between different modes of transport across borders and across sectors. In many of the countries in Asia, 5 to 10 per cent of the workforce is employed in inland transport sector.

Significant improvement of the occupational safety and health conditions in the transport sector requires intervention at international and national level as well as at the work places. The active involvement of government, employers and workers is a prerequisite. Safety culture can only be developed if it is participatory.

The Safety Convention 2005 has been a national effort to deal with safety issues, at the national level, in a comprehensive manner. Realizing the need to focus on a specific sector, the convention placed thrust on transportation sector.

Views of lifetime practitioners and experts, on a number of important issues relating to challenges of occupational safety and health are presented in this volume. Also the actions relevant at this stage are presented as Recommendations of the Safety Convention 2005.

The Book is a valuable reference document on present needs and optional approaches to tackle issues of occupational safety and health.

October 1, 2005
New Delhi

Pradeep Chaturvedi
Chairman
Safety & Quality Forum
The Institution of Engineers (India)

Introduction

The Safety Convention is being organized as an initiative towards strengthening the safety movement launched by the Safety & Quality Forum of the Institution of Engineers (India) two years back. During these two years the Safety & Quality Forum has carried the message of safety and quality to different geographic locations in the country and to different categories of stakeholders. Overwhelming participation in our first ever Safety Conference held in New Delhi during February 2005 and regional and sectoral conferences held at Udaipur, Pune, Visakhapatnam, Hyderabad, Kolkata, Chandigarh and Dehradoon were indicative to large acceptance of our approach.

The Forum is highly encouraged with the observation of Ministry of Labour and Employment as follows:

"The Ministry highly appreciates the successful organisation and overwhelming participation in the Safety Conference in February 2005. It is gratifying to note that this Forum provided a suitable opportunity for interaction to all the stakeholders on different aspects and concerns on safety. This Ministry also welcomes the proposal to continue the concept of holding similar Conference(s) with a view to continuing this fruitful exercise on an annual basis."

Strengthening our movement further, the Safety & Quality Forum decided to launch the Safety Convention during September 27-29, 2005 and then make it an annual event. The Safety Convention has the objective to bring together experts from different sectors, on a platform, to exchange ideas, research, best practices and experiences on highly topical issues in the areas of occupational safety and health. The Convention will deliberate on people's outlook towards consequences for safety in all walks of activity, especially at work places and transport. The globalised economy has created eagerness and opportunity to take proactive steps by involving all stakeholders. It has been realized that there is a dire need to ensure a fine balance between conflicting interests of the nation, organizations, industries and individual persons to achieve objectives of occupational safety and health. As achieving any such goal, needs interaction of all the stakeholders, the Forum ensuring participation of all stakeholders.

The Forum has also taken into consideration the resolution of the Engineers' Mobility Forum, EMF, at its meeting held in Hong Kong during June 13-17, 2005. EMF has approved change of statutes to make 'safety of common man' a concern of professional engineers.

The Safety Convention is being organized with the title "Challenges of Occupational Safety and Health"; and a thrust on "Safety in Transportation.

The Convention will cover all aspects of occupational safety and health across various economic and social sectors of development. As Delhi has been recently in the limelight in terms of fire safety and road safety, it has been decided to organize a special session on

"How Safe is Delhi". The Convention will be organized in Plenary and Technical Sessions as given below:

Plenary Sessions

- Safety in Transportation Sector
- Challenges of Occupation Safety and Health: Impact of Innovation in Safety Measures.
- Standards and Practices: Role of BIS
- How Safe is Delhi?

Technical Session

- Safety in Transportation Sector
- Safety in Energy Sector
- Safety in Industry including Fire Safety
- Safety in Construction
- Natural Hazard Mitigation and Safety Measures
- Perception of Engineering Students

Papers relating to above sessions have been printed in this volume and made available for permanent reference. A unique feature of this year's Volume of Technical Papers is the inclusion of Case Studies by all the Awardees of Safety Innovation Award this year. The innovative approaches present in these Case Studies can provide lead to distinctive approaches for developing safety culture.

The Technical Committee under the Chairmanship of Prof. G.S. Yadava, has made outstanding contribution in evaluating the papers and structuring the Convention.

Shri Ashok Mittal, Concept Publishing Company, has once again provided his support in publishing Technical Papers in appropriately edited form.

The Quality Council of India, especially Lt.Gen. H. Lal, Chairman, Expert Committee and Shri Girdhar J. Gyani, Secretary General, Quality Council of India has extended their partnership support for the Convention and publication. These papers are also available on a CD for quick retrieval and use by the individual.

September 27, 2005
New Delhi

Pradeep Chaturvedi
Chairman
Safety & Quality Forum
The Institution of Engineers (India)

Contents

Section–II
Safety in Energy Sector

Section–III
Safety in Industry

Section–IV
Occupational Safety, Health and Environment

Section–V
Safety in Construction

Section–VI

Innovative Safety Measure by the Award Winners

Plenary Session–1

Safety in Transportation Sector

1

CHALLENGES OF TERRORISM AND SECURITY

O.P. Tandon*

ABSTRACT

Terrorism continues to be India's primary and most immediate security concern. In this backdrop, this paper attempts to examine the new threats which are emerging in the present day. Taking stock of political, economic, military and non-military aspects of India's security, it addresses the issues pertaining to the cross-border terrorism, smuggling of arms and drugs, chemical warfare, nuclear terrorism and radiation hazards.

In the Twenty-first Century, there are two landmarks in International Terrorism—September 11, 2001, the attack on Twin Towers where more than 2,600 people died at the World Trade Centre, 125 at Pentagon; 256 died in the four planes. The death toll surpassed that at Pearl Harbour in December 1941. The second landmark is the serial bomb attacks that left a trail of deaths and destruction on July 7, 2005 on the London transport network, in which 33 were dead and 350 injured. Following the terrorist attack in Bali and Jakarta (the targets being Australians) the March 2004 Madrid bombings killed 190 people apparently to punish the Spanish Government for participating in the war against Iraq. Denmark, Italy and Britain were the next in line. The line adopted by the group of terrorists was to organise number of attacks in order to punish the common enemy.

Before I discuss the challenges of Terrorism in India, some useful findings of the 9/11 Commission Report of US may be looked into.Terrorism was not the overriding security concern for the US Government and the Intelligence community struggled throughout the 1990s and upto 9/11 to collect intelligence. An outmoded structure and bureaucratic rivalries resulted in an insufficient response to the new challenges. Surprisingly, there was no National Intelligence Estimate on Terrorism between 1995 and 9/11.

The Commission of 9/11 proposed a strategy with three dimensions: (1) attack terrorists and their organisation; (2) prevent the continued growth of Islamic terrorism; and (3) protect against and prepare for terrorist attacks.

*Former Secretary (Security), Cabinet Secretariat, Government of India, 46, Usha Colony, Malvia Nagar, Jaipur-302 017

Needless to mention the United States continues to be the target even today. A group of F.B.I. counter terrorist analysts warned (August 2005) of possible terrorist attacks in New York, Los Angeles and Chicago around September 11, 2005. Washington and Las Vegas are also considered to be the potential targets. US, authorities, however, are divided on this forecast.

Terrorism continued to be India's primary and most immediate security concern, according to the Annual report of the Ministry of Defence released in November 2004. The prime threat remained fundamentalism and terrorism and the "ingrained adventurism of a section of Pakistan military." The terrorism faced by India is typically cross-border and State sponsored. (*Hindu*-Nov. 6, 2004)

In this background we have to examine the new threats which are emerging in the present day security environment. Today, security is more inclusive and includes political, economic, military and non-military aspects besides the economic performance of the nation. Terrorism and insurgency—both the internal and the externally controlled varieties, drugs and smuggling of arms, are most obvious sources of destabilisation of a nation, particularly, for a developing nation. It requires coordinated efforts of nation's security agencies, desiring external and internal alertness.

These threats have a direct bearing on our not-so-friendly neighbourhood, which refuses to die even after 58 years of Independence, because of unresolved land and maritime disputes. Its international land border of 15,120 km is shared on the East with Bangladesh (4,096 km), China (3,917 km) in the North, and Pakistan (3,310) to the North and West. It also borders Nepal (1,752 km), Myanmar (1,458 km) and Bhutan (587 km). India's 6,100 km long coastline and 1400 km of island territories border Sri Lanka, Maldives, Indonesia, Thailand, Bangladesh, Pakistan and Myanmar. Because of arbitrary division of territory at the time of Independence, 125 km of the frontier with Bangladesh—comprising 78 km of West Bengal, 3 km of Assam, 8 km of Mizoram and 36 km of Tripura—remain disputed. (Rahul Bedi, *Indian Express,* May 15, 1997).

Shri I.K. Gujral, while he was the Prime Minister of India, in his talk at the Royal Institute of International Affairs in London in September 1996, had summed up the nation's security scenario as, "In the east, there is China, a full-fledged nuclear power. In the south, there is Diego Garcia, a major American base for the nuclear submarines as well as aircraft carriers. In the west, the Gulf region is nuclearised by the United States. Is it possible for any government in India to remain indifferent to this gigantic array of nuclear arms across these borders. (P.M. Kamath, *Indian Nuclear Strategy: A Perspective for 2020* - Strategic Analysis - IDSA - March 1999). Following are some of the major threats which pose serious problems for India. These are :

Pakistan poses the most serious threat to India today, which primarily, emanates from terrorism. India's unresolved issues with Pakistan are well known and, for the past over five decades, Kashmir has, however, remained the central and overriding concern for Pakistan. It has provided covert support for insurgent and terrorist movement in Jammu and Kashmir, Punjab and the North-Eastern states. Also, Pakistan's continued support—financial, training and sophisticated weaponry—to the terrorists/secessionists, has been an issue of grave concern for India. The bulk of Pakistan's defence preparedness is directed against India and the proxy war in Kashmir represents a direct and immediate threat to India's security. Various terrorist organisation funded by ISI resort to terrorist activities in Kashmir in order to destabilise the state. US Government figures point to a sharp rise in terrorist attacks in Jammu and Kashmir during 2004. Worldwide, the number of attacks more than tripled from last year's 175 and about 300 of the 650 "significant attacks" in 2004 related to Kashmir.

India has become vulnerable to Narco-Terrorism as it is placed between the two major narcotic growing centres of Asia—the Golden Crescent (GC) Pak-Afghan border comprising Pakistan, Afghanistan and Iran and the Golden Triangle (GT) Myanmar-Thailand border comprising Myanmar, Thailand and Laos. It is difficult to estimate the quantum of money involved in the narcotics in this regard. According to a World Drug Report, total revenue accruing may range from more than $300 to $400 billion, 8 per cent of the total international trade. Drug trafficking is the most lucrative business, with profits higher than of the entire oil industry and second only to arms trade. The terrorists and their sympathisers indulge in drug trafficking for financial gains to support their cause. Thus they develop a nexus with the drug barons and extend support and protection to them. They are also involved in supply of fire-arms. India is a major transit state for drugs. Drugs smuggled from Pakistan to Mumbai are shipped or sent by air to the us and Europe. From Mumbai, another route is by land, where it reaches Tamil Nadu coasts and then shipped to Colombo and further to the West. Another major route is through the northeastern states of India to either Kathmandu or Delhi, from where they are smuggled to the West.

Cannabis which is grown in the Rif moumains of Morocco, Kif—the word for Cannabis covers the valleys and hill side in profusion. The,illicit crop is Morocco's biggest foreign currency earner and, according to European Union estimates, worth £ 2 billion a year making Morocco the world's biggest hashish exporter. There are reports that some of the profit earned for cannabis cultivated in Morocco have found their way to fund Terrorism.

In the Twenty-first Century, the terrorists would make use of Weapons of Mass Destruction (WMD)—biological, chemical, nuclear and radiological.

Biological agents like 'toxins' such as botulinium toxin or 'ricin' or 'anthrax' are extremely dangerous as they multiply in their victims. It is estimated that a milligram of 'botulinal toxin' (BTX) is capable of killing millions of people, and 'anthrax' is even more deadly. Apart from high toxicity, undetected by traditional sensors weigh in its favour. Biological warfare agents can even attack crops and livestock.

Chemical Weapons/Agents

Chemical warfare made its first official appearance on April 22, 1915, when Germans discharged over the French lines 168 tonnes of chlorine contained in cylinders. Result: 5,000 soldiers out of 15,000 died. This was followed by "phosgen and yperite" (mustard gas) which attack the skin and the eyes as well as the pulmonary tissues meaning death for those in the trenches. In 1936, research carried out by the Germans discovered 'neurotoxic' war gas: 'Tabon' which became the source of 'Sarin' and 'Soman' in 1952. Agents V, of which the most famous today is 'VX' completed this range. (Jean Paul Dufour - Le Monde). In Japan, AUM SHRINRIKYO, released Sarin gas in Tokyo subway in 1995, killing 12 and injuring 122 persons.

The chemical weapons are easily available and easy to produce anywhere. Chemical weapon is, 'the nuclear weapon of the poor'. Easy to manufacture and cheaper than H-bomb. *The Telegraph* (London) had reported that a small phial of the nerve agent VX, would be sufficient to bring the whole of London underground to a halt. The Government of India and other Governments will have to seriously consider about steps to counter them.

Nuclear Terrorism

Terrorist's could "go nuclear" in a variety of ways; Stealing a bomb; stealing nuclear materials and building a bomb; sabotaging or holding for ransom a reactor and causing or threatening to cause a serious nuclear accident is another form of nuclear terrorism.

While the probability of nuclear terrorism remains low, the consequences for urban and industrial societies could be catastrophic. An International Task Force on Prevention of Nuclear Terrorism had made certain recommendations which were published in the USA:

— The task force recognised that there was an element of risk in drawing public attention to the possible means of nuclear terrorism, but we consider the far greater risk to be leaving basic concerns unspoken and needed improvements undone. The public should understand the nature of threat. It will then be in a position to decide on protective steps, we believe should be taken promptly. The approaches explored by the task force intended to reduce the threat are—Physical protection; Intelligence and Emergency management.

— Theft of nuclear materials and their threatened use in a crude, home made bomb and sabotage of a reactor, fuel facility, or fuel shipment could have grave consequences. Denial of access to nuclear facilities should be the basic consideration in protecting against sabotage. Guard forces should be thoroughly trained. National authorities should task their intelligence agencies to apply sufficient resources to the threat of nuclear terrorism. (Preventing Nuclear Terrorism—Paul Leventhal and Yonah Alexander, Report of the International Task Force).

Radiation Hazards

A large number of accidents involving radiation sources and radioactive materials have been reported during the past half century. People have died from causes attributed to excessive radiation exposure, and many more have suffered serious or disabling injuries. Many reported serious accidents are linked to breaches in the safety of radiation sources. It could occur because of unreliable equipment or because of human mistakes or regulatory oversight. It is here, in such cases, it is to be ensured that the human error is not deliberate or regulatory oversight intentional. In United States alone, the Nuclear Regulatory Commission (NRC) annually receives 200 reports of stolen or abandoned radioactive sources. However, the use by terrorists of chemical, biological and perhaps down the line—radioactive materials as weapons are no longer unimaginable crimes of the future. The World Customs Organisation (WCO) has reported 234 confirmed cases of seizures between 1993-98 (*Source:* International Atomic Energy, Vol. 41, No.3, 1999, Vienna, Austria) *(Journal of International Atomic Energy Agency No.* 3, 1999 published from Vienna).

Targeting Sydney Nuclear Reactor

Just short of opening of the OLYMPICS Games in Australia in September 2000, New Zealand Police uncovered a possible plot to target a "nuclear reactor" in Sydney claimed to have been hatched by Afghan "sympathisers" of the Saudi rebel, Osama bin Laden. This is one of the biggest security scare to hit New Zealand on the eve of Olympic Games.

In the back drop of galloping terrorist activities all over the world and in India, in particular, there is need for a fresh look on the collection and assessment of intelligence, in order to quell their growth.

Until 1997, the US intelligence community viewed Osama bin Laden as Financier of Terrorism and not as a terrorist leader. In Februrary 1998, Bin Laden issued a self-styled Fatwa, to kill Americans because of American occupation of Islam's holy places and aggression against Muslims. These "Fatwas" and such warnings have to be taken note of and not ignored.

Training— It is a matter of great relief that the attempt to storm the Ramjanam Bhumi in Ayodhya (5 July 2005) was frustrated by the alert CRPF jawans. This was the second time that terror attackers met their match in well-trained CRPF Commandos which eliminated the Jaish-e-Mohammed terrorists who stormed the Indian Parliament (December 13, 2001) Akshardham Temple (Gujarat, September 24, 2002) terrorist killed 44 and injured 70 by spraying bullets inside the temple. In Jammu and Kashmir, the terrorists are still at-large and their activities need to be contained.

Coordination

Among various agencies like civilians, military, fire fighters, police officers, emergency medical technicians and emergency management professions need to be fortified to improvise an effective homeland defence against an unprecedented challenge. Steady determination and resolve under horrifying overwhelming conditions will save lives and inspire the nation to join the crusade against terrorism.

The possibility of various threats has increased manifold, particularly the use of Weapons of Mass Destruction (WMD) by terrorists. The national preparedness on the 'National Security' front is rather fragile and none of us is safe from the threat of WMD. Therefore, this threat need to be attended to in right earnest and not left to chance.

MANAGING SAFETY
Human Dimensions

A. Gangopadhyay*

ABSTRACT

This paper attempts to explore human dimensions of managing safety in organisations. Focusing on two approaches to motivating change—behavioural and humanistic psychology, it discusses at length the Hawthrone studies, hierarchical structuring of human needs, levels of applying motivational constructs, hygiene theory of motivation, intrinsic and extrinsic motivation, socio-technical system model and socio-technical systems approach. Different approachs to the problem of accident prevention have also been dealt with.

Since safety practitioners do not have line responsibility to direct employees and supervisors to react safely, they must generally rely on motivation to promote the safety concept and practices in the organisation. First-line supervisors also generally consider motivation as an important method of improving safe performance in their department. Motivation is the actual initiating of safety action by company personnel. Its purpose is to fulfil a programme's objectives. Planning and organizing are necessary to find out if the objectives are being reached. Motivation is how we impel others to carry out these objectives.

There are two primary approached to psychologically motivate people to improve performance. Each approach includes ways of promoting and explaining personality changes, and has proponents who adapt the techniques in various ways. The two approaches to motivating change are known as behavioural and humanistic psychology. It is the viewpoint that each can be used to improve the overall safe performance of the entire organisation, from the safety practitioner's own performance to the performance of other employees.

Behavioural and Humanistic Perspectives

In discussing ways to change a person's behaviour, the behaviorists avoid talking about such mental constructs as motivates and concentrate instead on the behaviour they wish to change.

*Professor & Head, Department of Personnel Management and Industrial Relations, Tata Institute of Social Sciences, Deonar, Mumbai-400 088

They say that all that is really known about people is their behaviour in particular situations. In particular, a Skinnerian behavioural approach believes that the influence of peers, parents, and the immediate environment reinforces certain types of behaviour and punishes others. They speak of "conditioning a person's behaviour". Two primary proponents of this way of looking at things are B.F. Skinner and Joseph Wolpe. A variation of the behavioural viewpoint is that of social learning theory, which places a greater emphasis on the cognitive process and such concepts are imitation (modelling).

At the opposite end is the humanistic approach, which concentrates on the subjective, internal mental workings of individuals when trying to improve performance. Such techniques as sensitivity training investigate the potential for human growth (emotional) rather than describing personalities. When humanists do talk about behaviour, it is in a subjective manner. Primary proponents of this viewpoint include Abraham Maslow, Carl Rogers, and Sidney Jourard. Humanistic followers believe that people are born "O.K." and with a psychological balance that can be blocked by outside factors.

In comparing the two viewpoints the behaviorists stress the use of rigorous scientific methods, which require an objective analysis, while the humanist approach relies more on intuition in changing personalities. Behaviorists often are thought of as exerting more direct control, whereas the humanists approach is considered more indirect. Behaviorists concentrate on the means of changing behaviour, and followers of the humanistic approach are concerned with the end results of this change.

Hawthorne Studies: Redefining Human Dimensions

In fact, no study of motivation would be complete without a discussion of the famous Hawthorne studies. It is difficult to imagine any other single research work that affected the work perspective as much as did these studies. They consisted of a group of studies conducted at the Hawthorne works of the Western Electric Company over a period of 15 years, beginning in 1924. The original idea of the studies was to investigate the effects of illumination on productivity. The results, however, did not turn out as expected.

It was apparent, through research interviews of the employees involved that other factors were affecting the group besides external ones such as changes in lighting. It was later discovered that the employees were producing more, not because of physical surroundings, but because of psychological incentives -b because they were receiving special attention and believed that management was interested in their welfare. As a result, they developed a good attitude toward the company a desire to perform at peak efficiency. This was the beginning of behavioural research. As a result of the Hawthorne Studies, it was discovered that the attitudes of employees could be significantly changed by the development of a cooperative atmosphere between workers and supervisors.

Management's concern and a positive motivational technique can be used to develop a safety awareness and a reduction in the severity of accidents that do happen.

Hierarchical Structuring of Human Needs

A scale based on the urgency of the need involved can measure motivation. There are different styles of motivation for different levels of awareness by employees. The best known and one of the most respected theories about these levels is A.H. Maslow's *Hierarchy of Human Needs.* He has determined through his research that human needs can be classified. These priorities can be separated into the five categories listed here:

1. *Physiological Needs:* Survival involves the basic needs for food, water, air and rest. These requirements have to be satisfied before higher motivational or emotional needs can be taken care of.
2. *Safety Needs:* These include the need for security, both in the physical and psychological senses. They also involve needs for protection of body and position or status and reduction of physical and psychological hazards. Only after these requirements have been satisfied can an employee or person seek a higher level of motivation.
3. *The need to belong and be loved:* This involves the need for attention and social belonging. Basically, an individual is believed to seek such meaningful relationships with people after the first two needs listed above have been met.
4. *Esteem Needs:* These occur only after the three needs stated above have been met and come down basically to a desire for self-respect. This means striving for competence and achievement, and receiving respect from other people for those achievements.
5. *Self-actualisation Need:* After all other needs have been met, the last or highest form of motivation is self-actualisation. This is the desire for self-fulfilment or in other words, being as much as one can possible be.

According to Maslow each of the lower needs must be satisfied before others become predominant. He physiological needs (air, water) must be satisfied before an employee or person will be concerned with a higher level of motivation such as safety and security needs. Only after each lower level of needs is satisfied can an employee seek rewards and satisfaction at a higher level.

Applying Motivational Constructs by the Practicing Managers

Understanding these levels of need should help supervisors of safety specialists in understanding the needs of their people. If management recognises this as a framework for analysis, it should improve the supervisor's knowledge of human behaviour.

Obviously there are differences between individuals, but a supervisor should know the general personality of the group. Today, with the strong national unions and a more highly educated work force than ever before the needs (psychological) are different from what they were in the past. Most employees today are further up the ladder in regard to job security than were their predecessors. Therefore, a safety programme that tries the Heavy handed approach of creating a fear about lost jobs and livelihood if they (workers) do not adhere to safety rules is out of date. This approach, which is still often used in production facilities does not appeal to employees since most can survive without their present jobs. Workers today, unlike those in the depression era, are above this level of desire.

If we return to a survival economy, this "psychological safety approach" could be effective. Its technique, described in Fig. 38.1 is simple but so is survival. But at present most employees are not on this level, as already noted, and its procedures will have little effect on safety records.

The second level, which is used by most companies today, is the "safety and security approach". Many of these techniques, as described in Fig. 38.1 are the standard methods employed for current programme. The employee is looking for protection and security in his or her job. At this level workers will seek causes and policies which will improve their jobs and make them secure. The "blood and gore" advertisements are most effective for these individuals. However, it is the author's contention that most employees have progressed even above this stage. Most people feel fairly safe and secure in their jobs and positions just as they feel no need to worry about physiological needs being met. However, for the small number who are not secure, this approach to safety will be effective.

FIRST LEVEL : Physiological Safety Approach

1. Threat of termination for violation of safety rules
2. Hard-sell Safety (i.e. blood and gore advertisements)
3. Scare procedures

SECOND LEVEL: Safety and Security Safety Approach

1. Safety incentive plans (i.e. money)
2. Hard-sell Safety (i.e. blood and gore advertisements)
3. Negative safety audit (find something wrong)
4. Physical guarding rules and regulations
5. Scare procedures

THIRD LEVEL: Belongingness and Love Safety Approach

1. Soft-sell safety (i.e. identify safety with family and love)
2. Safety committees (rotating procedures)
3. Safety suggestion system
4. Recognition of individual safety awareness
5. Recognition of group safety records
6. Positive safety

FOURTH LEVEL: Esteem Safety Approach

1. Positive safety
2. Participative (by employees) safety
3. Safety suggestion system
4. Safety committees (employee-oriented)
5. Safety surveys (point out good and some fair points)

FIFTH LEVEL: Self-actualisation Safety Approach

1. Creative safety
2. Participative (by employees) safety
3. Positive safety

Fig. 2.1: Applying the Maslow need Hierarchy to safety

The next level of awareness is "Belongingness Safety". Employees in this group are safe, secure and comfortable in their positions. They are experienced and believe they have some value. They will now week companionship at work. To these individuals it is a motivating force to form some friendships and develop a closeness with peers. The vast majority of progressive companies generally adhere to applying the belongingness safety approach to this type of employee. If an organisation wants its safety programme to be effective, it will use most of the techniques listed in Fig. 38.1, as well as others, to generate a feeling of safety being identified with groups and recognition by groups of individual efforts. It can be assured that most blue collar workers fit this approach, however, much depends on the company. If it is strict and authoritarian, its employees will not develop or seek a higher level of motivation. If it is progressive and concentrates on developing the capacities of its personnel to their fullest, its employees will seek those higher levels of awareness. Therefore, to develop an effective accident prevention programme requires assessing the development of employees and the type of management programme in effect at a particular location.

Many employees at present, and probably more in the future, seek the fourth level. This is "esteem safety". Employees of this type are probably experienced and something of experts in their area. They are proud of their work and seek respect for accomplishments. When status is very important to an individual, the approach described in Fig. 38.1 under esteem safety can be effective. Appeals to the individual ego are effective. Positive safety and participative method may especially appeal to such a person. In this area management should specifically "talk to the ego".

The last level is "self-actualisation". In this area employees seek to completely fulfil themselves. Such an idea probably will never be reached, but people of this type are concerned with broadening themselves and seek abstract and fulfilling experiences. An approach has yet to be truly developed for satisfying this kind of individual. Except perhaps for a corporate president of safety or a famous safety person, few people could have the input into a safety programme needed to satisfy at this level. But areas of consideration to provide satisfaction would include creative safety as well as major participation in safety programmes.

As can be seen, many of the areas overlap. Some concepts can be applied to several different levels of the hierarchy, while others are strictly limited to one. And some techniques, such as positive safety, can even be used in all areas, although their effectiveness will be strongest at those levels programmes.

In conclusion, it should also be pointed out that many companies today use techniques that come from different approaches. This illustrated the need for proper planning. If most employees are at a motivational level above the current safety techniques at use where they work, adjustments should made taking this into account. A change in approach may produce dramatic results after it is implemented. Employers and employees must be communicating on the same level for the employee to appreciate safety efforts.

Motivation—Hygiene Theory

Frederick Herzberg is noted for developing the motivation hygiene theory of employee motivation. Through extensive research, Herzberg developed a "critical incident" technique. This consisted of conducting employee interviews, where the subjects were asked when they were satisfied in their work environment. They then were asked to be specific about what had caused them to feel that way. The subjects were next asked what made them very dissatisfied and what specifically had caused those feelings. These responses were all recorded by the interviewers. Through this method of analysing 1500 employees in 12 separate investigations, Herzberg collected information concerning (a) factors leading to great satisfaction and (b) factors leading to great dissatisfactions. He found that there are two different types of needs which people have. These needs are not basically affected by each other, and each type affects the personality in different ways. If persons were dissatisfied with their jobs, their main concern was with the environment in which they worked. However, whenever they felt good about their jobs, their satisfaction had to do with the actual work itself. Herzberg classified environmental factors as hygiene or extrinsic factors. He believed that by fulfilling these needs, one can keep employees from becoming dissatisfied with their work. Examples of such dissatisfaction factors include company policy and administration, job security, salary, personal life, and the type of supervision one is under.

Herzberg called the second group of needs motivators as opposed to the first group, these can lead to human growth and to developing the maturity of employees. He believed that these motivators were more effective in motivating improved performance. Often referred to as intrinsic factors, they include how one felt about one's own job achievement, advancement potential and the recognition and challenge that one received at work. Herzberg, therefore, concluded that increasing job satisfaction and decreasing dissatisfaction are two separate and unrelated functions. So complete elimination of dissatisfiers will not lead to positive satisfaction or motivation. If motivation is wanted, one must concentrate on satisfiers.

Intrinsic and Extrinsic Motivation

The question of what is a motivating climate for workers has been of concern to managers everywhere. In focusing on developing and promoting an effective work performance, there are two main viewpoints. The first sees workers as motivated by intrinsic rewards. In fact, evidence has indicated that positive attitudes and behaviour results from emphasis on intrinsic characteristics of the job. Some of these intrinsic factors include employees having feelings of achievement, variety, authority and promotional opportunities about their jobs. The primary proponent of this view has been Frederick Herzberg. Managers with this viewpoint have advocated a concern for work content factors such as job enrichment and design improvements. There is scientific evidence that positive attitudes towards work and related behaviour are the result of increased emphasis of fundamental characteristics in the design of jobs.

One of the most often used and effective intrinsic motivation techniques is job enrichment. Proponents of this philosophy believe that any time you can eliminate dull routines or boring techniques and replace these with meaningful work experience, it will have a positive influence on morale. Job enrichment is just the opposite of simplification, which is the breaking down of jobs into small components. Job enrichment characteristics include: (1) making employees responsible for the total function or line specialisation, (2) allowing employees to both start and complete a project or task, (3) diversification of duties so that no one person has all of a boring task (there should be interesting parts in all jobs), (4) delegating decision making down the hierarchy as much as possible, and (5) conducting rotating job assignments.

The second viewpoint holds that extrinsic factors or job context such as improving working conditions, pay and job security provide the main area where motivation can occur. Mitchell Fein is the most obvious supporter of this viewpoint. He has said, "They (workers) exchange their skills, efforts and time for what they can buy back outside the organisation."

Neither intrinsic nor extrinsic rewards are able to supply a complete motivational package. Extrinsic rewards are more useful to some types of employees (less educated, low paying jobs, older employees) and intrinsic rewards are more appropriate for others. When developing a motivational plan, those unique situational factors that exist in an organisation must be taken into consideration. A variety of rewards should be available since security or self-actualisation along cannot appeal to a large enough group. And because most individuals have money needs, which vary in strength for each person, it is appropriate to have several monetary ways of motivating performance.

In the discussion of motivation there is a need for identification of a term called motivational sincerity. Motivational sincerity means that motivation must come from the heart. In other words, there must be a sincere interest in the employee's welfare. If the interest is not sincere, it should not be shown at all.

Employees must be allowed to identify with safety and receive rewarding experiences from their safe work. Unless safety practitioners believe that better psychological working conditions can improve the meaningfulness of life for employees, they will fall into the category of efficiency engineers, whose behaviours will be seen as manipulative and whose use of motivational techniques will be counter-productive. Management also, not just safety specialists, must feel or be taught to be receptive to this idea, otherwise there will be much wasted efforts. If supervisors, safety specialists, and top management all believe this, they can seek an atmosphere that will help the growth and development that each individual must have before safety consciousness is a part of the organizational climate.

Socio-Technical System Model

The concept of "Socio-technical systems" emerged from studies and consulting work by the Tavistock Institute in U.K. "The essence of the emerging interventions (was) to try to develop a better fit between the technology, structure, and social interaction of the workplace so as to enhance desired organizational outcomes".

Generally, the Tevistock work and later a studies use socio-technical concept as a base for organization development projects in autonomous work group design, job enrichment and job enlargement. Many of these early studies involved social changes in work arrangements more so than changes in technology; changes in the technological aspects of "socio-technical" organizations were fewer and had loss profound results, perhaps because to some existing factories and machines were to costly and actually unnecessary for task completion.

Realizing the expense and difficulty if modifying existing plants, particularly those with large machinery, the socio-technical approach was applied to the design of new plants. The best known of these new plants designs, considering both technology and psycho-social aspects of the organizations, are the saab-scanio and volve plants in Sweden.

Socio-technical systems simply refers to the motion that organizations are more than task systems, or structural entities, or human systems. They are, in fact, a composite of task (Technical) subsystem factors, human processes and management subsystems.

To further elaborate, the socio-technical system view of organizations may be conceived of as comprising three interrelated subsystem: (a) technological, (b) human subsystem, and (c) management subsystem. These subsystems are defined in the following ways.

1. *Technological Subsystem* - includes the technology itself, work flow, information flow, job roles and relationships, task configurations and relationships, policies procedures and job feedback.
2. *Human Subsystem* - includes organization climate, level of motivation, communication, level of commitment, participation, cooperation, satisfaction with work environment, willingness to accept change, work group factors, human resources development, satisfaction with compensation and company policies and procedures.
3. *Management Subsystem* - include management technologies and structure used to control and direct technology and people.

The following diagram (Fig. 2.2) shows the interface between the subgroups (Margulics and Raia, 1978)

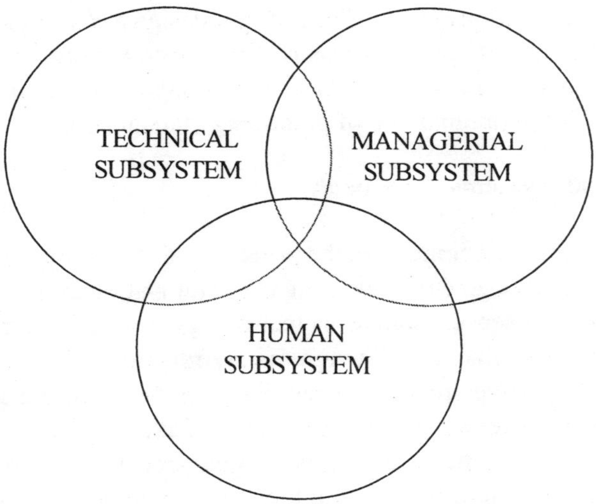

Fig. 2.2: Interface between the subgroups

Management's role and responsibility and perhaps its most difficult task is to dynamically manage the interface between the technology and human subsystem. Those interface, when managed properly, provides the appropriate compatibility, or "fit", between the two subsystems. Understanding both the technical and human subsystems through appropriate assessment is essential to managing this "Fit", even during times of static conditions. When a technological change is imminent, concern with the 'fit' becomes imperative because of the high probability that technical changes create a "misfit" between the two systems.

Traditionally, organizations have not balanced their concerns for technological and human subsystem change. The result has been "misfit", which has the apparent consequences of lowered morale, decreased productivity, lower quality of work, increased conflict and even sabotages, waste, job stress and duplication of efforts. The final result for many organizations has been increased costs of technological implementation and often, under-utilization or failure of the new technology.

Characteristics of Industrial Work Resulting in Psychological Problems

- Mechanically Controlled Work Pace;
- Standardized Motion Patterns;
- Detailed Predetermination of Working Methods and Tools;
- Constant Repetition of Short-Cycle Operations;
- Requirements on High Superficial Attentions;
- Utilizations of Solely or Mainly Motoric Functions in the Individuals;
- Combination of Foregoing Work Characteristics with Payment by Results;
- Inadequate Opportunities for Social Interaction;
- Authoritation Supervision of Work;

The Socio-technical Systems perspective, then, provides a guiding model, which is comprehensive, manageable and pragmatic for introducing technological transitions in the course of achieving organizational goals. The ability to design technological innovation, while at the same time maintaining an awareness of the human component, affords the socio-technical systems approach an opportunity to be rooted in the science of organizational behaviour as well as in the pragmatism of practicing management.

The Socio-Technical Systems Approach

Some theories of organization emphasize the principle of "technological determinism". As such, these theories view the goal of improving the organization as the more efficient use of human resources. In this sense, change in technology may be needed to affect human performance in an enduring manner. Although the environment cannot be ignored, the socio-technical systems perspective studies the relationships between the social and technical systems, considering the internal and external forces that act upon it.

Socio-technical systems theory provides the guidance and direction for the diagnosis and implementation of required change and takes into account the complex relationships between people, tasks, management and technology. The theory has evolved a set of fairly reasonable and recognizable propositions, which specify the following:

1. The design of organizations must fit its goals and takes into account the technology, the human system and the processes of management.
2. Employees directly affected by organizational change must be activity consulted in the design and implementation of new task systems and now managerial structure.
3. Subsystems must be designed around specific recognizable tasks and support systems (human factors) must be congruent with technological design.

4. Effective use of the now technology must include the development of a high quality of life.

Developing a Supporting Model for Participative Safety Management

The presence of protective equipment's, danger signs, safety suggestion boxes, contests etc. is a common sight in many industrial plants. The safety function has become an accepted part of today's modern and complex organization. There has been a marked progress in the principles of accident prevention.

Different Approaches to the Problem of Accident Prevention

Engineering Approach

The engineering approach concentrates on reducing technical hazards by redesigning machinery, equipment and work procedures. Although it was responsible for the greatest and most immediate reduction of accidents, the emphases of mechanical safety measures have reached a point of diminishing returns. In most cases, machine have been made as safe as technology will permit.

Human Factor Approach

Recognizing the limitation of engineering approach, emphasis was next placed on personal factors identifying accidents prone employees. The controversy can be summed as follows:

> "There may be some truly accident-prone, people, but they constitute only a small portion of the workforce. Trying to track these down in the hope that it will eliminate most accidents does not stack up as a highly fruitful endeavour."

Staff Approach

Another development in safety management was to establish separate department and a specialist is hired, who can handle the safety problem at workplace. But gradually, it became evident that safety is a liner function and not a staff one. Safety must be the responsibility of the whole organization and not a staff one.

Participative Approach

Safety is the responsibility of the whole organization and not of a specific department. Underlying the line approach is the assumption that increased participation by the supervisor and employee's result is an increased commitment towards safety objectives.

The purpose is to develop a scheme, which describe s and relates relevant variables that act upon both individual and organizational needs in the implementation of safety function through participative technique.

In many facts there is no general agreement in industry on the kind or degree of employees participation in safety programmes.

The following statements reflect a wide range of opinions in the area:

1. Safety cannot be imposed. It requires participation. One method is to form small groups from the shop-floor employees.
2. Why does workers-participation in safety programmes have such a favourable effect? Men are involved in something important to their security.
3. Safety rules, too, cannot be imposed without group participation. Employees may obey some of them to avoid punishment and accept other as group norms because they recognize their value. But in almost any department, safety practices can be made stronger by group norms if the supervisor enlists employees-participation in selecting them.
4. Since there is no doubt that the workers play an important role in accident prevention, variety methods have been tried to make him an active participator. Unfortunately, few, if any, have achieved success in and of themselves.
5. Participative safety forums are not acceptable, for they specifically emphasize the questions of safety and thus prevent it from being seen as a part of normal operating procedure.
6. Participative forums are for the birds. The committee defeats the primary purpose of the management approach. It places responsibility for the safety programme in the hands of a random group.

There is no common agreement on the impact of employees-participation on the reduction of accidents. There has been less research that has explored the "non-safety" benefits that might be gained through employee-participation in safety. Virtually no studies have been attempted to determine whether there are any important individual, group and organizational relationships those results from employee's participation in plant safety programme. There are intangible gains to be made from the prevention of accidents that do not yet appear susceptible of even approximate measurement. Since safety involves a considerable measure of human relations, it is so broad and complicated that it will probably always remain art as well as science. Goodwill and cooperation with respect to safety, which is a non-controversial subject between employer and employees, leads to goodwill and cooperation in more delicate problems of workplace, such as, wage, bonus, etc.

EXHIBIT-1

An Interaction - Influence System Via Safety Function Through Participation

A

Inherent Factors related to safety within organizations

↓

- Work-related factors
- Number of employees
- Technology of work
- Plant layout
- Employees' attitudes
- Management Considerations
- Philosophy towards safety
- Objectives of the safety programme
- External factors
- Government regulations
- Voluntary Safety Associations
- Economic conditions

B

Conditions basic to the Organization and Implementation of the safety process

↓

- Role Relationships
- Top management
- Supervisors
- Subordinates
- Organizational constrains
- Location of safety in organizational structure
- Horizontal relationships to other functions (Methods, production control etc.)
- Imposed safety controls
- Organizational education (extent of safety education)
- Policies and rules relating to safety
- Power structures
- Union
- Line-staff
- Departmental status
- Developed group norms
- Procedural process
- Techniques to implement safety management
- Auxiliary functions of safety (i.e. medical, service fire production, security)
- Development of safety standards
- Enforcement of safety codes

C

Management Approaches toward safety

↓

- Participative active use of both supervisors and employees
- Overdirective
- Centrally administered and controlled by either safety or line personnel.
- Underdirective little or no effort to implement safety.

- Requirements for individual participation
- Belief in importance of participating in safety
- Belief illegitimacy of participating in safety
- Psychological involvement
- Activation of independence needs
- Requirements for organisational participation
- Economic feasibility
- Urgency of decisions
- Employee experience
- Proper communication channel

EXHIBIT-1 (Contd.)

Safety Programmes

D

Component developed from participative safety teams

- Operational components communicative methods to disseminate safety information
- Essential communication network
- Developed safety rules
- Submitted safety reports
- Formulation of suggestions
- Social-psychological components
- Representation on different status levels interaction on other than technical or work-centred problems
- Reliance on individual competence and knowledge of the work environment
- Activation of need for dependency or autonomy recognition of super-influence
- Exercise of communicative skills
- Decision beyond routine issues multiple levels of decision-making

E

Supportive conditions arising from participating in safety programmes

- Positive supervisory factors
- Identification with company goals
- Increase feeling of job responsibility
- Personal recognition
- Identification with work
- Recognition of employee needs
- Improved interpersonal skill
- Empathy towards other supervisory problems
- Non-technical work-orientated interaction
- Negative supervisory factors
- Hostility and resentment towards management
- Reduced interaction
- With employees
- Reduced sensitivity to employee needs
- Disinterest in safety programme

- Positive employee factors
- Increased feeling management recognition
- Ability to exert influence
- Recognition of other rank and file employee problems
- Recognition of company goals
- Increased influence through group action
- Development of group building and maintenance roles
- Increased feeling of achievement
- Negative employee factors
- Reduced cooperative relationships with supervisors
- Reduced confidence and trust in supervisors
- No change in peer attitude towards representatives' role in safety
- Disinterest in safety programmes.

F

Potential effects on the total organisation

- Improved supervisory and employee attitudes
- Increased motivation towards safety
- Reduction in production costs
- Increased productive capacity
- Improved union-management relations
- Public-relations value
- Increased personal growth
- Reduction in accident-frequency and severity rates
- Increased flow of communication and information
- Improved control over accident hazards (The absence of all these factors)

Exhibit 1 is a conceptual scheme which describes some of the variable that are relevant in the development of an interaction—influence system through participating in plant-safety management. An interaction-influence system is a comprehensive tern, which includes the processes of an organization, it is management philosophy, and the motivation it taps. Every organization has structure, observational and measurement processes, communication processes, decision-making processes, influence processes, attitudinal dimensions, and motivation characteristics. This scheme attempt to look at the separate elements of a safety programme and to combine them into a system approach surrounding the concepts of employee participation.

Factors Related to Safety Within Organization (A)

Column A describes factors inherent within each industrial organization, which set parameters on the degree and extent of any safety programme. Work-related conditions or established certain conditions under which safety programmes function. For example, the physical layout of the workplace, the extent of automation, and the sheer size of the workplace dictate certain outcome of any accident-prevention programme.

Conditions Basic to the Organization of the Safety Function (B)

Conditions basic to the organization of the safety process operate to some degree in every organization. "Role relationships" refer to the behaviour of the principle participants in the safety policy, which forms guideline developed by the company, and, in turn, sets the direction and tone of the basic programe.

Management's Approaches to Safety (C)

The concepts outlined in columns A and B exits to some degree in all safety programmes and could be described as relevant environmental conditions. Column C represents what could be described as the decision-transformation stage and outlines the approaches that can be taken in the process of implementing safety programmes. Often this third stage is confused with the more visible techniques and methods are used. Thus, although safety committees exist, they may not represent a truly participative approach.

There are some very basic conditions, which must exist before individual; employee's participation in safety programmes becomes an effective technique. A person must be free to rearrange his particular goals in light of his experience. Involvement in safety programmes must provide experiences, which allow a person to alter this belief and sentiments concerning safety as it relates to his work.

Individual must also be willing to accept participational activity before it an become an effective managerial technique. People with low independence needs, people who work under safety programmes that are highly authorities and reflect central control, might well find opportunities for participation as signs of weakness in leadership. Thus, hopes for immediate acceptance and compliance via employee participation may be dysfunctional in light of the safety objectives.

The individual employee to be relevant and important must also view participation in safety programmes. His involvement in safety programmes should be perceived by him to increase positive goals and reduce negative ones. It might well be that these positive goals extend beyond the realm of safety and include such factors as recognition, status, power, and achievement.

Thus, in summary, the approach by line personnel; in implementing safety programme is dependent upon four basic assumptions, which must exist before participative techniques are useful. These are:

1. Psychological involvement.
2. Fulfilment of a need for independence.
3. The relevancy or importance of the area of decision making.
4. The extent to which the employee view the area as a legitimate one in which to make decisions.

The absence of anyone of these four factors makes the implementation of safety programmes through the line organization via employee participation an ineffective venture.

In addition, certain organizational requirements are necessary. For example, the use of participative techniques must be economically feasible. Admittedly, it is virtually impossible to determine the exact costs of participation. However, there are times when the costs of the participation's time, for example, would be prohibitive. There are practical restrictions in additions to time and sheer cost, including employee experience and adequate communication structures, which severally restrict the organization's use of participative techniques.

Components Developed from Participative Safety Programmes (D)

Column D describes two broad classifications of components that evolve from the use of employee's participation. Operational factors reflect the type of activities directly associated with the immediate objective of safety programmes: the reduction of accidents.

Supportive Conditions Arising from Participating in Safety Programmes (E)

Supportive conditions are assumed to exist when the experience of the employees and supervisors are viewed by them as relevant and contributory to, or maintaining their sense of personal worth. "Sense of personal worth" means experience, which individuals consider important and which satisfies individual; or group needs. Column E outlines both positive and negative supervisory and employee supportive conditions that may develop through involvement in particpational activities.

Potential Effects on the Total Organization (F)

Column F speculates on the effect that participation will have on organizational goals. These consequences relate to safety goals, attitudinal effects, and external environmental factors affecting the operations of an organization. Although there is some evidence to support not only relationships between employee participation and the development of improved safety practices, but also reduction in accident costs, reduction in accident severity and frequency rates, and improved public relation value—Column F suggests an area for further research.

Some Action Plans Promotion of Participative Safety Forums in the Organization

DON'T

1. Make example of people who violate safety rules.
2. Impose safety standards and procedures on employees.
3. Check to see if people are working safety.
4. Develop and install new safety procedure.
5. Ignore safe work practices stimulate safety by emotional appeals and elaborate campaigns.
6. Make safety exclusively the safety professionals.

DO

1. Help people see the need for rules and consequence of violation.
2. Involve employees in solving safety problems and setting standards.
3. Provides self-monitoring system.
4. Make improving safety procedures part of each person's job.
5. Recognize safe work practices.
6. Stimulate safety by making safe work as a part of everyone's job.
7. Share responsibility for safety with all employees.

3

SAFETY ON INDIAN RAILWAYS
An Overview

P.C. Sharma*

ABSTRACT

No railway system can survive by ignoring the vital aspect of safety as it is not only significant for passenger traffic but also for transportation of material. This paper presents an overview of safety on Indian Railways. Analysing the role of Indian Railways in nation-building after independence, it describes its mega-size network, periodically review of its safety preparedness, critical assessment of major inquiry committees, its corporate safety plans and special Railway Safety Fund to give much needed impetus to the renewal of assets and technology inputs.

Introduction

Railways have come to be recognized as the safest mode of mass transportation on account of inherent characteristics of the system. Railway managements have zealously guarded this image over the ages. Safety has thus come to be recognized as the key issue for the railways and one of its special attributes. No railway system can survive by ignoring this vital aspect as safe and timely transit is not only significant for passenger traffic but also for transportation of materials, in todays highly competitive environment. Safety ranks highest among the factors in selection of a mode of transportation—above the cost and transit time. Accidents hurt business not only in the immediate context but also in the long run and ultimately only those means of transport flourish which are perceived to be safe. It is, therefore, the key performance index which the top managements need to monitor along with productivity.

Indian Railways—Building the Nation

The foremost task for the Indian Railways after independence in 1947 was to strengthen and modernize its infrastructure not only to support the economic development in the planned

*Adviser (Safety), Ministry of Railways, Government of India

era but also to rid the system of obsolesce and poor maintenance. 1950s were primarily devoted to pulling up arrears in maintenance of track, signalling and rolling stock besides providing small rail links to the heavy industries like steel plants that were being set up. The inputs given by the nation for renewals and restoration not only brought about an improvement in productivity but also helped bring down the incidence of accidents on the Railways to a great extent. Number of accidents on Indian Railways came down from 8481 in 1950-51 to 2131 in 1960-61. Derailments which constituted bulk of the accidents came down from 7527 to 1415 over the same period. There was also appreciable reduction in the number of collisions, although there was increase in the categories of 'fire in trains' and level crossing accidents. The unrelenting efforts of Indian Railways in strengthening the infrastructure and modernizing it have led to substantial improvement in the safety performance in subsequent decades, which primarily forms the basis of this paper.

Indian Railways—Mega-size Railway System

Indian Railways operate on continental and gigantic dimensions with nearly:

- 63112 route kms
- 39900 Passenger vehicles
- 7681 locomotives
- 215 thousand wagons
- 6850 block stations
- 1472 thousand workforce
- 120 thousand bridges
- 16750 manned level crossings
- 20600 unmanned level crossings
- 97 loco sheds
- 50 workshops and production units
- 318 carriage and wagon maintenance depots
- Rs. 520 million expenditure on staff per day
- Rs.1 billion revenue expenditure daily.

And daily transport output is approximately:

- 2 million Train kms.
- 14 million Passengers
- 1.5 million tonnes freight loading
- 8700 Passenger trains run
- 5700 Freight trains run

What is Safety?

Safety is basically the product of good practices at all levels of functioning, i.e. design, manufacturing, maintenance and operations. Safety is compromised when the laid down standard practices are infringed. First symptoms of deterioration in the safety performance are evidenced by the increase in the number of failures. Overlooking these warning signals can be disastrous as each of these is an accident waiting to happen. Accidents and assets failures lead to hold ups, missing of schedules and disrepute amongst the customers, which affect the business adversely.

Although no technology is fail-proof, an error rate, howsoever small, being inherent in any man-machine system, reliability of the equipment is the most important factor in the efficiency and safety of a transport system. Objective of the various research organizations is to develop equipment and systems which have near zero level of failure rate. Side by side with the induction of advanced technology, it is imperative that the same is maintained properly and replacements, renewals made as and when needed. If the accidents have to be minimized, it is imperative that the equipment in use is always kept in fine fettle.

Over the last decade or so for which data is available, Indian Railways have laid special emphasis in improving health of the infrastructure and rolling stock. While steam traction, which was highly fault prone has been completely phased out from the main line operations, induction of modern technology and initiation of various measures have brought down the failures of diesel and electric locomotives on line to just about one-third. Four-wheeler freight wagons which were primarily equipped with screw couplings and vacuum brake system have been almost phased out. Small population of four-wheeler tank wagons that still exists is being moved in close circuits and under surveillance. All new rolling stock is being manufactured with more efficient air brake system and existing rolling stock with adequate residual life is being retrofitted to air brake in the workshops. These steps have also brought down cases of poor brake power, train partings and coach detachments very substantially. Failures of electric traction equipment have been contained despite substantial growth through induction of modern technologies. Failures of signalling and interlocking gear have been controlled by accelerated replacement of mechanical signalling with improved systems of panel interlocking and route relay interlocking. Over the period, signalling and

Table 3.1: Equipment Failure on Indian Railways

Year	*Rail*	*Diesel Loco-motives*	*Electric Loco-motives*	*Wagons*	*Poor Brake Power*	*Train Parting*	*Coaches Detach-ments*	*OHE*	*Signals*
91-92	7129	13609	8112	92798	3449	2142	573	1093	236289
92-93	6590	12753	7957	60983	4671	1989	579	778	206608
93-94	4184	11822	7047	71549	3314	1749	602	655	198445
94-95	2980	10220	6219	58415	3188	1583	604	382	161395
95-96	2810	8474	5891	44242	3458	1423	432	256	139730
96-97	2965	7625	4956	37150	3140	1313	361	169	125259
97-98	2672	7077	4686	23490	2367	1172	250	163	111014
98-99	3852	6347	4035	17631	1428	1135	200	144	100554
99-00	8416	5703	3806	13529	868	1034	169	166	118397
00-01	10660	5389	3685	8782	474	1081	230	503	132735
01-02	10678	4808	3211	7209	259	1013	194	309	121242
02-03	10374	4416	2976	5796	164	826	148	651	110640
03-04	8613	4301	2929	4586	60	688	156	475	117240
04-05	7162	4410	2639	3598	56	668	142	541	123752

interlocking aids like track circuiting, block proving axle counters, and solid state technology have been inducted into the system to minimize incidence of accidents due to failures on the part of Station staff. Almost all high density routes have been provided with modern signalling equipment. Bulk of the failures in signalling gear are, therefore, from the low density sections, which still have mechanical semaphore signalling systems. Number of failures of track is primarily attributable to substantial growth in the number of welds, more prone to failure, with progressive growth in long welded rails and aging effect on AT welds.

As a result of upgradation of the infrastructure, there has been consistent improvement in asset failures despite steady growth in volumes of business both on freight as well as passenger segments resulting in high growth in revenue as can be seen from (Fig. 3.3).

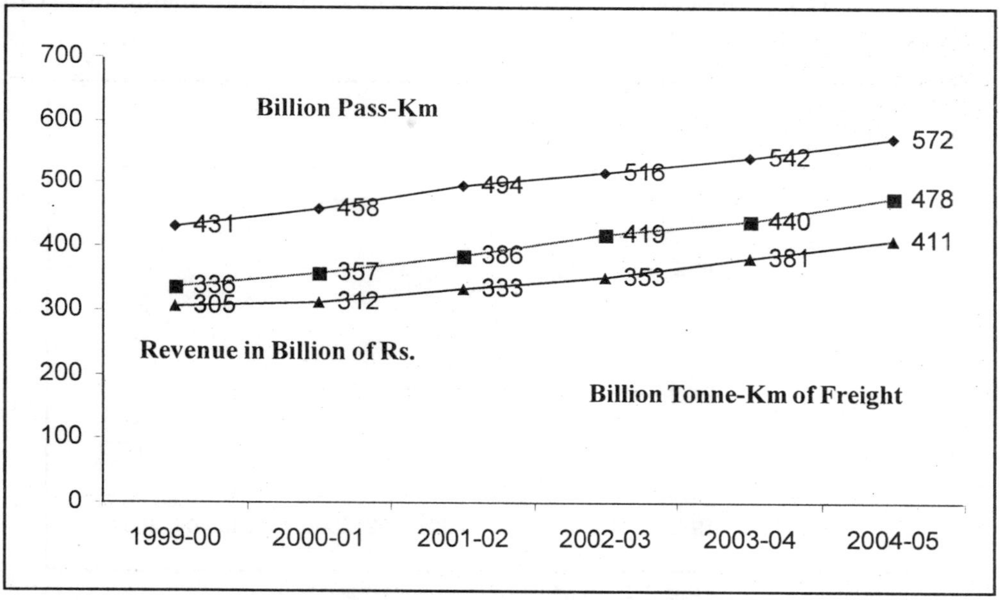

Fig. 3.1: Turnover Indices of IR

While freight in terms of billion tonne kms grew by 37.7 per cent, growth in the passenger traffic over the same period has been 32.5 per cent. Growth in the revenue has been of the order of 42.3 per cent.

Indian Railways—In Perennial Quest for Safety

With such a massive utilisation of assets, safety is of paramount importance not only for its sake, but also for operational efficiency. Hence, the highest priority is accorded to safety and the rail mode continues to be the safest means of transportation for public. Indian Railways have always considered Safety as the key business objective at all levels of management. No compromise is tolerated in Safety of Rail users and investment has been done in all possible areas which could enhance safety in working of Indian Railways. From time to time, Indian Railways have been periodically getting its safety preparedness reviewed by expert committees headed by eminent personalities, mostly retired judges/chief justice of

Supreme Court of India. Following 4 committees have scrutinised IR safety in last four decades:

- Railway Accident Committee-1962 (Kunzru Committee)
- Railway Accident Inquiry Committee-1968 (Wanchoo Committee)
- Railway Accident Inquiry Committee-1978 (Sikri Committee)
- Railway Safety Review Committee-1998 (Khanna Committee)

Major recommendations of first three committees have been implemented and implementation of recommendations of Khanna committee is underway.

As result of regular review and initiatives undertaken Safety on IR has improved considerably by way of continuous reduction in train accidents despite manifold increase in traffic as can be seen from (Fig. 3.2 and Fig 3.3):

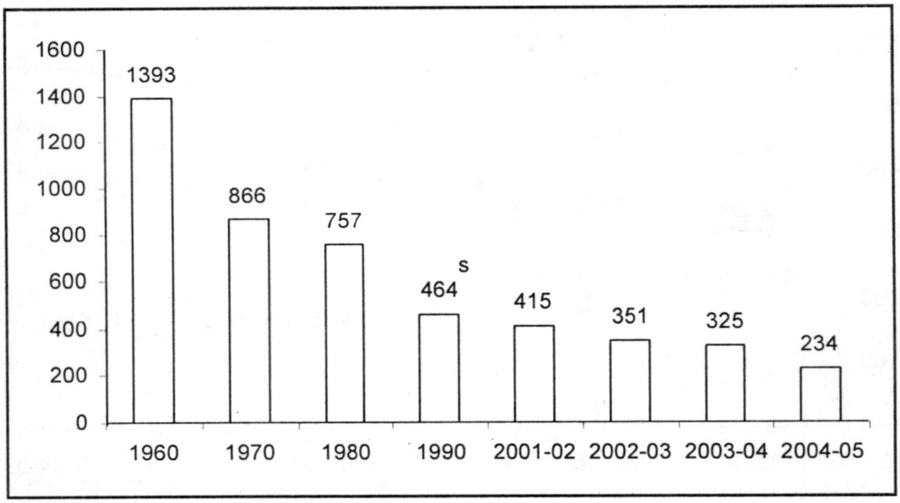

Fig. 3.2: Trend of Train Accidents per Year

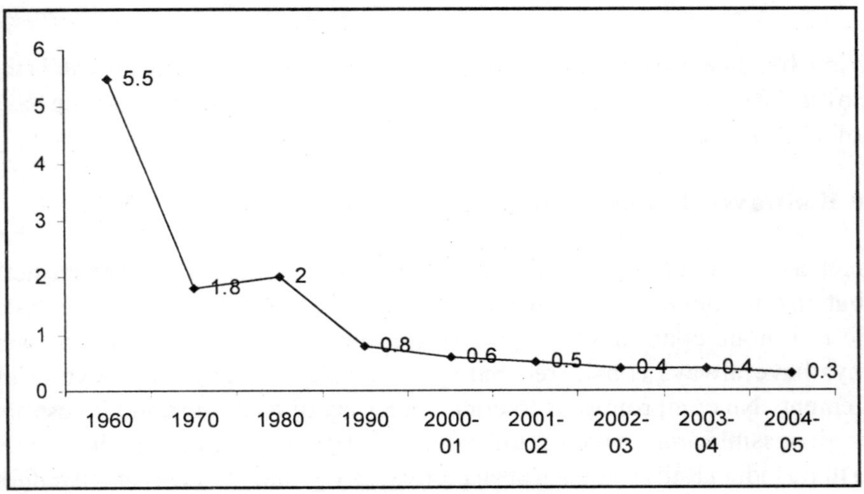

Fig. 3.3: Trend of Accidents per million Train Kms

As can be seen from the (Fig. 3.3), accidents per million train km., a composite index of safety, dropped sharply from 5.5 in 1960s to a figure of 0.31 in 2004-05. IR has thus taken long strides in improving safety of rail travel.

Category-wise share of consequential train accidents is important to comprehend potential hazards of different type of accidents and efforts made to contain each type separately. Looking at different categories of accidents in (Fig. 3.3), derailments which form bulk of them, have come down from 1066 per year in the sixties to 138 in 2004-05, collisions which are the most serious type have come down from 83 per year in the sixties to 13 in 2004-05. Fires in trains have also registered substantial decline from 104 to 10 per year in the period under study. Level crossing accidents have not shown much improvement. Railways have no control over the road users whose failures lead to almost all such accidents. Also, there has been phenomenal growth in the number of road vehicles in the period under study.

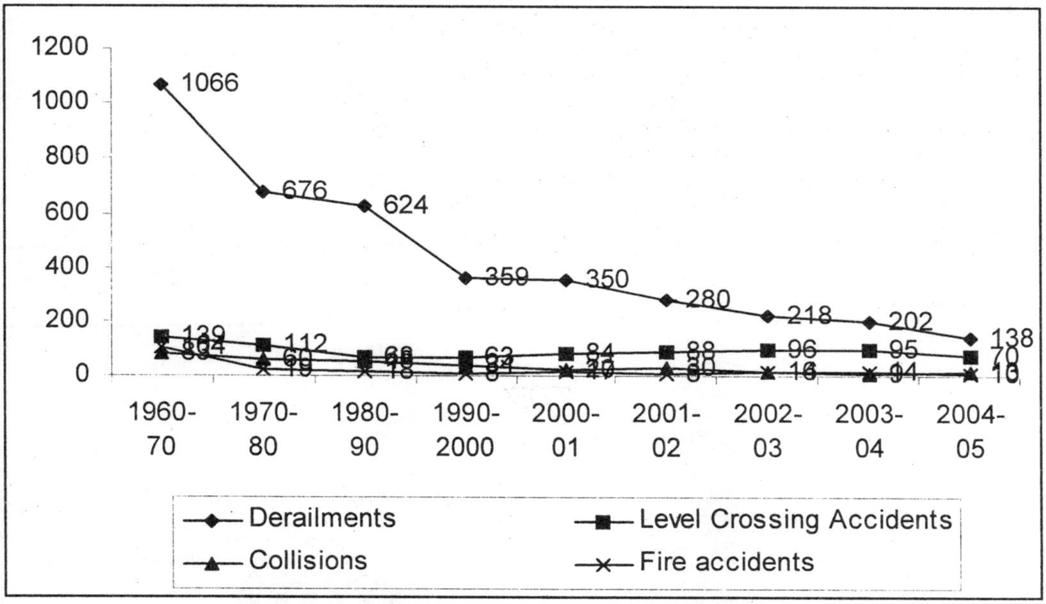

Fig. 3.4: Type wise Accidents on IR Since 1960 to 2005

During the last decade (1994-95 to 2004-05) derailments were a major contributor (67%) in the total tally. These were followed by level crossing accidents (21%). Collisions and 'fire in trains' accounted for 6per cent and 3per cent respectively.

In any plan for prevention of accidents and minimising their impact, saving of human lives must merit highest consideration. Collisions are the most dreaded accidents for any railwayman. On Indian Railways collisions accounted for 6 per cent of the accidents of which only 4 per cent involved passenger carrying trains, but these resulted in 33 per cent of the total casualties. Accidents at level crossings accounting for 21 per cent of the total accidents on IR were responsible for 49 per cent deaths. Derailments though constituted bulk (69%) of the train accidents, these accounted for 16 per cent of the casualties.

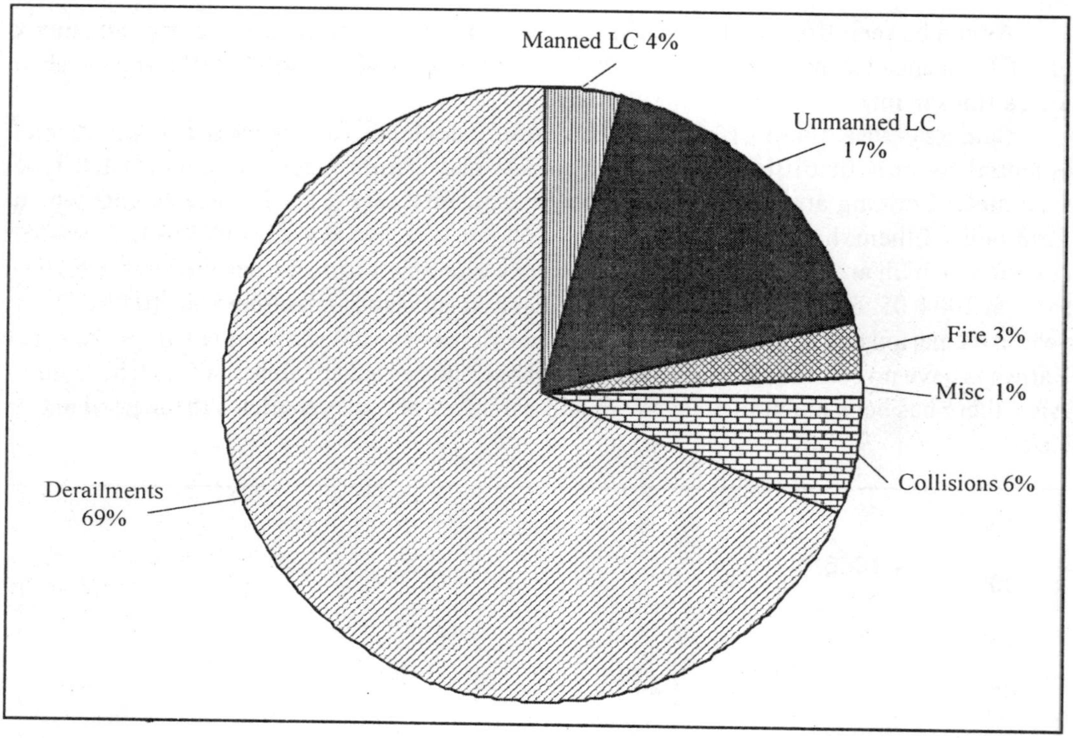

Fig. 3.5: Type-wise Accidents Year (1995 to 2004-05)

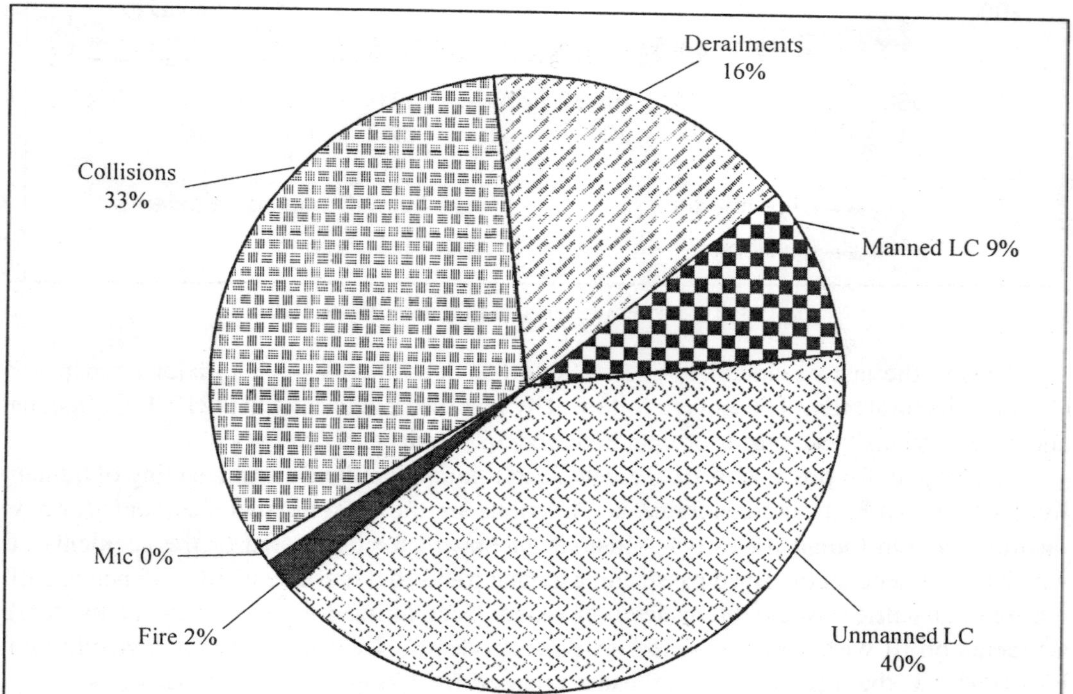

Fig. 3.6: Casualties in Train Accidents (1995-96 to 2004-05)

Corporate Safety Plan

With view to accelerating the pace of the safety drive Indian Railways formulated a 10-year Corporate Safety Plan, which was presented to the National Parliament in August, 2003. This plan envisages achieving following broad objectives:

(a) To achieve reduction in rate of accidents per million train kilometres from the present level of 0.44 to 0.17 by the year 2013.
(b) Implement measures to reduce chances of passenger fatalities substantially in consequential train accidents by 2013.
(c) Focus on development of manpower through major improvements in working environment and training to reduce the accidents attributable to human failure by 40per cent by 2013.
(d) Achieve safety culture on all fronts including maintenance depots, worksites, stations, controls etc.
(e) Progressively achieve an environment of "Fail-proof" from the present "Fail-safe" system of asset failures by upgrading the systems by 2013.
(f) Prioritization of safety related projects.
(g) Implementation of accepted recommendations of Railway Safety Review Committee 1998 at an accelerated pace.

To ensure that financial constraints do not inhibit the implementation of the Plan, fund requirement for the Safety enhancement works, as outlined in the Plan, has been identified as Rs.318 billion including Rs.170 billion Special Railway Safety Fund already committed.

Collisions being worst type of accidents are sought to be completely eliminated by 2013. Various initiatives taken to minimize chances of human error that can lead to collisions, include improving brake power of the trains through progressive use of air brake system, bogey-mounted brakes, disc brakes and improvements in signalling infrastructure—2500 of 4700 interlocked stations are fully track circuited. Auxiliary Warning System has also been installed over about 500 kms. including heavy urban area of Mumbai.

Biggest initiative underway in prevention of collisions is introduction of Anti Collision Device (ACD) developed indigenously. Anti Collision Device (ACD) works on a satellite based Global Positioning System (GPS) and Angular Deviation Count principle for identification of track layout. It is intelligent microprocessor based equipment, consisting of a central processing unit, a global positioning system and a digital modem for communication with other ACDs. There are two types of ACD equipments, viz. 'mobile ACDs' for locomotives and brakevans and 'stationary ACDs' for Stations and level crossing gates. All the ACDs interact with each other and exchange information within their radio zones upto 3 kms. and results of ACD interaction lead to a decision whether the loco ACD shall apply brakes or not.

There are about 38000 level crossings on Indian Railways, of which 16,750 are manned. All the accidents at un-manned level crossings take place on account of failure on the part of the road users as evidenced the world over. While manning does bring down the accidents, it has other limitations of causing unduly long detentions to road users apart from problems of finding resources for the purpose particularly when most of these level crossings exist in sparsely inhabited remote locations. Although massive inputs in providing grade separators (road-over-bridges/under passes), manning and interlocking with signals and telephones have been provided, the problem of negligence on the part of the road users is virtually beyond reasonable control. Since 1990-91, 203 grade separators have been provided, 1435 level crossing gates have been manned and 1747 level crossing gates have been interlocked. As most of these accidents lead to heavy casualties, involvement of Government agencies and NGOs is being resorted to educate and counsel road users.

Derailments which though account for proportionately less casualties have serious impact on operational efficiency and reliability. With the focused attention on human resource, technological upgradation and modern maintenance practices, the present level of human failures causing derailments is bound to improve substantially. With phasing out of derailment-prone and less reliable 4-wheeler tank wagons from the fleet, investments being made through SRSF/DRF into rehabilitation of track and rolling stock etc., and anticipated improvement in the skills and quality of manpower through enhancement of HRD, it is feasible to attain a substantial reduction in derailments.

Based on the above mentioned strategy the Corporate Safety Plan envisages reduction of accidents on Indian Railways by the year 2012-13 substantially. Collisions are targeted to be completely eliminated. Derailments will come down by 60 per cent and Fire accidents by 80 per cent. It has not been possible to project assured improvement in level crossing accidents as there is no control over the circumstances that lead to such accidents.

Imvestment in Safety

Indian Railways is placed in a peculiar situation in as much as it is not entirely free to revise fares and freight rates commensurate with the variations in cost of inputs. While rate of increase in cost is dictated by a multiplicity of factors beyond the control of the Railways, there is a strict parliamentary control over the fixation of tariff, which is generally done as an annual exercise with prior approval of Parliament. The compulsion of democratic pulls and pressures compel the Railway management to carry certain sections of traffic far below the cost of carriage resulting in the exercise of budget making a super balancing act. The special efforts have, therefore, to be made to ensure that adequate provision is made in the annual budget for replacement and renewals highlighting this inadequacy. From time to time special expedients have, therefore, to be resorted to wipe out the arrears in renewals. One such effort has been in the recent past to create Special Railway Safety Fund of Rs.170 billion giving much needed impetus to the renewal of assets and technology inputs. The level of expenditure on safety related works including renewals of assets during the last five years has more than doubled as may be seen from (Fig. 3.7).

Safety related inputs have been provided to Indian Railways steadily apart from inputs for capacity enhancement. While these have led to Indian Railways carry higher volume of traffic from year to year, there has been gradually decline in the number of accidents as would be clear from the (Fig. 3.8).

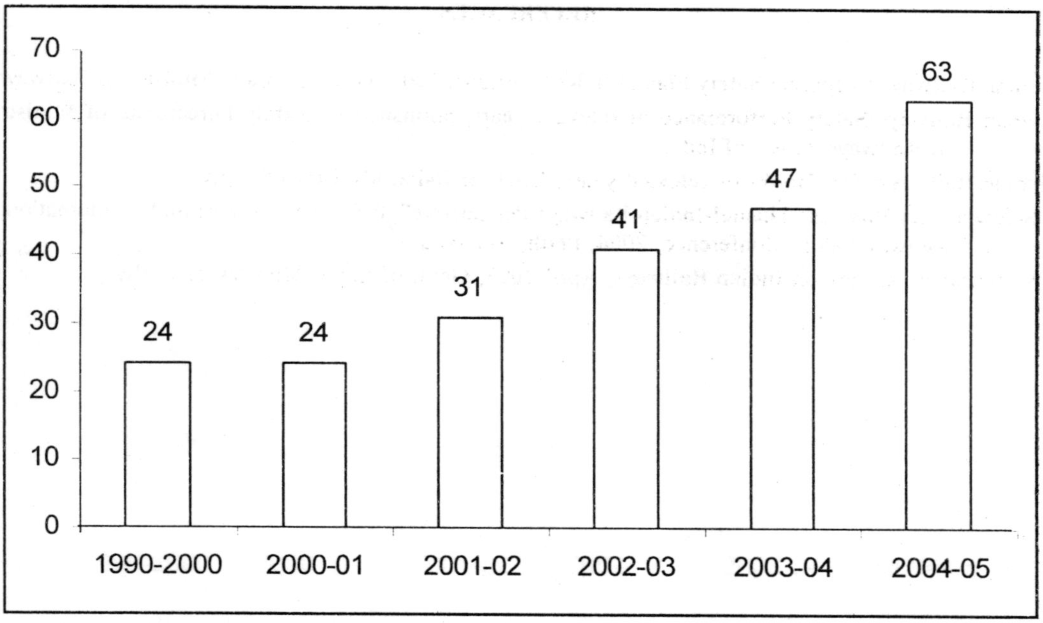

Fig. 3.7: Expenditure on Safety

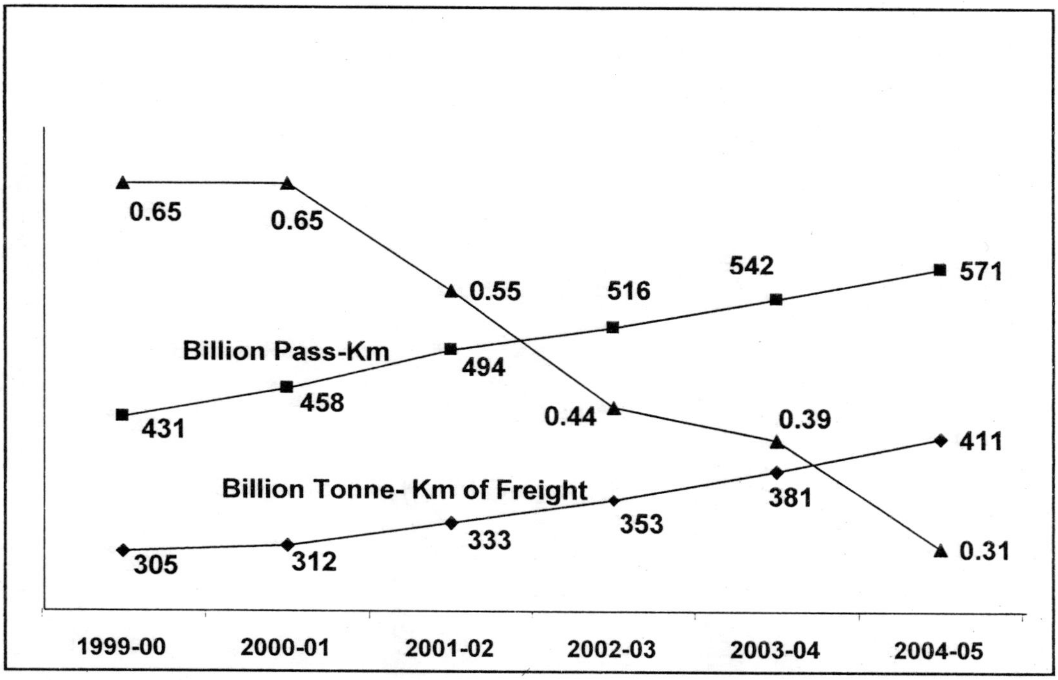

Fig. 3.8: IR Growth Vis-a-vis Safety Performance

REFERENCES

Indian Railways Corporate Safety Plan (2003-13), August 2003, Govt. of India, Ministry of Railways.

Indian Railways-Safety Performance of relevant years, published by Safety Directorate of Ministry of Railways, Govt. of India.

Indian Railways Year Books of relevant years, Govt. of India, Ministry of Railways.

'Safety as Key Business Theme!-Indian Railways Perspective", P. C. Sharma & Amitabh, International Railway Safety Conference, 2004, Perth, Australia.

White Paper—Safety on Indian Railways, April 2003, Govt. of India, Ministry of Railways.

4

OCCUPATIONAL SAFETY AND HEALTH IN THE TRANSPORT INDUSTRY

Ingrid Christensen*

ABSTRACT

The transport sector is considered as one of the hazardous industries in terms of occupational safety and health. This paper seeks to address a wide range of issues concerning safety and health of workers in transport industry. Presenting an overviews of the health and safety scenario in the transport sector, it deals with the measures for the improvement of occupational safety and health conditions such as international conventions and agreements, awareness raising, data collection and analyses, research, and implementation as well enforcement of occupational safety and health legislation.

Introduction

In the world of today transport ranges from the more simple forms of transport such as use of animals, pushcarts or bicycles for shorter distances over trucks, trains and boats for domestic use to the planes and ships making use of the latest and most sophisticated technology for transport of goods and people across the continents.

Rapid technological innovation has changed the transport sector over the last century, often enabling it to grow faster than overall economic activity. The globalization of production and changes in management strategies have increased the demand for mobility. Over the past two decades, the sector has been affected by deregulation and liberalization (including partial or total removal of rules on competition), privatization (including corporatization, contracting out, and partial or total change in ownership), and restructuring (increased consolidation, trans-national ownership and alliances as well as the spinning off of activities to independent subsidiaries). The sector is changing as a result of economic integration at regional and world level but also because of more stringent environmental regulations and the drive for sustainable, seamless transport systems. Further advancements in information and communication technologies, expanded satellite systems and e-commerce will promote increased integration between different modes of transport across borders and across sectors.

---

*Sr. Specialist on Occupational Safety and Health, ILO Subregional Office for South Asia, New Delhi

These trends, observed initially in the OECD and transition countries, are spreading rapidly to other parts of the world.

In many countries in Asia, 5 to 10 per cent of the workforce is employed in the inland transport sector. The relative absence of regulation in road freight and urban road transport, together with the inability of the formal sector to cope with social needs, has led to the fast growth of informal activity in these sub-sectors in many countries.

The Situation in Terms of Occupational Safety and Health

In Asia and the Pacific Region, the transport sector is considered as one of the hazardous industries in terms of occupational safety and health. However, while news headlines tend to focus on larger disasters such as plane crashes, capsizing of ferries or severe oil spills, the ordinary working conditions for women and men of this sector often receive much less attention.

For the road traffic, it is assumed that a high proportion of traffic accidents involve transport workers, although the exact proportion is not known. Factors related to poor road design, vehicles not adapted to local people, excessive hours of work, fatigue, and poor driving skills associated with inexperience and poor judgement are common causes of traffic accidents.

Additionally, loading and unloading heavy cargoes, noise, vibration, irregular meals, stress from long working hours and night driving are causing serious health problems for many transport workers. The poor maintenance of both roads and vehicles and the lack of proper roadside facilities are also common constraints. The transport of hazardous materials and wastes, the nature of which is often not known to workers, poses a special risk.

Seafarers are exposed to a number of risk factors in addition to the risk of boat accidents and drowning. Several job-functions on board include working in awkward positions and undertaking strenuous work, exposure to noise from the engines and whole-body vibrations, long working hours, often at times uncomfortable to the human body rhythm.

For many transport workers, the job entails staying away from family and communities for days or even weeks or months interrupting their family and social life.

Reported work-related injuries have generally declined in the transport sector, but new hazards and risks have been identified and recognized recently, including those from: exposure of cabin crews to radiation; the quality of air on board aircraft; handling of and exposure to asbestos by railway workers; exposure of women in advanced stages of pregnancy to stressful conditions in air traffic control work; exposure of lorry drivers to organized banditry in some countries, changes in employment and manning patterns in the shipping industry impacting the working conditions etc. Violence against transport workers has become a serious issue in all sub-sectors worldwide. Some ILO research has been devoted to these issues to provide information and general guidelines to constituents.

Underreporting of work-related accidents is assumed to be of a significant size and in a number of countries almost hardly any work-related diseases are notified.

HIV/AIDS is a disease that has become associated with the transport industry. In the earlier stages of the pandemia, the spreading of the diseases along the major transport corridors was noted. Rather than a direct occupational hazard, it is predominantly an indirect hazard as a result of the working and living conditions faced by many transport workers.

The Improvement of Occupational Safety and Health Conditions

Significant improvement of the occupational safety and health conditions in the transport sector requires intervention at international and national level as well as at the workplaces. The active involvement of government, employers and workers is also a prerequisite.

At the international level, international Conventions and agreements dealing with safety, environment as well as occupational safety and health of the workers engaged in the transport sector have been adopted. While some of the Conventions cover not only the transport industry, but workplaces in general, others are related to specific sectors of the industry. The ILO Occupational Safety and Health Convention No. 155 is an example of the former. This Convention requires ratifying States in collaboration with the social partners to formulate, carry out and periodically review a coherent national policy on safety and health. Those States are also obliged to maintain laws and regulations to ensure the safety and health of the workers. Moreover, the Convention outlines the responsibilities of employers and the duties and rights of workers.

The SOLAS Convention, — an example of an instrument targeting a specific sector, in this case the maritime sector — was prepared under the aegis of the International Maritime Organization, and provides for, among other things, the application of the International Safety Management (ISM) Code. The objectives of this Code are to ensure safety at sea, prevention of human injury or loss of life and avoidance of damage to the environment, in particular the marine environment and to property.

Also the aviation industry are governed by specific safety standards that are applied worldwide, of which some have direct impact on the safety and health of the crews.

The ILO Code of Practice on HIV/AIDS provides guidance on how countries can address the HIV/AIDS aspect in the World of Work.

At national level, governments in consultation with employers' and workers' organisations should establish the national framework for improvement of occupational safety and health in the transport sector. As is the case at international level, occupational safety and health within different sectors of the transport industry are often the responsibilities of more Ministries (Transport, Shipping, Aviation, Railways etc.); hence governed by different legislation.

Awareness-raising and promotion, data collection and analyses, research as well as preparation, implementation and enforcement of occupational safety and health legislation are among the frequently used instruments. In this regard, employers' and workers' organisations as well as institutions like the Institution of Engineers can play an important role in transforming the interventions at national level into applicable interventions at workplace level.

For enterprises interventions, the ILO Guidelines for Occupational Safety and Health Management Systems provide a systematic management approach to safety and health, not only for the transport industry, but for industries and sectors in general. It provides for (larger) enterprises to prepare a policy, establish an organisation to implement the policy, to plan and implement improvements in the working environment, to monitor and evaluate the measures and finally to assess the need for adjustment of the management system. For small and micro-sized transport companies, less formal instruments would usually be more appropriate. For both approaches the involvement of workers is a key factor.

The provision of adequate training and information is urgent in view of the special precautions required throughout the transport industry. Because of the constant movement of the workplace, it is more difficult to regulate the design and maintenance of roads and vehicles and to incorporate safety and health aspects in management.

The concrete prevention starts with the identification of the possible hazards and the assessment of the associated risk. Since most transport work takes place in shifting conditions, the identification can be more complicated than at permanent workplaces, and tools and procedures will need to cater for this. Simple checklists can be useful tools for such exercise, which should be repeated with regular intervals. Based on the assessment, corrective measures to eliminate or reduce the hazards should be introduced. However, in the transport industry not all hazards are preventable within the enterprise itself—the risk of traffic accidents or the design of cockpits, drivers' cabin etc. as examples; hence other means to supplement more conventional preventive measures should be considered to at least reduce the risk. Finally, checks should be carried out to ensure that the measures have the intended effect.

Further information on occupational safety and health in the transport industry and HIV/AIDS in the World of Work is available at the ILO websites

http://www.ilo.org/public/english/protection/safework/
http://www.ilo.org/public/english/dialogue/sector/sectors/transp.htm
http://www.ilo.org/public/english/dialogue/sector/sectors/mariti.htm
http://www.ilo.org/public/english/protection/trav/aids/index.htm

REFERENCE

The summary is based on the above website references.

5

SAFETY IN SHIPPING
Getting to the "Heart" of the Matter

Naomi Rewari* and Saujanya Sinha**

ABSTRACT

The implications and consequences of an accident at sea are significant particularly with the increasing global focus on protection of the marine environment in addition to safety of life, the ship and the cargo which she carries.

With stringent laws and extensive international media coverage, a major incident can spell disaster for those concerned, and the successful management of the crisis and its fallout will determine the ultimate outcome and long term effect of damage.

It is a paradox that although the stakes are higher than ever before, statistics demonstrate that actually the shipping industry is safer than it has ever been. The fact remains, however, that although incidents are less frequent, no one in the industry can afford the consequences of a major accident.

Even though it is now widely accepted that the human element has a key role to play in the majority of incidents at sea, there has been a slow and languid response to dealing with the fundamental issue of performance of an individual in a professional environment, particularly during a crisis or emergency situation. This paper sets out to describe significant contributing factors to the current scenario regarding safety in shipping and to recommend measures to meaningfully affect the performance of those individuals who are responsible for the safe operations of ships on the high seas.

How Shipping is Regulated Internationally

Regulations concerning shipping are developed at the global level. Because the shipping is inherently international, it is vital that shipping is subject to uniform regulations on matters such as construction standards, navigational rules and standards of crew competence. The alternative would be a plethora of conflicting national regulations resulting in commercial distortion and administrative confusion.

*Director, Training, Applied Research International

**Consultant, Applied Research International

The International Maritime Organisation (IMO), which is a specialised agency of UN, has been responsible to adopt a comprehensive framework of detailed technical regulations, in the form of international diplomatic conventions, which govern the safety of ships and protection of the marine environment.

National governments, which form the membership of IMO, are required to implement and enforce the international rules, and ensure that the ships, which are registered under their national flags, comply. The level of ratification and enforcement of IMO Conventions is generally very high in comparison with international rules adopted for shore-based industries.

The principal responsibility for enforcing IMO regulations concerning ship safety and environmental protection rests with the flag states (i.e. the countries in which merchant ships are registered—which may be different to the country in which they are owned). Flag states enforce IMO requirements through inspections of ships conducted by a network of international surveyors. Much of this work is delegated to bodies called classification societies.

The Principal Regulations Governing Maritime Safety

The introduction of the major conventions has to a large extent been a knee jerk reaction, with significant changes brought about after an incident has taken place. Even from the days of the infamous sinking of the "Titanic", the industry can map a predictable and concerning pattern of afterthought and post incident decision making and action. The latest example is the introduction of "The International Ship and Port Facility Security Code, 2002", which was ratified and enforced in record time post 9/11.

The following are the major international shipping conventions, adopted by the IMO concerning safety and pollution prevention.

Dealing with the Ship

SOLAS (International Convention for the Safety of Life at Sea, 1974) lays down a comprehensive range of minimum standards for the safe construction of ships and the basic safety equipment (e.g. fire protection, navigation, lifesaving and radio) to be carried on board. SOLAS also requires regular ship surveys and the issue by flag states of certificates of compliance.

MARPOL (International Convention for the Prevention of Pollution from Ships, 1973/1978) contains requirements to prevent pollution that may be caused both accidentally and in the course of routine operations. MARPOL concerns the prevention of pollution from oil, bulk chemicals, dangerous goods, sewage, garbage and atmospheric pollution, and includes provisions such as those, which require certain oil tankers to have double hulls.

COLREG (Convention on the International Regulations for Preventing Collisions at Sea, 1972) lays down the basic "rules of the road", such as rights of way and actions to avoid collisions.

LOADLINE (International Convention on Loadlines, 1966) sets the minimum permissible free board (this directly corresponds to the amount of cargo that is permissible to load), according to the season of the year and the ship's trading pattern.

ISPS (The International Ship and Port Facility Security Code, 2002) includes mandatory requirements to ensure ships and port facilities are secure at all stages during a voyage.

Dealing with the Shipping Company

ISM (The International Safety Management Code, 1993) effectively requires shipping companies to have a licence to operate. Companies and their ships must undergo regular audits to ensure that a safety management system is in place, inclllding adequate procedures and lines of communication between ships and their managers ashore.

Dealing with the Seafarer

STCW (International Convention on Standards of Training, Certification and Watch-keeping for Seafarers, 1978/1995) establishes uniform standards of competence for seafarers.

The Reduction in the Number of Ship Losses

Relatively few ships actually sink at sea. The vast majority of the following "losses" simply refer to ships which are damaged and "written off" by the hull insurers as being beyond economical repair—described by underwriters as "total constructive losses".

The figure below covers the entire global industry and indicate the steady improvement in safety performance since the 1990s.

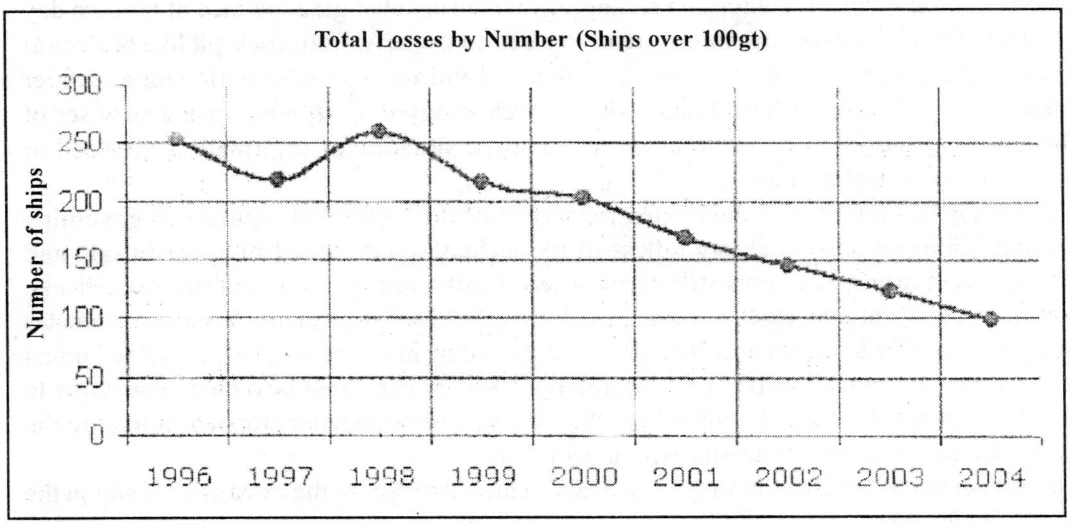

Source: Lloyd's Fairplay.

The Reduction in the Number of Accidents

The following figure concerns insurance claims for third party liability, such as incidents involving personal injury, cargo damage, pollution, or damage to property (e.g. other ships or port equipment).

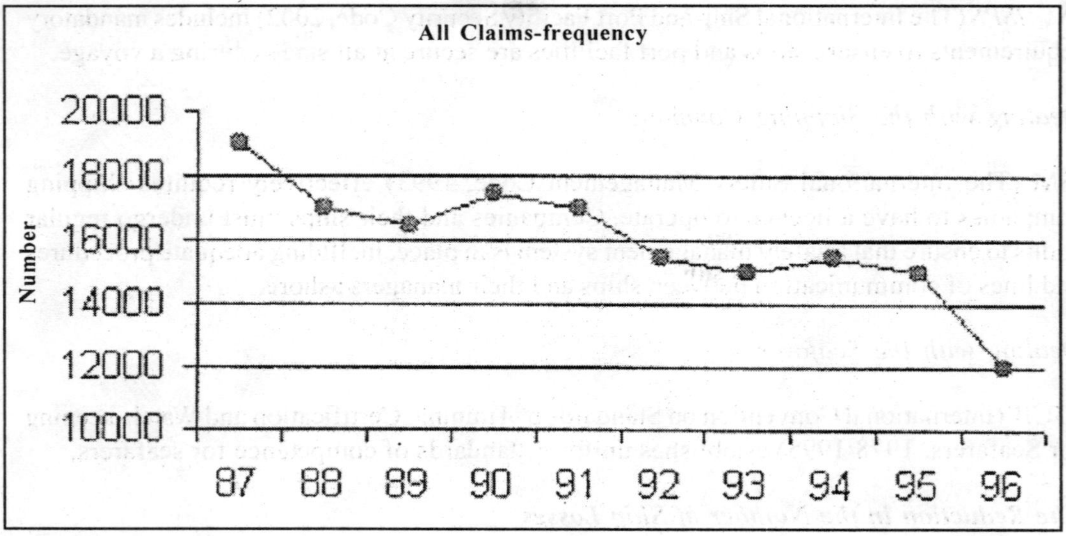

Source: UK P&I Club

How Safe is Modern Shipping?

The current level of safety in the industry can be attributed to a large extent to the comprehensive set of conventions in place. However, other significant factors include the recent developments in science and technology which are changing the face of modern day shipping. From global positioning systems to electronic charts, from cock-pit like bridges to unmanned engine rooms, the availability of detailed and accurate information for a seafarer today is vast. The rapid introduction of this technology does though pose a new set of challenges as the seafarer has to adapt to prevent a situation of information overload or over-reliance on technology.

Herein lies the crux of the matter, regardless of the number of regulations governing systems, ships or personnel; regardless of the technology designed to assist in safe and efficient operations, accidents still occur at sea. Analysis of the causes of these accidents, both immediate and root, reach the same conclusion: "the human element". Cautious estimates put the figure that 80 per cent of accidents are caused by human error, in reality, the human factor figure for maritime accidents is probably 100 per cent. Just how close you come to 100 per cent would depend upon where the investigator/researcher stopped in tracing the particular casual chain for any particular accident.

To get to the heart of the subject, we cannot afford to ignore the most crucial cog in the wheel: the human.

The Heart of the Matter

It is a challenge to investigate why an individual at a certain moment in time takes a decision which can lead to an accident. A cursory review of the situation in many case studies often leads to puzzling questions and answers:

Question: Was the individual competent to do the job?
Answer: All relevant professional qualifications and certifications were in place?
Question: Was s/he trained well enough in the task?
Answer: All stipulated training programmes had been completed.
Question: Did s/he know what s/he was supposed to do ?
Answer: S/he had the relevant amount of experience.

..........and yet..........

a wrong decision was made.

These puzzles lead to further questions... does the system then have fundamental flaws? Has there been a gross omission somewhere in the system? Is there any effective way to tackle these aspects?

The answer lies in the fact that to date, largely we have considered the individual at the "head" level and have failed to a great extent to account for the "heart". Experience as well as our understanding of human nature clearly shows that an individual in a moment of crisis, emergency or at an important moment in time is just as likely to:

Follow his instincts rather than training
Be as illogical as logical
Rush into action as to calmly weigh the options....................

.... unless........he has had the specific opportunity to try and do otherwise.

An important part of the solution therefore has to be to: acknowledge, educate and train, evaluate and be prepared to consider the individual as a whole; as a being with thoughts, feelings and emotions which are going to significantly effect performance and unless are accounted for, will continue to present problems.

A Holistic Approach to Safety Training and Professional Development

Since it is so important to consider all aspects of the individual, then this approach needs to be reflected in our selection, training and development of the professional seafarer.

Selection and Promotion

The Indian shipping industry has laid clear eligibility criteria for entry level selection which is based on academic performance and medical fitness. For the first time this year, the Indian administration is introducing the important additional dimension of personality evaluation: the screening of candidates to consider their potential not just to cope with the technical demands of the career, but also their suitability to face the psychological and emotional demands of the tough life on the high seas.

Similiarly in an industry in which promotion of personnel is often governed more on supply and demand rather than readiness for professional development, it is becoming increasing important to recognise the value of evaluating an individual's psychological profile to assess whether they are able to successfully assume a position of leadership across a range of varied situations. Suitability to effectively work at a junior level does not necessarily mean that the individual will have the capabilities to progress to more senior ranks. The industry needs to start to seriously track and monitor its personnel across a number of attributes from entry level through to most senior level.

Utilization of Simulation Technology

Simulation Training

The unique opportunity of facing a real life crisis in the safe environment of a simulator has a number of advantages for training professionals versus the typical classroom approach or even on the job training methodologies. The simulation environment allows the individual to be exposed to situations which he may rarely have the opportunity to face in real life, where he is able to combine his technical knowledge with practical application of skills, and perhaps most importantly where he has to "feel" the experience and employ the necessary soft skills to effectively handle the situation. Exposure in an immersive environment skilfully observed and guided by expert facilitators with detailed briefing and debriefing before and after the exercise can provide the ideal platform for the individual to build the necessary confidence to handle a similar real life experience.

To gain maximum benefit, it is important that the individual who "experiences" a simulation training programme has to move through an internal process of realization, acceptance and change. For this to occur, an environment of trust and understanding needs to be built up between facilitator and participant. The facilitator has to realise that the participant in simulation experience is not a subject being exposed to a set of discreet "experiments", but is a professional who needs to be handled sensitively and with due acknowledgement of the individual's past experience and competence. With this holistic approach to training, meaningful and individualized change can take place. In this way, the training becomes less a matter of "delivery" of a programme for the facilitator and rather a "journey" of discovery for the participant.

In terms of soft skills, a carefully designed simulation exercise can assist in evaluation of a range of human resource skills includmg : leadership, clarity of command, situational awareness, effective delegation, prioritization of tasks, ability to plan ahead, communication abilities, resource management, problem solving techniques, stress and anger management, response to unexpected situations and emergencies.

At both an individual and team level, the participants can be closely evaluated on a set of predetermined criteria. With the introduction of relatively affordable, flexible and customised simulation technology in India over the past few years, the capability exists for an individual to practice an operation, manoeuvre or series of actions in a specially created and customised environment. From the re-created engine control room of a vessel in which panels, gauges, dials and alarms replicate the actual panels on board, to a piece of machinery which is programmed to behave in the erratic manner in which the actual particular machinery has been doing on board, from recreating an incident in which the participant is juggling a range of responses including for instance, handling the aftermath of an oil spill including emergency response, communication with external parties, handling, or warding off the hungry media, the extent of meaningful application of simulation technology is vast and is actually a valuable resource which is waiting to be fully tapped.

PARAS (Post Accident Research and Study)

Experience has demonstrated how the accurate analysis and understanding of the chain of action which leads to an incident is important for effecting both short term and long term change to ensure that lessons are learned and a similiar incident does not happen again. Simulation technology has an important role to play in this.

PARAS involves the recreation of the incident with the potential for an individual to even re-enact the situation, attribute causes and to "practice" correct courses of action. PARAS has rich potential for use within the insurance and legal sectors.

Conclusion

If the shipping industry wants to move even further towards setting new levels in safety standards, it is crucial that it addresses fundamental issues at a micro as well as macro level. In terms of preventing accidents, accounting for the "human factor" is of primary importance. To date, taking into consideration the individual with all his/her attributes and personality profile has received little real attention in terms of training and development of the professional seafarer. This paper has highlighted some realistic and practical measures which can be introduced to assist in shaping a seafarer who has the necessary soft skills as well as technical competence to make the right decision at the right time and thereby hopefully preventing a disaster at sea.

REFERENCES

International Convention on Standards of Training, Certification and Watchkeeping for Seafarers 1978 as amended in 1995.

Lloyds Register-Fairplay January 2005.

UK P&I Club.

6

SAFETY CULTURE

Pradeep Chaturvedi*

The engineering fraternity has been given the responsibility to aim at safety and development at the heart of its professional activities so as to ensure service to the common man. This was reiterated by the World Federation of Engineering Organisations, WFEO, at its various meetings; and the Engineers Mobility Forum, EMF, at its meeting at Hong Kong during June 13-17, 2005. The intent was so strong that the member countries resolved to change the statutes to include safety of common man in the duties of professional engineer. It is thus, natural for the Safety and Quality Forum to evolve consensus on this aspect.

People's outlook towards consequence of safety, at work places and in all walks of activity in a changing world, has created eagerness to take proactive steps by involving all stakeholders. The global concern has established that there is a need to bring about a fine balance between conflicting interests of the nation, organizations, industries and individuals to achieve objectives of occupational safety and health. Such an approach calls for a much broader concern and interaction to arrive at a consensus.

Evolution of Integrated QHSE Management Systems

Quality management system and Environmental management system opened avenues for better understanding and culture of quality. This also encouraged experts to think in terms of developing integrated standards including occupational safety and health. It started off with separate manuals and separate dedicated staff for implementation. In the early part of this century, a brilliant idea emerged to consider all these in an integrated manner. Integrated QHSE Management System provides a systematic, structured and documented approach. The system includes, legal compliance in an auditable and globally recognized manner in line with the internationally agreed guidelines. It is now possible to introduce these integrated systems in a cost-effective manner. It is but logical that these four aspects are integrated to ensure better and effective legal compliance.

*Chairman, Safety and Quality Forum, The Institution of Engineers

The opening up of the trade under the WTO regime is further giving strength to work towards integrated management system as that is becoming a prerequisite to trade all over the globe, secure partnerships and maintain stakeholders' loyalty. Productivity of a production unit and the trade profitability in the global markets depends on ability to manage the quality issues of activities, product and services, coupled with environmental, health and safety risks. An effective system can lead to reduction of breakdowns and thereby the down time, which will result in improving the overall production and market cycle.

In international conventions and the ILO guidelines for occupational safety and health management systems provide a systematic management approach to safety and health, not only for the transport industry, but for industries and other sectors in general. It provides for (larger) enterprises to prepare a policy, establish a unit to implement the policy to plan and implement improvements in the working environment, to monitor and evaluate the measures and finally, to assess the need for adjustment of the management systems. For small and micro-sized enterprises, less formal instruments would usually be more appropriate. For both approaches, the involvement of workers is a key factor.

Occupational safety and health covers various activities such as Occupational Safety and Health Culture; Leadership and Management Breakthrough; Behavioural Safety Approach; Impact of Occupational Safety and Health Standards (OHSAS 18001) and Legal Framework; Integrating Occupational Health and Safety with Corporate Social Responsibility; Understanding of value of Occupational Health and Safety; Development of Management Practices; Measurement Standards and Indicators; Capacity Building Programmes and HRD; and Implementation of Health and Safety Core Practices.

In addition, the environmental aspects covering the following are also integrated with safety and occupation health:

- Management of Discharges and Emissions;
- Toxic Work; Hospital Waste; Impact of ISO 14001; and Urban and Municipal Waste.

At the international level, international conventions and agreements dealing with safety, environment as well as occupational safety and health of the workers engaged in the transport sector have been adopted. While some of the conventions cover not only the transport industry, but also workplaces in general, others are related to specific sectors of the industry. *The ILO Occupational Safety and Health Convention No.155* can be considered as an example of the same.

At national level, governments in consultation with employers' and workers' organizations should establish the national framework for improvement of occupational safety and health in the transport sector.

Safety and Transportation

Rapid technological innovation has changed the transport sector over the last century, often enabling it to grow faster than overall economic activity. But, this transformation has not totally shut doors on push carts; animal drawn vehicles or bicycles. The globalization of production and changes in management strategies have increased the demand for mobility.

Transport sector is an tool of economic development and employment provider. Even small sub-sectors have an important role to play. In some countries in Asia, upto 10 per cent of the workforce is employed in the inland transport sector demonstrating the utility of every form of transport. The inability of the formal sector to scope with social needs, has led to the fast growth of informal activity in these sub-sectors in many countries.

Why Road Safety?

Road accidents have been the worst offender in terms of fatal accidents. The scale of problem can be realized by a comparison with the number of deaths every year. This comes to over 100,000 people killed in road accident in the year 2002. Out of total loss of 120,000 due to all disasters, including natural disasters. This figure is almost 30 times that in developed countries. Accidents on highways; State roads and others have been equally responsible for this total tally. The government loses around Rs. 50,000 crore per year due to these accidents. The figure being about 2-3 per cent of the GDP.

A number of causes for road accidents have been identified and these include; poorly designed and tested vehicles; poorly maintained vehicles, poor discipline and no education of road users; poor design and maintenance of roads, poor enforcement of traffic rules, poor and primitive traffic management, mixed traffic (Non-motorised traffic dangerously mingles with fast traffic).

Major elements that can contribute in managing safety on roads include: lack of planning for safety; road traffic aspects (control devices, encroachment and parking, and road user behaviour and skills); traffic laws enforcement; accident recording and investigation; road safety research; and Road Safety Action Plan.

Occupational health and safety management system helps to improve safety and health of workers and also helps the organization improve the quality of their products and services as well as the environmental conditions at the work place. Development of safety culture management model in any organisation typically progresses from regular compliance driven approach (i.e. unsafe condition centric) to development of self-directed OH&S management systems (i.e. process centric) and finally, to the adoption of the behaviour based safety approach (i.e. individual behaviour management centric approach). This process of shaping positive safety culture in any organisation clearly requires both, the institutionalization of an effective PDCA based OHSMS process as well as individual behaviour based safety management systems.

Effectiveness of available safety hazard control and quality management system is dependent on interaction of engineers, policy makers, academicians, corporates and other civil society organizations and need a lot of social engineering work.

Major transport accidents can frequently be traced due to failure of behaviour-based safety management system. Use of behaviour modification strategies to promote the safe behaviour is an important aspect. If observations are done at regular intervals, trend data can be prepared. More recently, the emphasis in safety has shifted towards the behaviour and attitude of employees at all levels of organizations. Safe behaviour strategies have improved and changed culture which is far beyond the compliance of norms.

7

INTEGRATED QHSE MANAGEMENT SYSTEM IN ONGC

U.S. Hazra and K.K. Pandey*

ABSTRACT

Quality of product and services is one objective each manufacturer and service provider strives to achieve in order to outplay competing manufacturers and service providers to survive and excel in the very competitive present day market. The sustainability is another name of the success. Present definition of sustainability offers to perform financially, socially and environmentally. However, to ensure or to increase the probability of sustainability it is important to encompass some components, other than direct financial and environmental components, like loss prevention, accident prevention, total safety culture to ensure enhanced financial performance besides quality of product and services.

QHSE Management system wherein quality management system integrated with occupational health, safety and environmental management system appears to provide a near total management system which holds the key to sustainability.

Introduction

ONGC is a company involved in exploration and production of natural hydrocarbon across the country in onshore as well as offshore areas. Company also bids for exploration and production of hydrocarbon abroad through its subsidiary ONGC Videsh. Such a big company, where activities range from acquiring seismic data, processing, collation, exploratory drilling, production and work over operations to restore sick oil wells, requires an integrated management system to unite various units. The various activities have unique relationship and are interdependent for performance. In view of it Company's top management decided to develop a complete management system so that these units of the company perform cohesively to ensure overall performance of the company.

*Both at ONGC, Mumbai.

Background

It was decided during Vichar Vishleshan- II (2nd - 4th November, 2003) that all establishments of ONGC be certified on QHSE management system (Quality, Occupational Health Safety and Environment Management system) by 31st March, 2005.

It was proposed to have QHSE certification based on ISO-9001 for Quality Management, OHSAS-18001 for Occupational Health and Safety Management and ISO-14001 for Environment Management due to following reasons:

- Most widely accepted standards worldwide.
- OHSAS 18001, though not an ISO standard, is developed on the lines of ISO 14001.
- These standards are compatible and can be integrated.
- These standards provide freedom to develop management system as per industry requirement.
- ONGC already had number of installations certified based on these standards.

Significance of Integrated QHSE Management System

Integrated QHSE management system provides a systematic, structured and documented approach to address quality, health, safety and environmental issues on precautionary principles. The system includes legal compliance in an auditable and globally recognized manner in line with internationally agreed guidelines. This leads to institutionalizing the QHSE management system to provide a framework for spontaneous automatic growth including legal compliance.

Integrated management system is fast becoming a prerequisite to trade globally, secure partnerships and maintain stakeholders' loyalty. Integrated system apart from creating a total management solution help the organization to work as a complete unit with a common objective, while promoting development in a balanced and holistic way. Trade profitability hinges on ability to manage quality issues of activities, product and services coupled with environmental, health and safety risks. Genuinely effective management system helps reduce the amount of downtime and deliver highly motivated, effective and productive workforce and a healthy bottom line. Registration of a management system demonstrates the responsibility to employees, clients, stakeholders and society as a whole by ensuring that the organization does everything in its power to reduce the risks.

Size of Opportunity

- Onshore: 190 production facilities and 122 drilling and work over rigs.
- Offshore: 11 process complex with 124 associated well head platforms and 8 jack-up rigs.
- All 36 Geophysical field parties engaged in collecting geo seismic data across India including data processing centres.
- 2 survey vessels, a flotilla of support and survey vessels.
- All well logging systems.

- Entire system of seismic data collection and processing for offshore.

Road Map to Certification

It was decided to harness internal resources for developing QHSE management system in view of tight time frame, reduce outlay and develop an ownership of the system by the field personnel.

A four stage road map was developed to meet the target which was approved during the 4th Key Executive Meet on 30th - 31st May, 2004.

Stage I : Planning : EDC 30th September, 2004

— Framing site specific policy
— Hazard identification and risk analysis
— Environment Impact assessment
— Document preparation, vetting and dissemination
— QHSE training
— Identification of legal and other requirements

Stage II : Implementation: EDC December, 2004

Application of procedures, plans and programmes identified and documented in the management system.

Stage III : Internal Audit : EDC 15th January, 2005

Stage IV : First Stage/Pre-assessment and Certification Audit: EDC 31st March, 2005

Task Realizations

Formation of core QHSE team

A QHSE core team of officers having experience of field operations and development of management system was assigned the task to facilitate implementation of QHSE across ONGC.

Development and Dissemination of Model QHSE Document

Most important prerequisite for certification of QHSE management system is documentation of policy, principles and procedures in conformance to the requirements of standards, i.e. ISO 9001, OHSAS18001 and ISO14001. There was a need to develop an integrated model QHSE document which apart from satisfying the needs of standards, should also contain all the ingredients like HAZOP analysis, Environment impact assessment module to identify significant health, safety and environmental risks, procedures to act proactively to address these risks, legal requirements and compliance etc. Besides it the model QHSE documents should encompass the needs of all sort of operations ranging from geophysical survey, drilling, production and processing of crude, work over operations and others. An extra care was taken to underline the QHSE elements which underpinned the behavior based safety management so as to make users own the system and apply it with the sense of custodianship.

Management Sensitization

To develop awareness among the top managers, QHSE core team set out on the task of management sensitization at different Assets, Basins and Services. Top managers were made acquainted with the requirements of QHSE management system, significance and benefits.

QHSE Training

Course content for QHSE management system based on the requirements of ISO 9001, OHSAS18001 and ISO14001 and practical pursuits to develop and implement management system was designed. With the assistance of ONGC academy 24 courses were conducted concurrently in Assets and Basins during August to September, 2005 and 630 personnel were trained which in turn helped in developing the management system in their respective Assets, Basins and Services.

Interpretation of Model QHSE Document and Preparation of Installation Specific QHSE Document

Assistance was provided to all the Assets, Basins and Services to understand the system's requirements, its practical aspects, procedure to conduct HAZOP analysis and Environment impact assessment, working out policy, procedures plan, programme and emergency plans, and finally, to document all these.

Internal Audit of Management System

In order to facilitate internal audit, a mandatory requirement prior to certification, core team formulated a comprehensive checklist and helped in concurrently conducting internal QHSE audits in all the installations across the ONGC.

Identification of Certifying Body

Hiring of certifying bodies was carried out centrally at corporate HSE. This resulted in timely certification at very competitive price. Precautions were taken to ensure that certifying bodies across ONGC had accreditation with the members of International Accreditation Forum (IAF).

Third Party Certification

Agencies to conduct the certification audit were finalized by end December, 2004. Approximately 98 per cent of certification assignment was awarded to Indian Registrar of Quality System which is a subsidiary of Indian Registrar of Shipping an autonomous non-profit organisation. IRQS is accredited through RvA (Raad voor Accreditatie, the Dutch Council for Accreditation), Netherland and NABCB (National Accreditation Board for Certification Bodies), India. The balance was awarded to M/s ICL Certification Limited which is accredited through JAS-ANZ (Joint accreditation system of Australia and New Zealand).

Outcome

As on 31st March, 2005, 98.16 per cent of ONGC operations were successfully audited and recommended for certification on Quality, Occupational Health Safety and Environment Management System based on ISO-9001, OHSAS-18001 and ISO-14001.

Major Facilities Certified on QHSE

- 117 onshore drilling and work over rigs.
- 186 onshore production installations.
- All 11 offshore production complexes including 124 associated well head platforms.
- 06 offshore jack up rigs.
- All 36 Geophysical field parties engaged in collecting geo-seismic data across India including data processing centres.
- All well logging systems.
- Entire system of seismic data collection and processing for offshore.

Status on 31 August, 2005

Yearwise certification	*QMS*	*OHSAS*	*EMS*	*Total*
Certification on 31-3-04	31	07	118	156
Certification accomplished during 04-05	382	390	289	1058
Total Certification as on 15th August, 2005	413	397	407	1217

Abridgement of QHSE certificates is appended in Annexure A.

Saving Incurred

No external consultancy was taken for developing and implementing QHSE management system in spite of enormity of work. It resulted in saving of time and money other than developing ownership with the management system by line managers and operators as they were totally involved.

Substantial savings were also accrued on account of centrally identifying the certification bodies.

Benefits of QHSE Management System

- Secure partnerships and maintain stakeholders' loyalty.
- Integrated system apart from creating a total management solution help the organization to work as a complete unit with a common objective, while promoting development in a balanced and holistic way.
- Reduce the amount of downtime and deliver highly motivated, effective and productive workforce and a healthy bottom line.
- Registration of a management system demonstrates the responsibility to employees, clients, stakeholders and society as a whole by ensuring that the organization does everything in its power to reduce the risks.
- Demonstrate corporate commitment.
- Demonstrate corporate progress.
- Facilitate internal management decisions at strategic, operational and communication levels.
- QHSE investment and cost.
- QHSE certification will ensure continual improvement in quality, occupational health, safety and environment management due to regular surveillance audit.

Balance Certifications

Expected to be completed by 31/1/06

Installations	*Reason of delay*
Drilling Rig E-2000-IV Assam	Rig is under revamping and upgradation since July, 2004. Will be operational by August 05. Certification expected by December 05.Nos. of certificate—03
Rig Sagar Pragati Mumbai	Rig was under dry dock. First Stage Audit completed. The certification audit process to be commenced, once Rigresumes operation. The leg of the rig got damaged and to go again for dry dock. Operation is expected to resume by January 2006. The Certification Audit can commence only after operation is resumed. Nos. of certificate—03

Conclusion

- There has been a significant improvement in the awareness of the field employees towards QHSE.
- All ONGC operations today have similar management system, i.e. a rig at Assam and at Mehsana operates on the same management system.
- QHSE certification will ensure continual improvement in Quality, Occupational Health Safety and Environment Management due to regular surveillance audit.
- Presently each unit has its own QHSE objectives and targets. Some of the important objectives are not resolved as these require corporate policy to deal with. However during review at local management level QHSE issues of common nature will help

the Asset/Basin/Services management to plan broader objectives. These broader objectives help corporate management to take policy decisions. On the other hand corporate decisions are well implemented in the entire organization because of systematic QHSE management system in all its units. Thus, QHSE management system helps top management to make strategic planning and simultaneous implementation thereof for the benefit of entire organization.

- The system so established will be continuously upgraded by incorporating the best industry practices like DNV's customized ISRS protocol for benchmarking DuPont's 'STOP' etc.

8

ROAD TRANSPORT SAFETY AND ROLE OF INSURANCE

Deepak Godbole*

ABSTRACT

Safety of road transport is a products of the interactions between the driver, the roads, environmental elements and the vehicle. Advances in the design of vehicles and better road infrastructure have contributed to a significant level of road safety. Still, the vehicle accidents are a leading cause of death. This paper tries to examine the role of insurance industry in road transport safety, with special reference to India. Taking stock of losses due to road accidents, major areas of study pertaining to road safety and the impact of changing conditions, it describes the ways insurers contribute to improveroad safety—influencing road safety policy, financing safety initiatives, promoting research in road safety and use of new technology.

Transportation

Transportation basically involves carrying/moving goods and people. Air, Water, Rail and Road are some of the common modes of transportation. Safety conveys meanings such as security and protection. But the question is how safe are our transport operations?

Let us focus on road transport. It was in 1885, the first motor cycle was invented, soon followed by invention of a motor car in 1888. Today, world over, there are more than one billion vehicles and the sector is growing at 5.7 per cent a year. But the less known facts are to be also noted. The first pedestrian death due to vehicle accident was recorded in 1896 and the first accidental death of a car driver in 1898. Motor vehicle accidents are a significant cause of death in comparison with other transport accidents (air and space), homicides, wars and even diseases such as TB, Cancer, HIV infections, causes which are more often highlighted in the media.

In past, there had been two major studies of causes of death worldwide that have been published in the 'Global Burden of Disease' (World Health Organisation, World Bank and Harvard University) and in the 'World Health Report—Making a Difference' (WHO). These publications show that road accidents as a cause of death or disability were by no

*Deputy Manager, Management Secretariat, GIC of India, Mumbai
E-mail: deengee@gmail.com, deengee@rediffmail.com, deepakg@gicofindia.com

means insignificant, assuming ninth place out of a total of over 100 separately identified causes. Further, by the year 2020 forecasts suggest that as a cause of death, road crashes will move up to sixth place.

Motor vehicle accidents create significant delays, increase road maintenance costs, contribute to air pollution, consume energy and lower the productivity of the nation's industries and workforce. Despite substantial progress in improving vehicles and driver performance, significant challenges remain to assure that the roads make their fullest contribution to the safety of road travel. The assessment of driving risks is a common concern for road transportation safety and the automobile insurance industry. Elements concerned in one way or the other with road accidents are—vehicle manufacturers, drivers, pedestrians, institutional authorities, financial institutions, insurance companies, consultants, administrators, practitioners and researchers. Research is essential to counter this public health threat and to assess the success or failure of counter-measures taken from time to time.

Losses Due to Road Accidents

The death toll is highest and growing in developing countries, where pedestrians, motorcyclists, cyclists and passengers are especially vulnerable. In addition to human suffering, estimated costs of road traffic injuries are between 1 per cent and 2 per cent of GNP per annum in the developing countries. 85 per cent of all road accident deaths occur in developing countries and nearly half in the Asia-Pacific region. This represents a loss of approximately US$ 65 billion every year; almost twice the total development assistance received worldwide by developing countries. The developing countries like India are at the worst end of the spectrum. Where as the share of the vehicle population of developing countries constitutes hardly 40 per cent of the world vehicle population, but their share in the road fatalities is 86 per cent of the world road fatalities.

Speaking specifically of India, (Fact file—*source:* Institute of Road Traffic Education, a non-governmental organization based in New Delhi)

- India accounts for about 10 per cent of road accident fatalities worldwide.
- An estimated 1,275,000 persons are grievously injured on the road every year.
- Social cost of annual accidents in India has been estimated at $ 11,000.

Studies/Research/Analysis

Studying the problem of transportation safety necessarily involves studying:

— Human Behaviour
— Vehicle design and built quality
— Equipment design and quality—head lamps, horns, seat-belts, helmets
— Road infrastructure/construction
— Transport controllers—driving licence, parking rules, driving records, vehicle history
— Research and testing at laboratories
— Environmental concerns—emission control and use of cleaner fuels

— Driver education efforts
— Road safety equipments—lights, barricades etc.
— Educating cyclists, motorcyclists, pedestrians

Road safety is a result of deliberate efforts on the part of many sectors of society—Government and non-Government alike all of whom contribute to the reduction of the incidence, impact of road trauma and support safety efforts. As in case of many other ventures, in transport sector also, resolving safety and infrastructure problems has been limited by available resources. However, there are always opportunities to afford countermeasures, derived from emerging technologies, which could lead to significant safety improvements. However, lack of authentic data and non co-ordination between research institutes, road authorities, police and hospitals make the studies difficult.

Effect of Ever Evolving/Changing Conditions

Changing conditions make it necessary to continually monitor road safety. Changes in traffic, technology, environmental conditions and vehicle design characteristics influence safety problems and their solutions. Also, today's growing traffic volumes increase pedestrians' potential exposure to accidents. Increased truck loads and sizes, new brighter headlamps, aerodynamic vehicle designs and shapes are often incompatible with the existing road conditions and traffic regulations. Increasing number of young drivers, an aging driver population, non-compliance with traffic regulations, speedy and aggressive driving behaviour and driving while under liquor influence pose additional problems.

Countermeasures for addressing road safety problems take many forms ranging from educating the driver, enhancing the safety features of vehicles, to improving the roads and facilities. They fall into two major categories—those designed to prevent crashes from occurring and those which minimize the severity of accidents. Representative countermeasures for the major safety problems would include:

- Wide roads and areas of open space adjacent to the road, giving drivers room to safely regain control of their vehicles if and when they tend to loose control.
- Gradual side slopes lessening risk of vehicles over-turning.
- Effective use of signs and utility poles, barrier walls/guardrails.
- Traffic signs and reflective devices that help improve driver performance.
- Preventive maintenance of roads and timely repairs.
- Pedestrian/cyclist facilities

Necessity of Regional Classification

Because road conditions, vehicle condition, population, intensity of traffic, environmental conditions differ, it would be essential to group the globe into cohesive groups for studying and analysing road accidents. In order to aid interpretation of data, following major regional groups can be made:

- Africa
- Asia/Pacific

- Central and Eastern Europe
- Latin/Central America and the Caribbean
- Middle East and North Africa
- Heavily Motorised Countries, i.e. North America, Australia, New Zealand, Japan and Western Europe.

The first five regions can be collectively called as Less Motorised Countries.

As the world is divided into heavily motorised and less motorised countries, there are also differences found in terminologies/definitions. As defined by the Convention of Road Traffic (Vienna, 1968), a road death is deemed to have occurred when a person injured dies within 30 days of the accident (and as a result of the crash). However, not all countries use a 30-day definition with some countries using 'on the spot', within 24 hours, 3 days, etc. The standard international definition of an injury road crash involves a collision of a moving vehicle on a public road in which a road user (human or animal), is injured. Non-standardisation of terms and definitions acts as hindrance in data management and analysis.

Coming to the regional trends, researchers agree on the fact that the total number of people killed in road crashes in regions of the developing world continue to increase, whereas in the West there has been a steady decrease over the last fifteen years or so. For example, between 1987 and 1995 deaths in the Asia-Pacific rose by 40 per cent, in Africa by 26 per cent (excluding South Africa where the increase was minimal) and the Middle East/ North Africa region by over 36 per cent. Road deaths doubled in a few Latin American countries and rose by 16 per cent in Brazil. Central and Eastern Europe show wide variation. Conversely road deaths in highly motorized countries fell by about 10 per cent. Pedestrians are a particularly high-risk group throughout Africa and Asia as well as the Middle East. China and India dominate the Asia Pacific region and, in fact, lead the world in terms of road deaths. It is estimated that road fatalities are expected to continue to increase with a fatality toll between 900 thousand and 1.1 million in 2010 and reach between 1.1 million and 1.3 million in 2020.

Insurance

Insurance is a risk management tool. It helps rehabilitation after a loss has occurred. Though prevention is better than cure, every accident cannot be prevented. Sometimes the preventive measures are also not economically worthwhile. Such non-preventive misfortunes need curative measures. Therefore, some kind of relief measures like insurance have evolved. Motor insurance is a curative measure to deal with misfortunes of road accidents. Motor insurance is a typical insurance cover. It has two parts. One part covers 'damage to the vehicle' and the other part covers 'third party liability'. Road accidents do not necessarily affect the vehicle owner/driver/passengers. Some innocent persons or at times even property of an unconcerned party gets damaged due to a motor accident. The third party cover takes care of such situations. It is instrumental in reducing the hardship and trauma caused to the affected third party and the family members as the courts award financial compensation basing on the remaining useful working life of the affected person and the lost opportunity to earn because of the accident. Insurance relies on many small contributions providing sufficient revenue to compensate the minority who suffer loss. Most countries require those responsible for death or injury to compensate the victim, or the victim's family. Since few people can afford to do this from their own resources, they either choose—or are required by legislation—to insure against the risk of a claim. Most countries require compulsory third party motor insurance.

Share of Motor insurance in general insurance business: Motor insurance is a major portfolio for non-life insurers world over.

Country	*Auto/motor insurance as a per cent of tatal non-life insurance in the recent past*
U.S.A.	45
Canada	48
Malaysia	48
U.K.	33
Germany	34
France	46
Japan	52
China	70
Greece	50
India	38

Automobile insurance forms a major segment of general insurance business in India as well. High claims ratio, from year to year, affecting the viability of the business, in this segment, has been a cause of serious concern to insurers. Several factors have contributed to this situation. There has been a phenomenal growth in the number of motor vehicles, of various types, plying, without adequate road infrastructure. Steep rise in accidents, resulting in fatal and other grievous injuries to persons, has meant increase in the number and the aggregate quantum of liability claims. The Motor Vehicle Act stipulates that no motor vehicle can be used in a public place, unless it is covered by insurance against unlimited liability, for injuries caused to persons, arising out of its use. It is a fact that courts have been taking a liberal view in assessing quantum of damages in regard to such liability.

As against this trend the rates of premium charged are inadequate and not commensurate with the risk assumed. Motor premium rates in India are presently governed by statutory tariff and are inelastic. (Though, de-tariffing is being talked about for quite sometime now.) Ad hoc rate increases attempted in the past were resisted by user groups, particularly owners of commercial vehicles (truck/lorry) through their trade associations. Every year insurers in India settle claims worth Rs. 3000-3500 crores and above, for the benefit of road accident victims. Despite the efforts for prevention of accidents and expanding relief measure, the situation continues to be grim and deteriorating.

An understanding of the principles, underlying motor third party insurance, is a pre-requisite to appreciate, the relevant legal provisions and the procedures involved in third party insurance. The entire concept of third party liability/insurance revolves around this fundamental principle called principle of negligence. This principle, expects one to exercise care in one's daily social conduct. Every society has, not only acknowledged/recognized this duty of care, but also over a period of time, refined it to suit the needs of progressive societies. Many other Principles associated with third Party insurance are either variants or extensions of or limitations on this principle of negligence or duty of care. In the context of motor third party Insurance and claims the above principle can be defined as: "A person driving a Motor Vehicle on a Public Road must drive the vehicle carefully and cautiously, strictly in accordance with the prevailing rules and regulations, in such a way that he does not endanger the safety of others using the roads. If it is found that he was lacking, in the carefulness or cautiousness or in complying with the prescribed rules and regulations and thereby if any harm is done to other road users he is liable to compensate them for the harm suffered by them."

Some examples of the negligence in the use of vehicle may be found in the following:

- Driving in violation of traffic signals
- Turning the vehicle without giving signal
- Breaking no entry barrier in a one way road
- Poor maintenance of the vehicle
- Using a cellular phone while driving
- Driving the vehicles without fitness certificate
- Over-speeding
- Over-loading and improper loading
- Driving without wipers in a rainy season
- Absence of dippers in the vehicle
- Failure to use parking lights
- Ignoring "zebra crossings"
- Violating traffic rules while driving

Often the bursting of tyres, breaking of the axle, leafs, bolts, dislocation of ball joints are pleaded as the inevitable cause of accidents. But in fact they can be attributed to the poor maintenance of the vehicle. Variants of the basic concept of negligence are—Contributory Negligence, Composite Negligence and Vicarious Liability which is an extension of the principle of negligence.

There is well set machinery for dealing with motor third party claims in India. There are 'Lok Adalats', 'Motor Third Party Claims Tribunals', 'Workmen's Compensation Courts' and schemes like 'Jald Rahat Yojana', which try to speed up the claim settlement.

Conclusion

The safety of road transportation is a product of the interactions between the driver, the roads, environmental elements and the vehicle. Advances in the design of vehicles, improved driver performance and enhancement of the road infrastructure have all contributed to a significant level of safety. Motor vehicle accidents are still a leading cause of death, even if the trend has somewhat declined over the years. In India, minimising the number of accidents and consequent fatal and grievous injuries to persons lies mostly in the domain of Central and State Governments particularly in the matter of:

(a) Adequate development of road infrastructure to cope with the growing traffic density.
(b) Creating road safety awareness among the users of motor vehicles and the general public.
(c) Effective enforcement of statutory regulations relating to licensing of drivers and vehicles maintenance of vehicles, speeding and other traffic discipline, safety precautions like seat belts and helmets for riders of two wheelers etc.

But the insurance industry also plays a role in these matters. Insurers lend money for setting up vehicle manufacturing units as also for infrastructure development projects involving building of roads, flyovers, bridges etc. Insurers influence careful driving by incorporating restrictive conditions in the policy and providing incentives in the rating structure. They spend on publicity towards safe driving practices. And most importantly, they provide compensation not only to the owner of the vehicle but also to the affected third party as a social commitment, especially, when premium received is grossly inadequate compared to third party claims payments. Since insurance premiums are generally related to road accident costs, motor vehicle insurers have an incentive to reduce road crashes to help reduce claims and hence the premiums charged. That is why, insurers try to promote better driving habits by using a 'bonus/malus' system that offers discounts for claim-free driving and imposes penalties when claims are made. Typically, these apply to next year's premium that may be reduced by a 'no claims bonus,' or increased in view of bad claims record. Insurers can improve road safety by way of participating in shaping road safety policy; financing road safety initiatives; promoting more research on road safety, promoting use of technology to improve transportation safety and security.

IS INDIAN TRANSPORT SYSTEM SAFE

N. Jayaraman*

ABSTRACT

Air travel enables us travelling long distance in short time, Rail travel is to cover long distance covered by railway track and bus travel is to reach all nook and corners of covered by bus route. In all modes of transport safety is an important aspect. In Air Travel safety responsibility is taken care of by airport authority providing data regarding climate, availability of runway, lighting of runway, instrument landing system, radar and also by communication with aircraft and good aero plane maintenance. With regard to safety while travel Major Repair Overhaul and Pilot training and deployment responsibility lies with the individual airlines. Safety in rail travel is governed by signalling system, maintenance of track, bridges and communication. Safety in Road travel is governed by Motor Vehicles Act. By adopting traffic rules and regulations speed of vehicle and good driving habits and judgment to act in emergency will make the road transport system safe.

Introduction

The industrial development of a country leads to movement of mankind and goods to distant parts of world. The movement of mankind is in pursuit of Education, Employment, Business Development, Industrial Promotion and on Government Assignments. Similarly, goods has to be moved internally by trucks, trains, ships, planes and also beyond the boundaries of countries. In this respect the safety of movement with all types of movement assumes a primary responsibility.

Air Safety

The operation of Aeroplanes require good design, construction, manoeuvrability, safety during takeoff and landing. The following are the identified reasons that affect Air Safety:

> Fog , Icing, reduced visibility, weather related , ground collision, explosion, structural failure, cockpit fire, in-flight fire, mid-air collision, crashed on landing, mechanical failure, engine failure, onboard fire, stalling, broke up in flight, captain becoming ill,

*Consultant/N.J.Consultancy, No. 1, Second Main Road, I.I.T. Colony, Narayanapuram, Pallikaranai, Chennai-601 302

navigational error, electrical failure, air frame failure, striking power lines, hitting mountains, ran out of fuel, instrument failure etc.

All Airport Authorities to provide the following service:

(i) Prevention of collision between aircrafts.
(ii) Prevention of collision between aircraft and obstruction in manoeuvre area.
(iii) Orderly flow of air traffic.
(iv) Safe and efficient conduct of flight.
(v) Initiation and co-ordination of search and rescue activity.

The following are the statutory requirements for safe operation of an aircraft: implemented by airport authority. To ascertain whether aircraft engine and radio has been maintained periodically, providing certificate regarding maintenance of aircraft approved maintenance schedule during last 12 months and Certificate of Radio Transmitters, CMR (Certificate of Maintenance Review), Proof loading result, Rigging Data, Compression Check, Bonding Test Result and Compass Setting Result.

The Airports has to install Instrument Landing System CAT I/CAT II for operation of flights under difficult weather conditions.

With regard to safety of Helicopters, Helicopters used by private and military are great causes of concern in Air Safety. The helicopters having single engine is not found to be reliable at several instances of helicopter crash occur in single engine fitted helicopters. It is most of the crashes are due to fuel shortage, engine failure and also training needed by private helicopters requires further improvement.

Air Miss

The flights are expected to follow certain routes and fly at designated heights. It has been found that flying at heights above 35000 ft improves fuel efficiency. However, if two planes flying at the same height and at a distance of 3.9 miles, there is a possibility of air miss. This happens either one of the planes changes its altitude with the result that two planes will be travelling at the same height with minimum distance separation. Traffic controllers provide latest data regarding the feasibility of air miss. Most of the advanced aircrafts are provided with anti-collision device.

Escalator Accident

Passengers use escalators for speedy movement in and out of airport. However while utilizing the escalator occasional injury/death occurs. Poor maintenance/closure of escalator companies, lack of instructions/supervision by escalator staff during passenger usage are real causes of escalator accident.

Similarly, aprons are damaged during positioning of aircraft for passenger use due to wrong communication between ground staff and pilot.

The following are the modern developments for air safety:

Electronic Flight Bag(EFB): It is e-enabled gives flights all information they need to fly an aeroplane (aeronautical maps/charts) a manual for fault reporting and operations—logbooks are in digital format and puts them at crew's finger tips.

Landing systems: Global Positioning Landing System (GPS) integrates data from Global Navigational Satellite System, ground stations. It also accurately pinpoints planning position relative to runway and surrounding terrain, with wireless emergency lighting system. Clearances/Instructions are transmitted to arriving aircraft so that pilots and controllers don't have to engage in multiple voice communication by Tailored Arrivals.

The maintenance time/cost also reduced by removing four small windows (eyebrows) above front windshield. It reduces weight by 20 lb and 300 hours of periodic inspection.

Bursting of tires during landing is one of the major causes of Air Safety. In modern aircrafts there will be 10 tires, 2 nos for nose landing gear and 8 nos for manual landing gear. The tires are of high elasticity, high strength cord, weight efficient and wear resistant.

Is Indian Railways Safe?

Introduction

Indian Railways is striving to achieve maximum safety by adopting latest technology in Signalling,Telecommunication, Introduction of new Engines, Coach Building, Track Renewal, Track and Level crossing inspection which pertains to the machinery impact on smooth operation of train.

With regard to man-machine interface, drivers are trained in watching the track for correct signals, condition of track, real-time condition of track where caution is required and also with manuals for correct running of train.

Safety in Train

Train Separation: are based on speed and breaking performance of the trains: For trains travelling at 120 km/hr the breaking distance is 1 km. Service Break is set at 1 km and Emergency breaking at a distance of 650 metre. Blocks in which trains are allowed to travel without a second train being allowed in the same block is divided into 3 km, 8 km or 32 km. For Rajdhani express 3 block sections are cleared. For the trains to move safely the routing signal must have a max of 5 lamps glowing and a minimum of 3 lamps to glow. If no signal is visible the train has to travel at 15 km/hr and also to stop. A chronograph is provided which records Speed/Time from which distance travelled can be interpolated.

The movement of trains follow absolute work and permissible work system. When train moves under signal it is absolute work and when signal is faulty or not available permissible work is adopted with the approval of competent authority.

By interlocking: the movement of trains along points is followed with indication, of track occupancy. By this conflicting train movement, signal aspects, enforced speed at junction are the norms adopted to control regular movement of train.

Axle and Wheel Counter: The axle wheel counter system ensures that entire train from engine at the front to last bogies, namely Guard van has passed through the designated signal block. As a wheel enters a block short-circuits the current as track at either end are electrically insulated and a small electric passes through relay with track circuit becomes de-energized (Signal changes from Green to Red). From Axle counter solid state technology is adopted.

In a fail safe mode of operation if the relay accidentally gets energized, circuit is disturbed, track circuit status becomes a potential for danger for presence of train is unknown. In that case all track signals drops to red. By manual intervention after identifying the real cause manual clearance is given. The signals are lit by LED which gives more colour output, useless power and working life is around 10 years.

Failsafe Multiples Specification: In a failsafe condition if the signal fails it will lead to complete stoppage of train and will not be a cause for accident. The details of specification of power supply to batteries is as follows. The details of electrical requirements for signalling is as follows:

Working Voltage: 24 V or 48 V DC. No of input /output per card 32, Max number of input/output 64- Output relays—miniature or Q series/plug in relays. Transmission error checking —Detection upto 2 bits, Medium Radio, Optical fibre, Copper, Max Operating temp +55°C.

When several lines meet or diverge any train allowed to go into an area is *protected by one signal at rear.*

Interlocking: The Green Signal allow the train to move, yellow allows the train to move with caution and red makes the train stop. Signal units translate into Train move/stop move action. Point machines sets routes for moving portions of railway track. Track circuit monitor indicates the occupation of railway track and the point detector indicates the direction in which track is set at diverging point of track.

For the drivers a Benchworking manual is circulated with alterations along with caution order (Caution diary) which indicates km Speed at which the train has to be run, due to factors like work in progress in the track or sleeper may be defective. Gangmen are provided with detonator which will be detonated 600 metres before the defective area. Drivers are trained in the topography of the area. The drivers recruited first work as driver in Goods train as Assistant Driver in the passenger before being upgraded as Driver in the passenger drain. During the entire course of the service the drivers will be mostly working in the same topographical area. However, when necessity comes to transfer they will be trained in load running with engine with the guidance of a Traffic Inspector. The drivers undergo alcoholic test and if found guilty recorded and booked. Drivers with colour blindness, sugar problems and refractory problems are tested every three years. Drivers wearing specs has to carry additional set of specs while reporting for duty.

The change points are operated either manually or electrically. The signal points have coding life of 20 years and wear and tear can be predicted during preventive maintenance or during annual inspection and overhaul for every 3 years.

Track Protection: Warning Man Machine Interface TPW is designed for 120km/h. The breaking curve for running at the speed will be decided Pneumatically and Safety Ring Belt is second line of defence. For Safety integration level (Electronic solution)—a Software Critical application is adopted.

At track shifting it has been usually found that if the gap more than 3mm (between regular track and change track) exists its is due to presence of foreign particle like stone, passenger articles like chapel, ladies handbag which cause the more gap than mm which has to be removed for the signal to work.

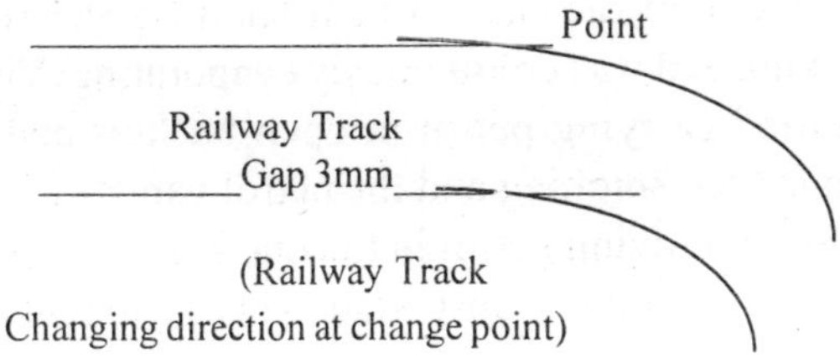

The short-circuiting of current through railway track is always essential to maintain a continuous signal. The presence of rust due to rain, water flooding the tracks, cracks in railway track are the main reason due to which signal points will fail to function and the train has to be brought to halt before proceeding.

Train protection and Warning System, Automatic Managing System are provided for safe running of the trains. If the driver observes an obstacle at 1 km distance, he has to start breaking and if this breaking does not start at 200m distance the automatic Train Protection and Warning System will automatically act and bring the train to rest. In European countries driverless system are in use with European Control System Level I and Level III.

In Indian Railways accidents occur during monsoon rains when rivers are flooded causing damages to tracks on bridges crossing the flooded rivers. Most of the train accidents during night time between 2 a.m. to 5 p.m.

Train accidents occur due to express or mail train hits a stationary train or goods train. Since the material of construction is steel the sleeping passengers normally are killed and remain stuck up in mangled/broken compartments. Anti-collision devices are yet to be provided to all trains.

Indian Trains have to cross a number of level crossing both manned and unmanned. Accidents occur both in manned level crossings as people/cattle continually cross tracks even if the gates are closed and also in unmanned level crossing. Poor visibility in curves, public decision to cross when the train has reached a distance of non-breakable distance also contributes to accidents. Normally, school buses with children, marriage parties in tractors are the victims of train accidents at unmanned level crossing.

Safety in Surface Transport: In US and other countries goods movement by trucks cover an average distance of 600 km/day. But in India we cannot think of such speedy delivery due to poor infrastructure.

In Indian Industry the movement of goods and passengers by surface transport is essential for the speedy movement. The Indian Roads are clogged and the Government of India are introducing Golden Quadrilateral connecting four metropolitan cities. However, where the traffic is peak: due to volume of passengers during peak hours of school and office and business travels, the vehicle move in close proximity and not maintaining a safe distance of 20 m between two vehicles.

The rate of Insurance Claim by motor vehicle accidents are bleeding the Insurance companies as the compensation payable by them are based upon the earning capacity of the diseased in the balance of life.

Further the movement of lorries for the transport of water/petroleum products , school buses/vans carrying children and office staff creating a chaotic transport which results in more accidents. In one of the cases a Petroleum Carrier with Petrol broke down and the petrol started leaking and was continuously evaporating. Villagers who are unaware of the risk of petrol started carrying petrol in open buckets endlessly. A person who was at a distance of 1 km started smoking and the petrol vapour got fire and in turn all the persons who were stealthily removing petrol in the one km intervening distance got burns. Similar instance occur during Chemical Transportation like Hydrochloric Acid, Sulfuric Acid, Alkalies, Benzene etc.

The lorry transport heavy loads of industrial use like machinery, iron rods which protrude beyond the length of vehicle, heavy materials like rock for granite factories, timber which are not properly fastened and during travel or turning in curves or hills cause havoc. The licensing system requires strengthening and driving schools run by reputed institutions must be considered for deployment.

Conclusion

Out of large number of trains carrying million passengers the number of accidents is very negligible. Even these accidents are due mainly due to small human error in judgment and also due to public crossing the level crossing in unsafe condition. In the long term when train passengers are wooed by airlines the speeding of train from the present 120 km/hr to 300 km/hr as in China or Japan will always maintain the supremacy in Indian Railways with one of the safest mode of transport. Indian Railways is one of the safest transport system considering the volume of passengers and goods movements carried everyday.

Air travel is safer and quicker and the number of accidents are more in the case of private helicopter compared to commercial planes. In the case of private helicopters, human judgment, maintenance may be cause for helicopter accidents. But in case of commercial flights the accidents are in the range of six sigma limits (less than 3 instances for million passengers travelling) and also offering greatest safety. But in adverse climatic conditions like torrential rain, fog, instrument landing system are the safest method for takeoff and landing. Even with advance technology Engine failure during flight, human error in manoeuvring the aircraft causes a few accidents.

In the case of road accidents the congestion due to number of vehicles on the road are increasing which forces traffic jam and to makeup the time lost speedy rash driving is resorted. With the introduction of four lane and Golden quadrilateral connecting state headquarters congestion in highways can be reduced. Good maintenance of vehicles, normal loading, use of quality spare parts and also strict inspection by Traffic Controlling authorities will prevent loss in road transport to a great extent.

The public has to be made aware of the harmful effects of chemical transports. Also distribution of petroleum products and water by tankers has to be minimized and laying of petroleum distribution lines and water supply through pipelines wherever possible has to be implemented. Some state governments are putting restriction of lorry movement during

8-10 a.m and 5-7 p.m. to reduce risk of accidents when office goers and children move from house and return home in the evening.

REFERENCES

Boeing —Technical Specification. *News Letter*.

Department of Signal & Telecommunication, Southern Railway, Chennai.

Internet.

10

TRANSPORTATION SAFETY IN THE CONTEXT OF WEATHER RELATED NATURAL HAZARDS

Medha Khole and D.M. Rase*

ABSTRACT

Safety in transportation, to a great extent, hinges upon the prevailing weather conditions, with the background of the importance of meteorological elements in the transportation sector, this paper examines and analyses the standard climatological patterns over India pertaining to weather related natural hazards such as heavy rainfall, heavy snowfall, landslides and fog.

Introduction

Weather related natural Hazards cause tremendous loss of lives and property all over the globe. Disastrous earthquake of Gujarat, Tsunami and recent floods and heavy rains over western India in 2005 are some of the most recent examples. About 60 per cent of the global disasters occur in Asia Pacific region and about 90 per cent of the deaths result from natural disasters occurring in the developing countries owing to various socio-economic factors (Khole and De, 2001). Over 80 per cent of damage from the natural disasters is of meteorological and hydrological origin (Obassi, 2001). The total global economic losses due to rainfall hazards range from US$ 3-4 billion per year in mid-1960 to US$ 25-35 billion per year in early 1990s (De and Joshi,1998). In the modern age all the human activities, particularly the economic activities strongly depend on the transportation sectors.

Safety in transportation (Road, Rail, Water and Air), to a great extent, hinges upon the prevailing weather/meteorological conditions. Various meteorological elements such as heavy rains, fog, etc., affect the visibility and this, in turn, affects the transportation safety in an adverse manner.

The poor visibility conditions due to fog and rain (particularly in the winter season, in the northern parts of India) lead to delays and/or diversion of Aircraft routes in the aviation sector resulting in serious economic loss. Also, poor visibility conditions often pose a threat

*Both at India Meteorological Department, Pune

of accidents affecting the security of human life. During winter season, heavy snowfall in the northern states of Jammu and Kashmir, Punjab, Himachal Pradesh and Uttar Pradesh, etc., often disrupts the transportation leading to a great loss, economically as well as in terms of human life. In the northern and northeastern India, occurrences of landslides, particularly during the monsoon season, constitute a major class of natural hazards and this too has an adverse impact on the safety of transportation.

Realizing this importance of meteorological elements in the transportation sector, it is worthwhile to examine and analyse the standard climatological pattern over India pertaining to these hazards. Such patterns, would serve as a basis in order to plan various transportation activities, additionally considering the changes for a particular epoch of time. With this background, climatology of following meteorological elements/natural hazards based on various publications of Indian Meteorological Department, has been prepared and presented in the paper.

(1) Heavy Rainfall
(2) Heavy Snowfall
(3) Landslides
(4) Fog

Data and Methodology

The data on the number of occasions of landslides/snowfall and number days affected due to heavy rains/landslides/snowfall for various relevant states in the northen/north eastern parts of India are extracted for the period of 21 years (1981-2001) from the Disaster Weather Events published annually by India Meteorological Department. Also, the data of number of fog days over nine stations in north/northeastern India were obtained from National Data Centre, IMD Pune.

Based on these data the mean frequencies of landslides, heavy rains (≥12.5 cm) and snowfall during the study period are computed and analysed. Also inter annual variability of these events has been studied.

Results and Discussion

Mean Climatological Frequencies

Fig. 10.1 depicts the mean frequency of landslides, snowfall and heavy rainfall for the period 1981-2001. It is observed that number of occasions of snowfall is more in Jammu and Kashmir and Himachal Pradesh as compared to number of occasions of landslides with frequency of one.

Inter Annual Variability

Table 10.1 depicts the years of highest frequencies of snowfall, landslides and number of days affected by heavy rainfall .

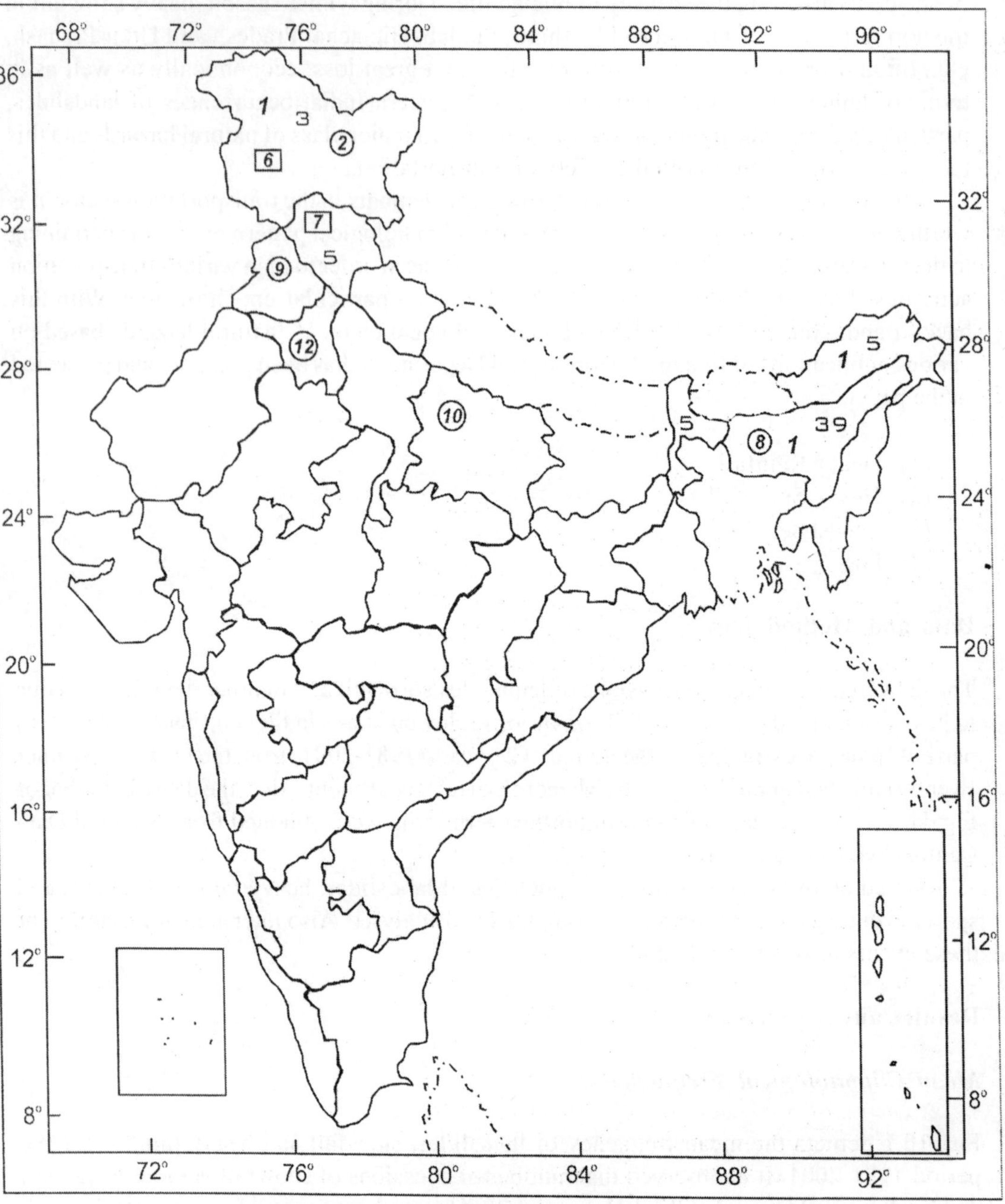

A Annual Frequency of Landslide

A No. of Days Affected Due to Heavy Rainfall

○ Annual Frequency of fog

□ Annual frequency of Snowfall

Fig. 10.1: Mean Annual Frequency of Weather Related Natural Hazards (1981-2001)

Table 10.1: Year and frequency of Snowfall, landslides and heavy rainfalls in North and Erastern States

Name of the State	Events		
	Year and Frequency of Snowfall	*Year and Frequency of Landslides*	*Year and No. of days offected by heavy rainfall*
Jammu and Kashmir	1982(34), 1991(14)	1981(04), 1983(02), 1986(02)	1996(11), 1993(9) 1987 (9)
Himachal Pradesh	1990(21),1991(19)	1985(03) 2000(03), 2001(03)	2000(20),1990 (14)
Assam	--	2000(3)1998(02), 2001(02)	1998(75)2000(68)
Arunachal Pradesh	--	1981-90 Nil except 1983(01)	1990(21),1982(18)

(a) *Landslides :* From the Fig. 10.2(a), it is observed that the highest frequency of the landslides during the study period occurred over Jammu and Kashmir during the period 1981 followed by in the year 1983 and 1986. In Himachal Pradesh the highest frequency is observed in the years 1985, 2000, 2001. During the years 1987 to 1990, no landslides were reported in the subdivisions under study except in Himachal Pradesh. It is also seen that there has been a general increase in the events of landslides during the recent years, i.e. 1998-2001. This may be attributed to increasing anthropogenic influence being one of the major reasons.

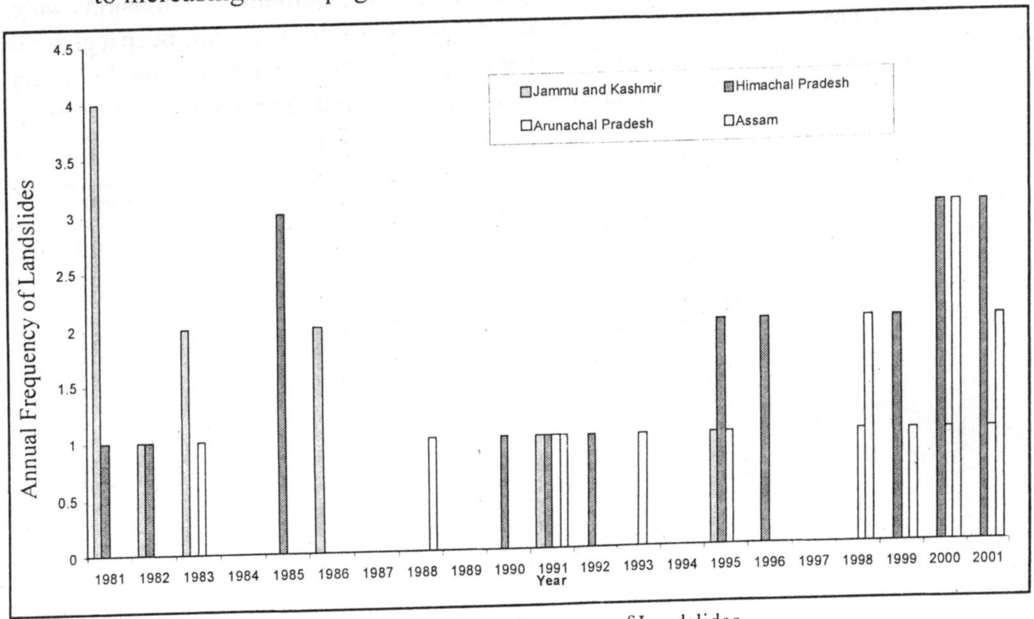

Fig. 10.2 (a): Annual Frequency of Landslides

(b) *Snowfall :* Fig. 10.2(b) depicts the inter annual variability of the events of annual snowfall over Jammu and Kashmir and Himachal Pradesh. It is observed that the highest frequency of the snowfall occurred over Jammu and Kashmir in the year 1982 followed by in the year 1991and in Himachal Pradesh in 1990 followed by in

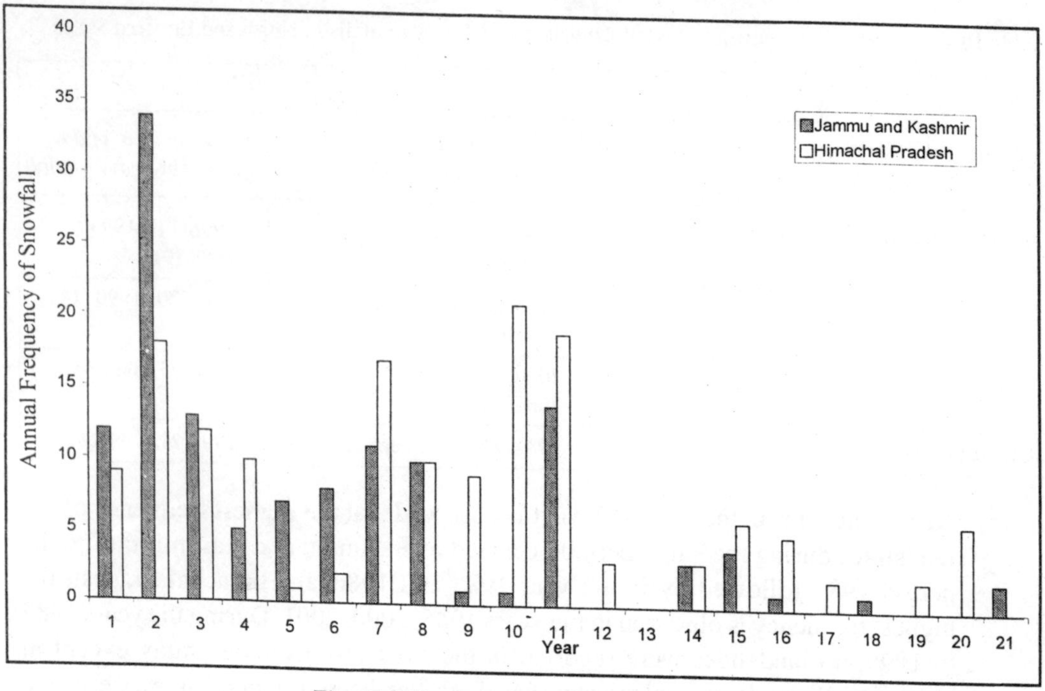

Fig. 10.2 (b): Annual Frequency of Snowfall

the year 1991. It is to be mentioned that during 1991 both the sub-divisions were affected by snowfall for a long period. It is also seen that there has been a general decrease in the events of snowfall over both of the subdivisions during the recent period (1992-2001) as compared to that during the initial part of the study (1981-1991).

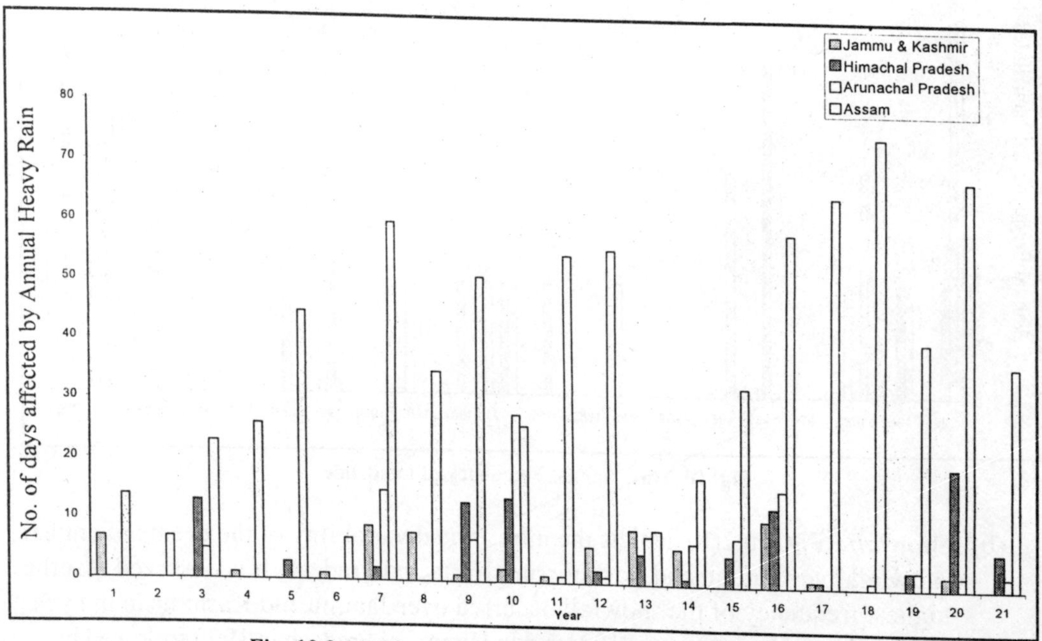

Fig. 10.2 (c): No. of days affected by Annual Heavy Rain

(c) *Heavy rainfall :* Fig. 10.2(c) depicts the inter annual variability of the events of annual heavy rainfall over Arunachal Pradesh, Assam, Jammu and Kashmir and Himachal Pradesh. It is observed that the highest number of days affected by the heavy rainfall occurred in Assam in 1998 followed by in 2000 whereas in Arunachal Pradesh the highest frequency occurred in 1990 followed by the year 1996. In Himachal Pradesh, highest frequency and the number of days affected by the heavy rainfall occurred in the year 2000 followed by in year1991. It is also seen that there has been a general decrease in the events of rainfall over all the subdivisions during the period compared to that during the initial part of the study.

Table 10.2: Year and number of fog days in North and Eastern States

Sr. No.	*Name of the Station*	*Year and No. of fog days*
1.	Agartala	2000(22), 1998(20)
2.	Gangtok	1996(13), 1984(12)
3.	Gulmarg	1982(10), 1981(09)
4.	Delhi	1997(30), 1996(26)
5.	Ambala	2001(14), 1999(14),1997(12)
6.	Amritsar	2001(20), 1986(15)
7.	Allahabad	1998(20) 1983(17)
8.	Gorkhpur	1990(26),1983(18)
9.	Guwahati	1990 (16), 1986(15)

(d) *Fog :* It is observed that the frequency of number of fog days is the highest in the month of January for all the stations and years under study (Table 10.2). It is also seen that in North eastern India, the highest frequency of fog days over Agartala is observed in the year 2000 followed by the year 1998, whereas in Ncrthen India, Delhi has the highest frequency in 1997 followed by 1996.

Conclusions

The following conclusions emerge from the study :

(1) The states of Himachal Pradesh and Assam are highly prone to the occasions of landslides. There has been a general increase in the number occurrences of landslides in recent years (1999 to 2001), which may be attributed to the increasing anthropogenic influences.

(2) There is a general decrease in the annual frequency of snowfall over Jammu and Kashmir and Himachal Pradesh in recent years.

(3) For the stations in Northern and Northeastern India, the frequency of number of fog days is the highest in the month of January.

4) During the period 1981 to 2001, the highest mean number of landslides is 1 for Jammu and Kashmir, snowfall are 7 for Assam and Meghalaya and highest mean number of fog days are 12 for Delhi. The highest mean number of days affected by heavy rain in Assam are 39.

Acknowledgements

The authors are grateful to Dy. Dir. General of Meteorology (WF) for providing facilities for completion of paper and to National Data Centre O/o Additional Director General of Meteorology (Research), Pune for supplying necessary data.

REFERENCES

De, U.S. and Joshi K.S., 1998, Natural disaster and their impacts on developing countries, *WMO Bulletin* 47 (4),336-343.

Khole Medha and De, U.S., 2001, Socio-economic impacts of natural disasters, *WMO Bulletin* 50 (1), 35-40.

Obsassi, G.O.P. 2001, WMO's Contribution to the development of Meteorology, Keynote lecture on the occasion of Sir Gilbert Walker Memorial Lecture and Award Ceremony, New Delhi, India, 20 March 2001.

11

SAFETY AND HAZARDS IN AMMONIA HANDLING, STORAGE AND TRANSPORTATION

I.D. Mall, V.C. Srivastava, A.K. Sahu and Bhawna Singh*

ABSTRACT

The paper presents a critical appraisal characteristics of ammonia; hazards of ammonia; risk and hazard associated with ammonia manufacture, storage, handling and transportation; and safety measures to be taken in handling and transportation of ammonia.

Introduction

Modern life is made possible by carefully controlled use of potentially deadly chemicals, fuels, solvents, purifiers, cleaners, pesticides, herbicides etc. Ammonia is one of the most versatile inorganic chemical which is not only used as major feed stock in fertilizer industry but also finds application in various organic and inorganic chemical industries. Total ammonia production in India and world is 13.5 and 1090 million tonnes, respectively[1]. Ammonia and its products as fertilizers are responsible for the green revolution. Some of the major applications of ammonia in chemical industry are: in the manufacture of nitric acid, pesticides explosives (TNT), ammonium nitrate, picric acid, teryl (RDX), acrylonitrile, resins (TDI), melamine formaldehyde, urea formaldehyde, methyl methacrylate (nylons), dyes and intermediates, adiponitrile, ethylene dichloride, hexamethylene diamine, cyclodedocanone oxime, hydrogen cyanide etc., pestivides (monouran, benomyl, picloram, captan, oxadiazon, orthone, glyphosphate, dimethoate, triflurulin); in the extraction of metals, as corrosion inhibitor in refinery and petrochemicals; and as industrial refrigerant[2,3].

Ammonia is a colourless, flammable gas with a pungent, suffocating odour. Its odour is perceptible even at very small concentration of 4 ppm. With rising temperature ammonia expands rapidly, increasing the internal pressure in vessels and pipes. Liquid anhydrous ammonia expands 850 times when released to ambient air and can form large vapour clouds.

*All at Department of Chemical Engineering, Indian Institute of Technology, Roorkee-247667, INDIA

These clouds are normally lighter than air and will rise. However, liquid anhydrous ammonia may also form an aerosol (small liquid droplets) which are heavy and sink towards ground.

Because of their widespread use, ammonia and its products are involved in many hazardous incidents. Major sources of ammonia leakages are either because of equipment failure or unsafe handling during transportation. Some of major equipments responsible for ammonia leakages are piping system, storage tanks, valves, compressors, urea synthesis section, and refrigeration systems. The US Occupational Safety and Health Administration (OSHA) have classified anhydrous ammonia as a hazardous material[4]. The Environmental Protection Agency (EPA)[5] has also classified it as an Extremely Hazardous Substance (EHS). In India, as per the Manufacture Storage and Import of Hazardous Chemical Rules, 1989 and as amended to date, ammonia has been classified as hazardous and toxic chemical[6].

Transportation, storage and handling of ammonia triggers serious safety, and environmental regulatory requirements for risk management plans, accident prevention programmes, emergency response plans and release analysis. Risk and hazard associated with ammonia manufacture, storage, handling and transportation can be classified as explosions, fire, physical injury like burns etc. due to physical contact with liquid ammonia, chemical poisoning due to long and short-term exposures to ammonia gas during leakages etc. Leakages of ammonia during storage, processing and transportation can be avoided and accidents can be minimized through proper safety precaution, good house keeping measures. A sound safety culture must be established throughout the organization and the most appropriate tools of safety assurance may be used at all stages in design, operation, handling and transportation.

The paper presents a critical appraisal characteristics of ammonia; hazards of ammonia; risk and hazard associated with ammonia manufacture, storage, handling and transportation; and safety measures to be taken in handling and transportation of ammonia.

Ammonia Production

Anhydrous ammonia is produced in about 80 countries. In 1974, the developing countries accounted for 27 per cent of ammonia capacity; however, today their share is more than 52 per cent. Total production of ammonia worldwide was 109.11 million tones in 2003: West Europe (8.94%), Central Europe (4.06%), East Europe and Central Asia (14.13%), North America (11.25%), Latin America (5.98%), Africa (1.03%), West Asia (Middle East) (7.37%), Asia (46.39%), Oceania (0.84%)[1].

Applications

Approximately 85 per cent of the world's ammonia production is used for fertilizer. Most of it is converted to solid form, i.e. into urea, ammonium phosphate, ammonium sulphate or ammonium nitrate[2,3]. Ammonia is a most useful chemical intermediate and has many important industrial applications in refrigeration, metals extraction, effluent treatment, hydrogen cyanide and detergents. It is also used in the fibres and plastics industry for the production of caprolactum, acrylonitrile, hexamethylene diamine and melamine.

Characteristics

Ammonia, NH_3, is a colourless alkaline gas, is lighter than air and possesses a pungent, suffocating odour. It is highly water-soluble and is soluble in chloroform and ether. It is easily liquefied under pressure. Characteristics of ammonia are given in Table 11.1[2-4, 7-9]. Comparison of ranking of ammonia with other toxic gases is given in Table 11.2[10].

Table 11.1: Characteristics of Ammonia

State: Colourless alkaline gas. **Odour:** Pungent and suffocating, **Taste:** Bitter. **Corrosivity:** corrosive. **B.P:** -33.35 °C. **M.P:** -77.7 °C. **Sp gravity:** 0.680 at -33°C. **Vapour pressure at 15°C:** 630 KPa. **Auto ignition temperature:** 651°C. **Flammable limits:** LEL 15 per cent, UEL 28 per cent (in air 15-18 per cent by volume). **Odour threshold:** 0.6 - 53 ppm (detection). **Solubility in water:** Very highly.

Short-term exposure limit (STEL): 35ppm; 24 mg/m³. **Fatal dose:** 5000 ppm. **Time-weighted-average (TWA) exposure limit:** 17 mg/m³ (25 ppm).

National Institute for Occupational Safety and Health (NOISH) REL: 25 ppm TWA; 35 ppm STEL. **American Conference of Governmental Industrial Hygienists (ACGIH):** 25 ppm TLV, 35 ppm STEL.

OSHA permissible exposure limit: 35 ppm STEL, 300 ppm DLH. **Reactivity:** Ammonia is alkaline and caustic, and is a powerful irritant. It is incompatible or reactive with strong Oxidisers, acids, halogens, salts of silver and zinc. It is corrosive to copper and galvanised surfaces. Liquid ammonia attacks some forms of plastics, rubber and coatings.

Warning properties: Odour and dense appearance of gas are distinctive warning properties. **Fire extinguishing materials:** Water spray, CO_2, Foam, Halon, Dry Chemical, any "ABC" class.

Table 11.2: Comparison of Ranking of Ammonia with Other Toxic Gases

Factor	*Ammonia*	*Chlorine*	*Mustard gas*	*Dioxin*
Dispersive energy	2	2	0	0
Persistence	1	1	2	2
Industrial Significance	4	4	0	0
Mode of Attack*	2	2	1	1
Inherent toxicity**	0	4	8	8
Injury of Fatality ratio	1	1	2	2
Long term effects	0	0	2	2
Total	10	14	15	15

*Based on the assumption that attack through the lungs is more likely to kill before medical treatment can be brought to bear than would ingestion.

**Inherent toxicity is given a maximum weighting of 8, industrial significance a weighting of 4, and rest of the factors 2. Within each factor the ranking is in the ratio of 2:1.

The concept of mortality index was introduced by Marshall[10] as a means of determining the potential for fatality in the chemical and process industries. This index was the mean number of the fatalities per ton resulting from the realization of the potential of given hazards. Based on major hazards the mortality index for ammonia works out at 0.02. For two store tanks and the ship storage connection it works out at 0.052[10].

Storage of Ammonia

Material of construction for storage of ammonia mainly depends on the operating temperature. Mild Steel may be used at ambient temperature[11], however, special steels are necessary at low temperatures to avoid embrittlement. Carbon steel has been successfully used for many years in both refrigeration and agricultural ammonia service[12]. However, after the use of anhydrous ammonia as a fertilizer in soils, stress corrosion failure of carbon steel storage and transport vessels have occurred. Only wrought iron or steel may be used for ammonia containers, piping, fitting and equipment[13]. Tanks, piping and electrical equipment must be bonded and grounded. Piping must be schedule 40 with welded joints and welded fittings or schedule 80 with either threaded or welded joints and welded fittings. Aluminium is highly resistant to dry ammonia gas even at elevated temperatures, however, when ammonia is moist or in solution as ammonia hydroxide, the rate of attack is low for all concentrations at temperature up to 120 °F. Due to this property, aluminium equipments have given excellent service in refrigerating systems handling liquid ammonia containing up to 50 per cent water. Nickel is not attacked by anhydrous ammonia and is resistant to aqueous ammonia or ammonium hydroxide in concentration up to 1 per cent. Nickel is not affected by mixtures of ammonia and air encountered in ammonia oxidation processes and has been used for piping carrying the mixed gases[12]. Nickel based alloys are resistant to anhydrous ammonia. Nickel-Chromium alloys perform well in pure ammonia atmosphere.

Storage Cylinder: Ammonia pressure storage vessels are usually horizontal cylinder of small capacities. An ammonia pressure storage cylinder should be designed for a pressure of at least 15.5 bars[11]. Cylinders must not be filled to more than 85 per cent of capacity by tare weight as marked on the cylinder by the manufacturer[14]. Cylinders must be protected from excessive heat. Storage areas must be kept free from all combustible material. Storage of ammonia cylinders near cylinders of hydrogen, acetylene, fuel gases, ether, turpentine, hydrocarbons, organic matter, or finely-divided metals should be avoided.

Spherical Vessels: Spherical vessels are used for larger capacities as compared to cylindrical vessels, 4000 tonnes. Operating conditions are 3 - 6 kg/cm^2 and –3 to 5 °C[15]. It is designed to withstand the stresses imposed when the vessel is full and the pressure of ammonia at the maximum design temperature plus the partial pressure of any inert gases, which may be present[11].

Refrigerated Storage: The refrigerated storage tanks are useful for higher capacities of 5000-45000 tonnes. The operating temperature is –33°C and pressure is atmospheric[15]. To maintain atmospheric pressure, in the ammonia tank corresponding to a temperature of –33 °C the gas vapourising in the storage tanks are to be withdrawn.

Handling

Similar to all chemicals, getting ammonia *on you* or *in you* should be avoided. Breathing the gas or sprays or mists generated by ammonia should be avoided. While handling ammonia eating or drinking is very dangerous to health. All work practices should minimize the release of ammonia. Awareness should be there for any signs of effects of exposure, exposures to fatal concentrations of ammonia could occur rapidly[4].

All employees who handle this material should be trained to handle it safely and without causing overexposure. Ammonia emergency equipment should be available near the point of use. Workers who handle it should wear protective clothing. Instant-acting showers should be available in the event of emergency, special eye-wash fountains or similar equipment should be available for eye irritation, proper respiratory protection equipment must be provided, and workers using such equipment must be carefully trained in its operation and limitations.

Cool, dry location, away from direct sunlight, sources of intense heat, or freezing availability are the factors, which are considered in the proper storing of ammonia. Only compatible materials are used for cylinders, process lines, and other ammonia-handling equipment. Lines should be purged with dry nitrogen both before and after maintenance activity. Copper, tin, zinc and their alloys are not suitable materials for use with ammonia. Cylinders should be kept away from incompatible material. Periodic inspections of process equipment by knowledgeable persons should be made to ensure that the equipment is used appropriately and the system is kept in suitable operating condition.

Special precautions need to be taken into account while handling gas cylinders. Cylinders should be protected against physical damage. Temperature of the area where cylinders are stored should not be allowed to exceed 52°C (125°F). There should not be any tampering with pressure relief devices in valves and cylinders. Electrical equipment should be non-sparking or explosion proof. During use, designated regulators, CGA fittings, and other support equipment must be applied. Use of oil or grease on gas handling fittings or equipment must be prohibited. All equipment must be properly grounded and bonded. Ammonia cylinders should never be directly connected to a vessel containing a liquid since suck-back may occur causing a violent reaction within the cylinder. To prevent suck-back, a trap, check valve or vacuum break should be inserted into the line. The trap should be of adequate size to take the total liquid volume sucked-back. Improving air circulation about the container or can increase the rate of gas flow by increasing the temperature of the room if it is below normal. Cylinders are not manifolded to increase output unless check valves have been inserted at the cylinder outputs to prevent exchange of material from one cylinder to another, causing a cylinder to become over-full.

If leaks develop in ammonia lines, those must be given prompt attention because they will become progressively worse. Containers are kept tightly-closed when not in use. Replace valve protection cap (where provided), mark empty cylinders "EMPTY"[4].

Transportations

All anhydrous tank wagons must be securely attached to the towing vehicle by a drawbar, hitch pin and safety clip, and suitable safety chains[16]. These parts should be checked to see that they are secure before each trip on the highway. It is highly recommended that the vehicle used to tow the nurse tank be at least equal in weight to the gross weight of the nurse tank. This will assist the operator in maintaining control of all the vehicles, at 25 mile/h and minimize the potential for an accident.

Regulations require: An SMV (Slow Moving Vehicle) emblem should be attached to the rear of the vehicle and visible from at least 500 feet. The words Anhydrous Ammonia (4 inch high) in contrasting colours should be placed on both sides and the rear of the tank[14].

The words Inhalation Hazard (3 inch high) in association with the anhydrous ammonia label must be on both sides of the tank.

Hazards

After a large release of chlorine, a release of ammonia is potentially one of the worst toxic hazard from the common bulk chemicals in the chemical industries. A concentration of 1700 ppm can be fatal for a half hour exposure. Its odour is detected by most people at about 5 ppm[11]. The hazards which are encountered in an ammonia plant depend somewhat on the process, including operating condition temperature, pressure etc.

Fire

Ammonia gas has an explosive range of 16 to 25 per cent by volume in air. Although it is classified as non-flammable under WHMIS (Workplace Hazardous Materials Information System) and its flammability potential is only slightly less than that of some gases that does meet the official flammability criteria.

Explosion

Contact of ammonia with certain chemicals including fluorine, chlorine, bromine, iodine, mercury and silver oxide can create explosive compounds. Moist ammonia vigorously attacks silver, copper, źinc and many of their alloys. Aluminium is attacked to a lesser extent. Iron and steel are inert to ammonia.

Health

Anhydrous ammonia is very corrosive, and exposure to it may result in burns to skin, eyes and respiratory tract. Ammonia produces intense congestion of the upper respiratory passages and immediate death from laryngeal spasm or oedema. For ammonia, there is a considerable amount of experimental data on animal; for mice L50 of 10150 ppm for a 10 min exposure and L50 of 4230 ppm for 60 min exposure[17]. Ammonia is hygroscopic; it migrates to moist areas like eyes, nose, mouth, throat and moist skin. For ammonia, the TWA exposure value is 17 mg/m^3 (25 ppm). The STEL value is 24 mg/m^3 (35 ppm). The OSHA has set a Permissible Exposure Limit (PEL) for the workplace of 50 ppm (35 mg/m^3) TWA. The Immediately Dangerous to Life and Health (IDLH) level has been set at 300 ppm[13]. Following symptoms are observed according to the exposed ammonia concentration:

Concentration	*Symptoms*
0.6 - 53 ppm	Odour threshold
25 - 50 ppm	Irritation of the eyes and mucous membranes, which can be tolerated for several hours
100 - 150 ppm	Immediate irritation of the throat, may be tolerable for an hour
400 - 700 ppm	Immediate, severe irritation of the respiratory system and eyes
> 5000 ppm	Result in rapid death due to suffocation or fluid in the lungs

Common Routes of Ammonia Exposure[7,9]

Inhalation

The most common way for ammonia to enter the body is through the respiratory system. Inhalation of ammonia vapours may lead to irritation of the nose and throat. Exposures to high concentration of ammonia gas can lead to coughing, laboured breathing, sore throat and in some instances, chemical pneumonitis and pulmonary odema. High concentrations of ammonia may cause unconsciousness and even death.

Contact with Skin or Eyes

Contact of ammonia (liquid/mist/gas) with the skin can lead to severe burns or dermatitis (red, cracked, irritated skin). Contact with undiluted liquid will cause frostbite, ulceration of the skin. Contact with the eyes causes pain, redness and prolonged exposure could cause blindness.

Ingestion

Ingestion of ammonia can damage the tissues of the mouth, throat, oesophagus and other tissues of digestive system. Ingestion of ammonia content can be fatal.

Environment

Nitrogen is essential for all life forms and ammonia is one of the many forms in which nitrogen exists in the environment. Ammonia is transported via numerous pathways in air, water, soil, plants and animals. This occurs by both direct transport and through the numerous inter-conversions between inorganic (such as nitrate and nitrite) and organic (such as proteins) forms. However, at unusually high levels, acute toxic effects can be observed which include the death of animals, birds or fish, and death or low growth rate in plants. Ammonia is more toxic under conditions of decreased oxygen concentrations. Chronic toxic effects may cause reproductive problems, lower fertility and changes in appearance or behaviour.

Ammonia Releases and Accidents

The three cause levels of any accident are represented by Fig. 9.1. Accidental release of ammonia can create a dangerous situation for both the handler and bystanders. Accidental ammonia release can occur due to overfilling of tank; handling the hose by the valve handle or hand wheel; due to faulty hitch pin or weakened tongue, weakened undercarriage structure; moving the tank before disconnecting the hose; faulty valves and deteriorated or out-of-date hoses; failure to bleed pressurized NH_3 from the hose before connecting or disconnecting, overturning an applicator tank; external overheating of the storage container etc.[14]. Accidental ammonia releases cause injuries and death to employees, emergency response personnel and people in surrounding communities.

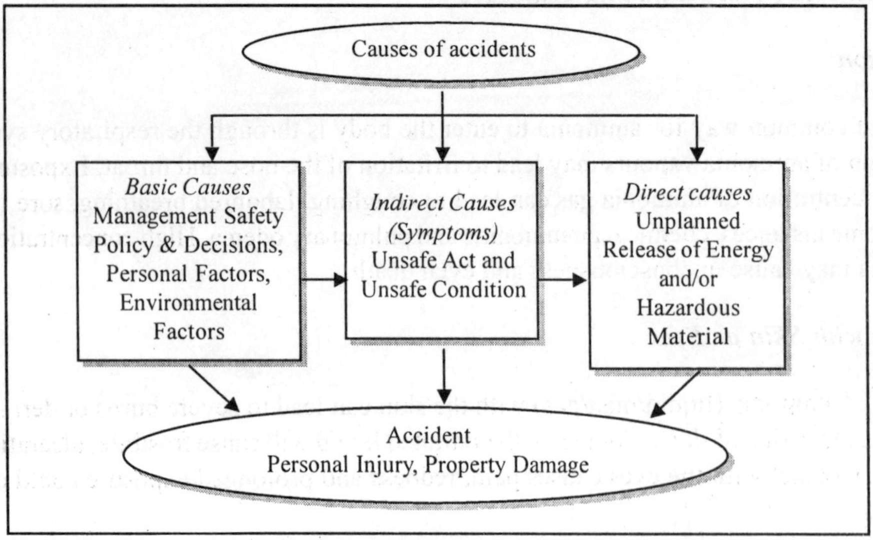

Fig. 11.1: Cause levels of accidents

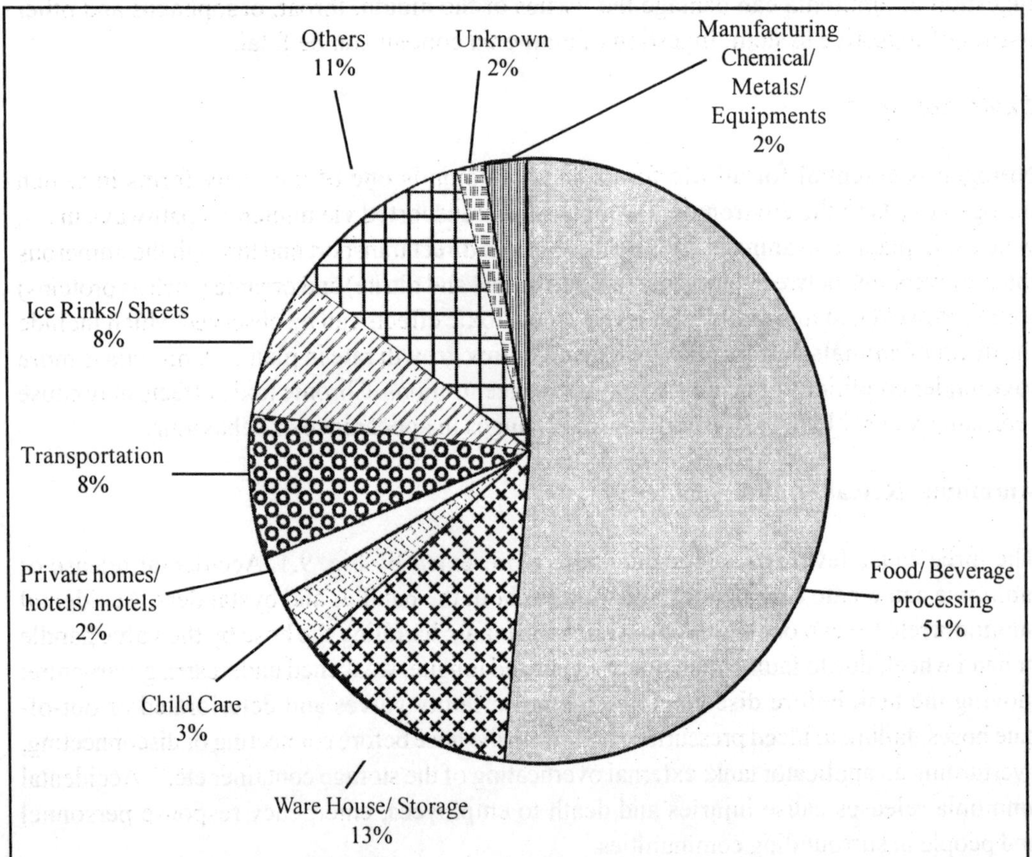

Fig. 11.2: Injured people following accidental ammonia releases[19]

In 1984, an explosion at a cold storage warehouse at Iowa, Kansas, Missouri and Nebraska in U.S.A. caused death of a firefighter and another seriously injury. In 1992, an explosion caused extensively damage due to rupturing of a refrigeration pipe struck by a forklift at a meet packaging plant. At North Carolina, 82 ammonia releases occurred between 1993-1997, most of the spills resulted from unintended releases from process vessels, storage above ground, equipment failure and human error[5, 14, 18]. An ammonia release at Cartegena, Columbia, in 1977, is reported to have caused a death toll of 21. In 1973, 18 people were killed at Potchefstroom, South Africa in an ammonia storage tank failure[11]. Fig. 9.2 is a typical statistical display of injured people following accidental ammonia releases in New York state of U.S.A.[19]

Safety Measures

The General Safety Measures Recommended are:[4, 13, 15, 20]

- Always use safety equipment, and check and recheck the condition of the equipment at the field and control room.
- Good work practices minimize the release of ammonia.
- Do not eat or drink while handling chemicals.
- Store containers in cool, dry location, away from direct sunlight, sources of intense heat, or where freezing is possible.
- Use only compatible materials for cylinders, process lines and other ammonia handling equipments.
- Workers must be thoroughly trained to handle ammonia without overexposure.
- Ammonia emergency equipment should be available near the point of use.
- Workers who handle ammonia should wear protective clothing.
- Hoisting of ammonia cylinders is not recommended.
- Never mix ammonia with other gases in the cylinder.
- Use only ASME code containers designed for storage.
- Tank cars carrying ammonia should be loaded and unloaded in strict accordance with tank-car manufactures recommendations and all established on-site safety procedures.
- Make certain that application equipment is locked and tagged-out safely.
- Purge gas handling equipments with inert gas before attempting repairs.
- Automatic monitoring devices for ammonia and oxygen should be installed.
- Portable fire extinguishers must be available for extinguishing fires in electrical motors.

Steps for Protection and Exposure Control During Emergency[4, 13]

- Trained personnel using pre-planned procedure should respond to uncontrolled releases. In case of large release, clear the affected area, protect people and respond with trained personnel.
- Protection of all personnel and the area.

- Use supplied air respirations protection during emergency response procedure to release.
- For eye protection, use splash goggles or safety glasses and face shield.
- Use triple-gloves (neoprene rubber gloves and nitrile gloves, over latex or N-DeXTM gloves), fully encapsulating chemical resistant suit and boots, hard hat, and self contained breathing apparatus for spill response.
- Rescuers should not attempt to retrieve victims of exposure to ammonia without adequate personal protective equipment.
- Supply maximum possible ventilation with explosion-proof equipment.
- Attempt to close the main source valve prior to entering the area.
- A water fog or mist can be used to control the vapour cloud resulting from the release.

First-aid Measures[2, 4]

- Remove victim to fresh air as quickly as possible.
- Trained personnel should administer supplemental oxygen and/or cardio-pulmonary resuscitation, if necessary.
- In case of frost-bite, place the frost bitten part in warm water or wrap the effected parts gently in blanket.
- Encourage victim to gently exercise the affected part while being warmed.
- If ammonia contaminates the skin, immediately begin decontamination with running water.
- Remove exposed or contaminated clothing, taking care not to contaminate eyes.
- If liquid is slashed into eyes, or if irritation of eyes develops, open victim's eyes while under gentle running water.
- Seek immediate medical attention.
- Precautions must always be taken to prevent suck-back of foreign materials into the cylinder by using a check valve, vacuum break, or trap, since suck-back may cause dangerous pressure changes within the cylinder.

Management must be co-operative and interactive on the issue of safety. It should insure that provision is made in company procedures and project programme for the necessary safety analysis. Study of historical release incidents play an important part in the development of a good safety culture as they provide routes and the corrective measures taken after any hazard that has occurred.

Disaster management plan for ammonia handling and processing is given in Fig. 9.3. The plan to be divided into two parts; on-site emergency plan and off-site emergency plan. In case of incidents, which might affect people and the environment inside the works only, on-site emergency plan is referred. Off-site emergency plan is referred in incidents that affect the people and the environment outside the work place only.

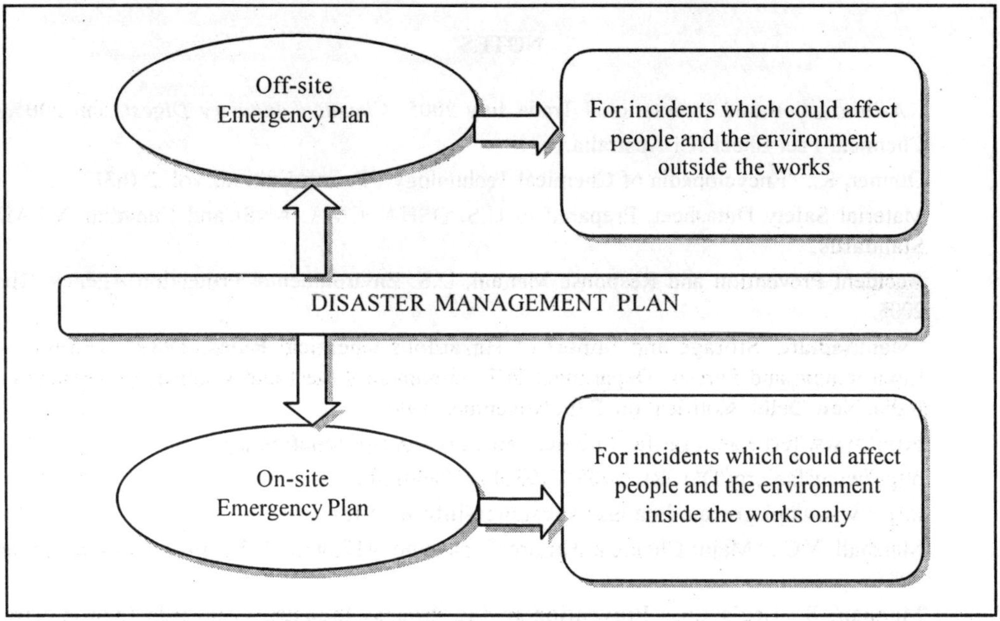

Fig. 11.3: Disaster management plan for ammonia handling and processing

Safety Appliances and Awareness

It is evident that there is always room for improvements in all aspects of safety. Each employee should be trained about the hazards of ammonia and importance of use of the safety appliances. Explosion and rupture incidents should be reviewed and then the exposure in their own plants should be assessed. Regular in-house safety bulletins are very useful in mitigation of accidents and increasing the awareness amongst the people. The safety bulletin may include hazards alert posters, signs and stickers, accident alerts, work safe bulletins, start safety tips, incident bulletins, constructive ideas, recall and product safety notice, investigation bulletin and preventive bulletin etc.

Conclusion

Ammonia, an intermediate in many chemical industries and refrigerant, is one of the most hazardous chemicals if not handled properly. A sound safety culture must first be established throughout the organization and use made of the most appropriate tools of safety assurance at all stages in design, operation, handling and transportation of ammonia. It can be handled safely when one follow proper procedures by use of proper safety and protective appliances and first aid. Although it has been never possible to remove risk from human activity, however, we can take steps to control the risks associated with the handling of ammonia at various steps of its use. Let us remember that "safety first" is "safety always" and remarks of William Shakespeare "Out of this nettle, danger, we pluck this flower, safety".

NOTES

1. IFA-Production and International Trade July 2005, *Chemical Industry Digest,* Jan 2005.
2. Chemical Fact Sheet ICI Australia.
3. Othmer, K., "Encyclopedia of Chemical Technology", Fourth Edition, Vol. 2 (638).
4. Material Safety Datasheet, Prepared to U.S. OSHA, CMA, ANSI and Canadian WHMIS Standards.
5. Accident Prevention and Response Manual, U.S. Environmental Protection Agency, April 2005.
6. "Manufacture, Storage and Import of Hazardous Chemical Rules, 1989". Ministry of Environment and Forests, Department of Environment, Forest and Wildlife, Governemnt of India, New Delhi, Notification, 27th November, 1989.
7. http://www.fertilizer.org/ifa/statistics/indicators/ind-production.asp
8. http://www.fao.org//DOCREP/005/Y1936E/y19360f.htm
9. http://www.npi.gov.au/database/substance-info/profiles
10. Marshall, V.C., "Major Chemical Hazards", page no. 417, 433, 325, Ellis Horwood Limited, 1987.
11. Mannan, S., Lee's Loss Prevention in the Process Industries Hazards Identification, Assessment and Control, Third Edition, Elsevier Butterworth-Heinemann, 2005.
12. Laque, F.L., Copson, H.R., "Corrosion Resistance of Metals and Alloys". Second Edition, The International Nickel Company, New York.
13. Health and Safety Guidelines, Engineering Data Sheet 4-04: Anhydrous Ammonia, May 1996.
14. Lawrence, T.J., Carpenter, T.G., Bean, T.L., *Safe Handling of Anhydrous Ammonium,* Ohio State University.
15. Mittal, M., "Storage and Handling of Ammonia and Safety Measures", *Chemical Industry Digest,* 3rd Quarter '97.
16. Shutske, J.M. 19 *Using Anhydrous Safety on the Farm*, 2005.
17. Carol J. Lehtola, Charles M. Brown, and William J. Becker, *Storage and Handling of Anhydrous Ammonia—OSHA Standards 1910.*111.
18. Occupational Illness & Injuries, NC Department of Health And Human Services Hazardous Substances Emergency Events Surveillance Program, http://www.dhhs.state.nc.us/dph
19. http://www.health.state.ny.us/nysdoh/environ/hsees/ammonia.htm
20. Williams, G.P., "Safety performance in ammonia plants: survey VI", *Process Safety Progress,* Vol.18, No. 1.

12

SAFETY MEASURES—AUTOMOTIVE VEHICLES
Standards and Practices

R.R. Singh*

ABSTRACT

This paper deals with the role vehicle standards play in helping road safety, with particular reference to India. Highlighting the major standards relating to transportation system, including product standards, methods of tests, code of practices terminology, dimensions, symbols, sampling and grading, it describes vehicle safety features, both active and passive, pertaining to vehicle design, tyres, brakes, anti-lock braking system, windows, mirrors, wipers and demisters, lights, audible warning devices, steering wheel, air-conditioning, safety belts, seats and so on.

What is a Standard?

A Standard is a minimum level of quality, performance, safety and reliability. If something complies with a Standard accepted in India, you can be confident it will work.

Each Standard is developed, maintained and regularly reviewed by a technical committee whose members represent the different interest groups that will be affected by it. For example, a technical committee developing a Standard for a vehicle part may have representatives of the vehicle manufacturing industry, trade organizations, consumer groups, researchers, government organizations and other interest groups.

In general, vehicle Standards lag slightly behind the latest technological innovations. This is because the experts have to be sure the technology works and is practical for manufacturers to produce on a large scale.

How do Standards Help Road Safety?

As Standards are introduced and vehicle manufacturers are required to comply with them, the cars on our roads become better at protecting the people inside them.

---

*Deputy Director, Transport Engineering Department, Bureau of Indian Standard, New Delhi-100 002

New cars comply with new or updated vehicle safety tandards and must meet many more tandards than in the past. This means that in general newer cars are safer than older cars. And many vehicle manufacturers make cars that comply with a particular tandard long before it becomes compulsory.

Examples of tandards that make a big contribution to vehicle safety are frontal impact, side impact and tyre standards.

Vehicle Standards

Vehicle tandards play an important part in helping road safety, but rarely get the recognition they deserve.

Vehicle Standards in India

In India Bureau of Indian t andards is actively involved in the formulation of standards for vehicle equipments and systems in association with Automotive Research Association of India, Central Institute for Road Transport, various manufacturers, users etc. A complete division council, i.e TRAN PORT ENGINEERING DIVI ION COUNCIL under the Chairmanship of h. B. Bhanot, Chairman ARAI is dedicated to the work of formulating standards related to transport system.

There are about 1000 standards relating to transportation system which include product standards, methods of tests, code of practices, terminology, dimensions, symbols, sampling and grading etc. in addition to this various I O standards are also recognized by BI under dual numbering system.

Recognizing International tandards and requiring vehicles to comply with them allows Indian consumers to enjoy the advantage of a wide selection of cars built to the highest a fety ta ndards in the world.

In Bureau of Indian tandards, commitee-wise programme of Transport Engineering Division on 1April, 2005 are as follows:

Committee-wise Programme of Work of Transport Engineering Division (TED) (As on 1 April 2005)

Sl.No.	*Committee Number*	*Title*
1.	TED 2	Automotive Prime-movers, Transmission and teering systems and Internal Combustion Engine
2.	TED 4	Automotive Braking ystems, Vehicle Testing and Performance Evaluation
3.	TED 6	Automotive Body, Chassis, Accessories and Garage Equipment
4.	TED 7	Automotive Tyres, Tubes and Rims
5.	TED 11	Automotive Electrical Equipment and Instruments
6.	TED 12	Freight Container and Pallets
7.	TED 14	Aircraft, pace Vehicles, Air Cargo Handling and Aircraft Electrical Equipment
8.	TED 16	Bicycles
9.	TED 17	hipbuilding
10.	TED 18	Inland, Harbour Crafts and Fishing Vessels
11.	TED 19	Marine Engineering and a fety Aids
12.	TED 21	Automotive p rings and u spension ystems
13.	TED 22	Transport Tractors, Trailers and Industrial Trucks
14.	TED 24	Transport Packages and Packaging Codes
15.	TED 26	Automotive Vehicles Running on Non-Conventional Energy o urces

The tests used to set Standards can be different in different countries. For example, frontal impact crash tests differ from country to country depending on the Standard. The American, Japanese and Australian Standards require the full width of the front of the car to be crashed into a solid barrier at a speed of 48 km/h (United States, Australia) or 50 km/h (Japan). This test is particularly effective for evaluating the performance of occupant restraints like seatbelts and air bags.

The European Standard requires an offset frontal impact crash test instead of a full frontal test. In an offset crash test, 40 per cent of the front of the car (on the driver's side) strikes a deformable barrier at 56 km/h. This test is used because offset frontal collisions are more common in real world crashes than full frontal collisions.

The tests are different but both sorts of test will result in a minimum setting that the vehicles must comply with. It is then up to manufacturers to decide how they will design their cars to comply with a frontal impact Standard.

Standards Don't Mean you can Take Road Safety for Granted

Although Standards cover many areas of car safety, some important aspects are not covered. These include vehicle size, weight and stability, and newer safety features like traction control. It may be some time before Standards catch up with the latest vehicle technology. It should be remembered that Standards do not represent the highest level of safety currently achievable, but the minimum or optimum that vehicles, their components and systems must meet. This means that not all new cars are created equal. Safer vehicles do not just meet Standards, they surpass them.

Vehicle Safety Features

Active and Passive Safety Features

Active safety features can help prevent crashes by providing the driver with better means of controlling the vehicle and avoiding hazards.

In the event of a crash, passive safety features are designed to prevent or minimize injury to the vehicle's occupants. Some features help absorb crash forces, some restrain occupants from colliding with the vehicle interior, and others prevent objects inside the vehicle from striking the driver or passengers.

Active safety	*Passive safety*
• Tyres	• Crush zone and safety cages
• Brakes	• Frontal impact protection
• Anti-lock Braking Systems (ABS)	• Side impact protection
• Traction control and four wheel drive	• Safety belts
• Handling and stability	• Air bags
• Audible warning devices	• Seats
• Seeing and being seen	• Head restraints
• Adjustability and comfort	• Safe vehicle interiors
• Minimizing driver distraction	• Load restraint
	• Fire prevention and protection

Bigger can be Better

All else being equal, a large new vehicle will protect occupants better than a small vehicle in a crash. Large 'crush zones' at either end of big vehicles make them better able to absorb crash energy. The crush zones and longer stopping distance help reduce deceleration forces reaching the occupant compartment and lessens the risk of occupants being injured.

Active Safety

Active safety features can help prevent crashes by providing the driver with better means of controlling the vehicle and avoiding hazards. Many vehicle features make an obvious contribution to active safety, such as well maintained tyres, brakes, lights and steering. Others are not so obvious, like seats and air-conditioning.

While active safety features can help reduce the chance of a crash, unfortunately not all crashes are avoidable.

Tyres

Tyres are one of a vehicle's most important active safety features. Whenever a vehicle accelerates, corners or brakes, it imposes forces on its tyres. Good tyres grip the road to allow a vehicle to stop, go and corner safely, which can be crucial in avoiding crashes.

Choosing the right tyre depends on the type of vehicle it will be fitted to and how the vehicle will be used. Tyres that may be suitable for a passenger car used for city driving could be unsuitable for a vehicle carrying heavy loads or used off-road.

Here are some tyre safety issues you should consider:

- e lect tyres with an approved standards mark. It is illegal to use or sell tyres that do not comply with an approved standard.
- Every tyre has a maximum speed rating. A tyre with a high speed rating will generally grip better and probably perform better at higher speeds than one with a low speed rating.
- A tyre has a load index that indicates its maximum carrying capacity. Ensure vehicles carrying heavy loads have tyres capable of safely bearing the weight.
- Check the tyre inflation pressure often to ensure it is correct. This will optimise the tyre's performance, load-bearing and wear and fuel economy.
- Tyres made of hard compounds usually last longer, but may not grip as well as softer tyres.
- Tyre specialists recommend all tyres on a vehicle should be of the same construction. Mixing tyre types on a vehicle compromises driving control. At the very least, the same type of tyre should be fitted to the same axle (front or rear).
- Having tyres of different construction or size fitted to the same axle is dangerous and illegal.
- Car manufacturers recommend that replacement tyres be the same type as those originally fitted, to maintain all-round driving performance.

- Ideally, tyres should be replaced in sets of four (or at least in pairs on the same axle) to maintain consistent vehicle handling and responses.
- If the size of a replacement tyre differs from that of the original tyre, ensure the new size is compatible with the wheel rim size and that adequate clearance exists in all conditions (including when snow chains may need to be fitted).

BIS has formulated following standards on tyres which are properly aligned with prevailing international norms:

1. *Automotive Vehicles - pneumatic tyres for two and three-wheeled motor vehicles - Specification,* [Supersedes IS 10914 (Part 1), IS 10914 (Part 4), IS 10914 (Part 5), IS 11157 and IS 12151].
2. *Automotive Vehicles - Pneumatic tyres for passenger car vehicles - Diagonal and Radial ply - Specification,* [Supersedes IS 10914 (Part 1), IS 10914 (Part 3), IS 10914 (Part 5), and IS 12441 (Part 1)].
3. *Automotive vehicles - Pneumatic tyres for commercial vehicles - Diagonal and Radial ply - Specification*, [Supersedes IS 10914 (Part 1), IS 10914 (Part 2) and IS 10914 (Part 5)].

Brakes

A good brake system has the following features:

- The vehicle should be able to stop in the shortest possible distance and still maintain stability.
- The vehicle can be easily controlled during braking.
- Braking performance should be consistent with brake pedal pressure.

Drum brakes offer very good braking on the rear axle, but are not as effective on the front axle where better heat dissipation is needed. When disc brakes are used on the rear axle, a drum handbrake is usually incorporated.

Disc brakes offer better performance than drum brakes during long braking, repeated braking, and when braking at high speed, because disc brakes are better able to dissipate the heat generated. Braking hard or frequently can create high temperatures where the brake surface and brake frictional material touch each other. The heat causes the frictional material to be less effective, reducing the brake's efficiency. This is called 'brake fade'. Fluid in the brake callipers can also boil under repeated braking, causing the brakes to fail altogether. The best disc brakes are ventilated—this cools the brakes and increases their resistance to brake fade.

Good brakes are essential for controlling a vehicle and so must be well maintained. Brake fluid may absorb water over time, which reduces its boiling temperature, so regular brake fluid replacement is essential. To help with maintenance, it is preferable to have brakes with an open design so that wear of the brake lining and pads can be easily checked. Any repairs or replacements must be made to the whole axle set, not just one end of the axle, otherwise braking may be uneven and unsafe. It is also necessary to regularly check the condition of brake hoses and replace them as necessary.

BI has formulated many standards on brake linings and braking systems such as:

1. I 2742 (Part 1 to Part 5) on brake linings,
2. I 11852 (Part 1 to Part 9), I 12821, 12831, 12851, 12884, 13207,13224,13225,13249,13363,13453,13603 are on brakes and braking systems.

Anti-lock Braking System (ABS)

The main safety benefit of AB is vehicle stability and steering during hard braking. With better control of the vehicle, the driver of an AB -e quipped vehicle has a better chance of avoiding a crash.

An AB -equipped vehicle has a computer with sensors that monitor wheel speed and detect when a wheel is about to lock up under hard braking. The computer sends a signal to valves that reduce and reapply brake pressure on that wheel several times a second, allowing the wheel to keep rolling. By preventing wheel lock-up, AB lets the driver brake hard and still retain steering control.

I 11852 (Part 9) is on Brakes and braking systems—Requirements for vehicles equipped with anti-lock braking systems.

Traction Control and Four-wheel Drive

o me new vehicles are equipped with a traction control system. These systems optimise grip and stability on the road during acceleration by measuring wheel rotation. If one wheel loses traction and starts spinning faster than the others, the system stops wheel-spin by reducing engine power or temporarily applying the brakes to that wheel. The wheel regains traction and the vehicle can accelerate smoothly, even on slippery surfaces. Limited slip differentials also help provide a more even distribution of traction forces when the vehicle is on a slippery surface.

Four-wheel drive can also enhance traction. The power of the engine is delivered to all the vehicle's wheels, so traction and steering forces are spread more evenly among the tyres. This improves grip and stability, particularly on surfaces two-wheel drive vehicles may not be able to handle safely.

Handling and Stability

A vehicle with good handling and stability is not only a pleasure to drive, it is also safer to drive. The vehicle holds the road and is more responsive to the driver's commands. This is achieved through a combination of the vehicle's steering, suspension, braking, acceleration, body characteristics and weight distribution.

Each manufacturer has its own system of improving vehicle stability, road-holding ability and handling performance. It is worth test-driving several vehicles to get a better idea of which one best offers you safe, efficient and comprehensive control on the road.

IS 11948 specifies the method of evaluation of steering effort of automotive vehicles and IS 5423 is specification on Shock absorbers for Automotive vehicles.

Seeing and Being Seen

A safe car should have features that enable the driver to see clearly and identify possible risks, and also allow other road users to see the vehicle and judge what it is doing. Some of these features are described below.

Windows

Windows should be of good size and allow clear vision in all directions, including a view of the vehicle's extremities. The rake of some windows may cause glare or reflections. Blind spots created by front and rear pillars should be minimal. Tinting windows can reduce driver vision, particularly at night or in poor weather conditions, and the level and position of tinting is restricted by law, so one must think carefully before buying a car with tinted windows. Consideration should also be given to how one's ability to see could be affected by loads carried in the vehicle.

IS 2553 (Part 2) is on Safety Glass for Road Transport and IS 14141 is on Electrical windshield washing system.

Mirrors

Good mirrors help in minimizing the car's blind spots and enable the driver to see potential hazards when reversing, parking or changing lanes. Mirrors should be easily adjustable from inside the car (preferably electrically) to suit different driving positions. Some vehicles have heated mirrors to stop them fogging up in bad weather. Dipping interior day/night mirrors can be quickly adjusted to prevent the driver from being dazzled by the lights of vehicles behind. One should be familiar with the characteristics of the mirrors fitted to the vehicle one is driving. For example, convex mirrors that reduce blind spots make vehicles appear more distant than a flat mirror would.

IS 14210 is specification on rear view mirrors of automobiles and IS 14346(Part 1 & 2) are on verification of driver's direct field of view. IS 4060 deals with flashers for direction indicators.

As an additional safety feature many cars also have turn signals in their side mirrors. This is a better spot for the turn signals because if a car is in your blind spot, its driver might not be able to see the back of your car.

These mirrors contain high-intensity light-emitting diodes (LEDs), usually arranged to form an arrow that can point either left or right. The LEDs are positioned behind the mirror glass so that from inside the car, the driver sees only a dimly lit arrow, but outside the car, other drivers see a very bright arrow.

Since the lights in these mirrors are LEDs, the system has an additional benefit: LEDs light up about a fifth of a second quicker than incandescent light bulbs. That may not sound like much, but at 65 miles per hour (105 kph), your car covers 19 feet (5.8 m) in a fifth of a second. LEDs could give someone the extra time and space needed to avoid hitting you.

Wipers and Demisters

Wipers and demisters help maintain clear visibility, and should be on the front and rear windscreens. Multiple-speed wipers are most effective in variable weather conditions. The blades should be replaced regularly to ensure the wipers work well both in bad weather and when cleaning the windscreens. Demisters for side windows can further enhance visibility. The most efficient demister of all is a good air-conditioning system. I 7827(Part 1 to Part 3) are specifications electrical windscreen wiper system.

Lights

Good headlights are an important safety feature for driving at night and in poor light conditions. They should give a good view of the road and roadside without distracting oncoming traffic. Fog lamps are designed to improve visibility in fog but should only be used in foggy conditions. I 5545 deals with fog lights for automobiles.

ome cars have wipers or water jets to keep the headlights clean. Large, well-positioned tail lights, brake lights, high-mounted stop lamps, reverse lights and indicators let other road users know that the vehicle is there and what it is doing, so crashes are less likely to happen.

Audible Warning Devices

Horns are fitted to all cars as warning devices, but it is better to use a vehicle fitted with a warning beeper that sounds when the vehicle is reversing. These beepers can prevent many crashes in areas with a lot of pedestrians and cyclists, such as city streets and car parks.

Adjustability and Comfort

Vehicle fittings must be able to be easily adjusted so you are comfortable and in control on the road. All adjustments should 'lock' in place until the occupant wants to change them. Mirrors have already been mentioned; here are some other features.

The Steering Wheel

Adjustable steering wheels can be moved forward and backward, up and down, and tilted to different angles. A correctly positioned steering wheel can help maximize your control of the vehicle.

BI has formulated following standards on teering control systems:

1. I 7065 is specification on steering wheels,
2. I 11939 is on Impact protection requirements and methods of measurement of automotive vehicles,
3. I 12159 is on method of steering geometry of automotive vehicles,

4. IS 13476 is on method of test for mechanical steering gear of automotive vehicles.

Air-conditioning

One should be able to adjust the temperature and ventilation to suit climatic conditions. Air conditioning helps prevent driver fatigue by keeping you comfortable and also demists the windows in wet or humid weather. Air-conditioning is not just a comfort feature; it provides real safety benefits.

IS 14618 specifies method of measurement of thermal performance of Air-conditioning system of Automotive vehicles.

Minimizing Driver Distraction

If a vehicle's instruments and controls are difficult to see and use, the attention is taken away from the road. This is when crashes can happen.

Well-designed instrument panels allow to see the speedometer, fuel gauge, clock and warning lights at a glance. Items that require adjusting, such as the stereo or air-conditioning should be placed within easy reach, and their controls need to be simple enough to be used without looking. Some people prefer cars that have knobs and dials as controls because they are easier to locate by feel than push buttons. Placing controls on the steering wheel is another way to minimize driver distraction.

Passive Safety Features

Crush Zones and Safety Cages

In frontal, rear and offset crashes, modern vehicles protect occupants by absorbing crash energy and reducing the forces transmitted to the driver and passengers.

The front and rear of a vehicle act as crush zones designed to crumple in a controlled and progressive manner, allowing the occupant compartment to decelerate more slowly.

Slower deceleration means less force reaches the occupants, so there is a lower chance of injury.

Safety Cages

The occupant compartment is a robust safety cage that diverts and dissipates crash energy and preserves the occupant's 'survival space' as much as possible.

The safety cage should include strong pillars to stop the roof from collapsing in roll-over crashes, as well as barriers to prevent the wheels, bonnet or engine intruding into the occupant compartment.

Door locks and hinges also form part of the safety cage and should remain intact so occupants are not thrown from the vehicle.

IS 13389/ISO3984 is a specification on Moving Barrier Rear Collision test method for passenger cars.

Frontal Impact Protection

Every new car should comply with one of the approved performance-based frontal impact protection standards. These standards ensure the front crush zone, safety cage and other occupant protection features work together to provide a prescribed level of driver and passenger safety in a frontal crash.

I 13390/I O3560 is a specification on Road Vehicle-Frontal fixed barrier collision test method.

Side Impact Protection

i de impact crashes can be particularly dangerous because there is no room for large crush zones to protect an occupant from crash forces. The door can be smashed into an occupant or the occupant's head may be struck, causing potentially fatal injuries.

Side Intrusion Beams

To minimize this danger, most new cars have sturdy side intrusion beams, cells, bars or other protection within the door structure. This provides a solid energy-absorbing barrier, while the safety cage helps to divert forces away from the occupant.

I 14682 is specification on lateral protection (side guards) and I 14247 /I O 2958 is on performance requirements of exterior protection for passenger cars.

Padding

Most vehicles also have padding on the inside door panels. This padding should be effective at absorbing impact energy, rather than merely cosmetic. ome vehicles have crushable structures in the doors to further absorb energy.

Side Air Bags

An increasing number of manufacturers install side air bags in some of their models. The side air bags deploy from the seat, the roof, the door or a pillar and protect an occupant's head and upper body.

They must deploy faster than air bags designed for frontal impact protection, as forces will reach the occupant sooner in a side on crash.

Although side air bags are currently uncommon, the significant protection they offer to an occupant's head, chest, abdomen and pelvis in a side on crash makes a car equipped with this technology worth considering.

In addition to this most heavy vehicles are required to be fitted with Rear Under run protective device as per I 14812 and the standard for front under run protective device is going to be formulated very shortly.

Safety Belts

Three-point lap and shoulder safety belts are the best means of occupant protection in almost all types of crash. Wearing a safety belt can halve the risk of being injured or killed in a crash. IS 15139 is on Safety Belt anchorage and IS 15140 is on Safety belt assembly.

How Safety Belts Protect You

Safety belts offer protection in three key ways:

- They prevent or minimize the 'second impact' in a crash—the impact of the occupants against the vehicle interior and each other. It is this second impact that causes injuries and fatalities.
- They stop the wearer from being thrown from the vehicle.
- They help absorb the wearer's inertial energy, allowing the wearer to slow at a rate similar to the occupant compartment.

In a 50 km/h crash, an average-sized occupant not wearing a safety belt would hit the vehicle interior with a force of three and a half tonnes.

Three-point Lap and Shoulder Safety Belts Preferred

It is preferable that three-point lap and shoulder safety belts are installed in every seating position in a vehicle, although this is not always possible in the centre rear seat of some cars. While two-point lap safety belts are much better than wearing no restraint, they do not provide the same protection as three-point safety belts.

Fitting Safety Belts

Safety belts should fit the wearer's body without any slack. The safety belt retractor usually makes this adjustment for the wearer and locks the safety belt in place if there is sudden movement.

Adjustable safety belt anchorages allow the shoulder sash to be moved higher or lower so that the safety belt can be worn more comfortably and safely.

New Safety Belt Technology

Although safety belts cannot stop forward movement completely, new technology designed as part of the vehicle's frontal impact protection system can further improve their effectiveness at restraining occupants.

- Webbing clamps grab the safety belt webbing to prevent more of the safety belt reeling out as it tightens on the spool in a crash.
- Pretensioners use a spring, compressed gas or small explosive charge that activates on impact to pull the safety belt tight before the occupant starts to move.

- Unloaders allow the safety belt to slacken in a controlled way so the wearer can slow progressively while still being safely restrained.

Remember that individual safety belt features by themselves do not necessarily make a vehicle safer. af ety belt design and technology has to be integrated into a vehicle's overall occupant protection system to be effective.

Importance of Complying with a Standard

It is a vehicle's compliance with a performance-based frontal impact occupant protection standard that is important, not the presence or absence of particular safety belt features.

However, within the range of a single model of vehicle, the addition of these items will increase the level of protection over the base model.

Air bags

An air bag is a supplementary restraint that works in combination with a safety belt to protect an occupant in a frontal crash. It is stored in the steering wheel hub or dashboard and inflates very rapidly in a split second after impact.

As an occupant moves forward in a crash, the air bag protects them in several ways:

- It slows the occupant's deceleration by deflating at a controlled rate. This does not mean it takes a long time for an air bag to deflate; inflation and deflation occurs faster than a person can blink.
- It prevents the occupant's head and chest from striking the steering wheel, dashboard or top of the windscreen.
- There is a large surface area to restrain the occupant's forward movement compared to a safety belt alone, so compression of the occupant/s chest is reduced.

Air bags work with Safety Belts

Air bags are designed to supplement safety belts, not replace them. o me air bags will only deploy in crashes severe enough to threaten safety-belted occupants.

Air bags are of limited benefit in side impact (unless they are side air bags), rear impact or roll-over crashes, or in crashes where there is more than one collision. A safety belt can protect an occupant in all of these crashes.

Any driver or passenger in an air bag-equipped vehicle must still wear a safety belt for their own protection.

Potential for Injury

Because of this deployment speed, there is the potential of injury from an air bag. New types of air bag deploy less aggressively, and air bags under development detect the size and position of occupants or take account of different crash speeds and adjust air bag deployment to offer the best protection.

There are several simple ways that occupants can avoid the risk of injury from an air bag:

- Always wear a safety belt.
- Do not sit too close to the steering wheel or slide the seat a long way forward. Shorter drivers should sit back from the steering wheel, but in a position that still gives them full control of the vehicle.
- Do not rest anything over the air bag cover, nor place anything between the air bag and an occupant (except a safety belt).
- Never put a rear-facing infant restraint in the front seat of a vehicle with a passenger air bag.
- Read the vehicle manual for any other information regarding the vehicle's air bag or air bags.
- If you buy an air bag-equipped vehicle, follow these precautions as well as any advice printed on warning labels in the vehicle.

Despite some risks, air bags are passive safety features that have reduced the injuries and saved the lives of thousands of people in serious crashes.

Seats

A vehicle's seats can be very effective passive safety features. A seat should be comfortable, but must also be sturdy enough to withstand a lot of force and protect its occupant in a crash.

The driver's seat should be able to be moved forward and backward, up and down, and the seat back reclined at different angles. This helps to be comfortable, to see well and to operate all controls easily. Many seats have an adjustable lumbar region to further improve comfort.

'Anti-submarining' Seats

Most new cars have front 'anti-submarining' seats designed to prevent the occupants slipping under the lap section of the safety belt and suffering abdominal injuries in a crash.
Strength of the seat

Strength of Seats

The seat back is often the only thing between the occupant and other passengers and objects in a crash: most new vehicles have seats built to comply with an international strength and safety standard.

A seat should be strong enough to act as a barrier against impact coming from behind the occupant. Some new vehicles have seats with deformable crossbars and padding to absorb crash energy.

Many seats also provide an element of side impact protection.

BI has recently formulated I 15546 on e ats, their anchorages head restraints for category M1 automotive vehicles.

Head Restraints

Head restraints are important safety features and should preferably be fitted to all seats—front and back. Head restraints must be permanently fixed or adjustable to a position high enough and close enough to the back of the head to provide support and protection for taller occupants. a fety belts should be comfortable and supportive to wear—adjustable safety belt anchorages will help them fit properly.

Preventing Whiplash

Head restraint position is critical for preventing whiplash in rear impact crashes. Whiplash is caused by the head extending backward from the torso in the initial stage of rear impact, then being thrown forward.

To prevent whiplash the head restraint should be at least as high as the head's centre of gravity (eye level and higher) and as close to the back of the head as possible.

Risks Associated with Poorly Positioned Head Restraints

A low head restraint can act as a pivot that will cause the head to extend even further backwards. If the head restraint is too far away from the back of the head it won't minimize movement. A poorly positioned head restraint may present up to six times the risk of whiplash injury of a safely positioned one.

Unfortunately, many head restraints cannot be adjusted high enough or close enough to the back of the head to protect taller drivers. Even adjustable head restraints that can be positioned correctly are often left at their lowest setting where their protection is minimal.

Fixed head restraints tend to be safer because they do not depend on the occupant to adjust them to the correct position.

Safest Adjustable Head Restraints

The safest adjustable head restraints offer protection to most drivers even at their lowest setting, can be tilted forward and lock into place when adjusted.

New Developments

A new development is a head restraint that moves forward and up if a pressure plate in the seat is pushed back by the torso in a crash.

Other vehicles position the head restraint automatically when the seat is adjusted by different sized occupants.

Safe Interiors

Most injuries in a crash occur when an occupant collides with the steering wheel, dashboard, door or other interior surface. Improving the safety of the interior of the occupant compartment, can reduce injury severity.

IS 152223 is a specification on Interior fittings of automotive vehicles.

Steering Wheel and Steering Column

Hitting the steering wheel can give drivers serious head injuries. An air bag and safety belt can help prevent these injuries, but the steering wheel can also be designed to lessen the impact.

The steering column should collapse to absorb crash energy and to minimize its intrusion into the passenger space. The steering wheel should deform to absorb energy if it is hit by the driver's head or chest in a crash.

Any other hard components the driver might hit should be covered by energy-absorbing padding.

Padding and Edges

All surfaces an occupant may strike in a crash should be as safe as possible. For example, the dashboard, door panels, roof, sun visors, mirrors, header rail and pillars should have thick energy-absorbing padding with no sharp edges underneath.

Jutting objects and hard edges must be kept to a minimum. All interior fittings, controls and surfaces must comply with safety standards from the major vehicle manufacturing countries.

Load Restraints

Load restraint is an essential consideration if the vehicle you are buying is likely to be heavily laden with parcels, equipment or baggage, as these can cause serious injuries in a crash if they are not restrained. Even a small object left on the parcel tray behind the back seat can become a dangerous projectile.

Vehicle Boot (or Trunk)

The vehicle boot should have tie-down hooks or eyes so heavy objects can be secured with rope. Ratchet-tensioned straps can also provide excellent load restraint. Station wagons in particular may have a luggage cover or net, which can offer reasonable load restraint, provided the material can be securely fastened and does not stretch too much.

Stowage Boxes and Compartments

Plenty of stowage boxes or pockets are a real advantage. Loose items can be safely held or locked up so that they do not pose a risk. Some stowage compartments may be located under the load floor, so occupant space is not reduced.

Fire Prevention and Protection

Fire prevention and protection is an important factor in vehicle safety. The fuel tank and pipe connections should be designed and positioned to minimize any likely damage in a crash.

Excess Flow Valves

An excess flow valve in a fuel injection system can prevent the fuel pressure pump discharging petrol if there is a rupture 'downstream' from the fuel tank. om e vehicles automatically shut off fuel flow in a crash.

Flame Resistant Materials

In the occupant compartment, flame resistant materials slow the spread of a fire. I 15061 specifies flammability requirements for automotive vehicles.

Cargo Barriers

A cargo barrier can be a vital vehicle feature for protecting people from loose loads in a crash. Cargo barriers usually have a steel tube frame, with steel mesh or polycarbonate preventing items from striking occupants. They are placed between the cargo and the front or back seat.

Conclusion

India is gradually harmonizing its automotive standards with global norms, giving due cognizance to the prevailing national concerns over safety requirements of the automobiles. This step is a move forward as far as the safety of the new vehicles is concerned.

But the fitness of the vehicles and consequently the safety of their occupants depend on the quality of the vehicles available in the market. Awareness of keeping the vehicle road worthy, the age of vehicle and the number of years the vehicle is allowed to be on the road. Due to easy finance scheme it is easy to purchase a vehicle but regular running and maintenance requires money, time, education and good mechanics. Non-availability of any one factor results in fast deteriorating condition of the vehicles which become prone to accident due to brake failures, engine failures or other unscheduled and un predictable break downs. The useful age of the vehicle is also dependent on road condition. It is high time that the concerned government does something worthwhile to address these issues.

13

HUMAN FACTORS IN TRANSPORTATION SAFETY

D. Majumdar*

ABSTRACT

It is predicted that death and disability resulting from road accidents in comparison with other diseases will rise from 9th to 3rd rank between 1990 to 2020. Road accidents as cause of death and disability would rank below heart disease and clinical depression and ahead of stroke and all infectious diseases.

Researchers have suggested that accidents can be the consequence of 'Conflict' situation involving the driver, the environment and the vehicle. Several investigations involving 'on-site' and 'in depth' analysis of accidents show that human behaviour was clearly the dominant cause. Human factors research deals these three categories, i.e. personal factors, the driving environment and vehicle characteristics, in accident causation for the prevention of injuries and improvement of safety standards.

Personal factors addresses the issues of personal characteristics, driver behaviour patterns, driver related capabilities and temporary impairment. The driving environment on the other hand includes the roads and highways, road markings, road signs, lighting and traffic as well as the natural features of the ambient environment. Visual displays and control, seating and space considerations, vibration and motion, vehicle mass, physiological effects of the use of normal and automatic transmissions, response time etc. are some of the important vehicle characteristics that are increasingly receiving greater importance in the promotion of road safety.

Key Words : Human Factors, Road Accidents, Transportation Safety

Introduction

There are approximately 700 million passenger cars, trucks and buses and more than 300 million motorcycles and mopeds (motorized two wheeler) worldwide. Nobody really knows how many bicycles are there. Along with food, shelter, security and health individual mobility is a basic human need; one on which our other basic needs are in part dependent.

*Defence Institute of Physiology and Allied Sciences, Lucknow Road, Timarpur, Delhi-110 054
E-mail: majum55@yahoo.com

A major side effect of increased mobility is increasing exposure to injury in traffic crashes. Murray and Lopez (1996) predicted that death and disability resulting from road accidents in comparison with other diseases will rise from 9th to 3rd rank between 1990 and 2020. Road accidents as cause of death and disability would rank below heart disease and clinical depression and ahead of stroke and all infectious diseases.

The causes of accidents can be characterized in many ways and at various levels of specificity. In general accidents can be attributed to human behaviour, the environment and the vehicle with some interaction among these factors. Older and Spicer (1976) suggested that accidents can be the consequence of conflict situation involving the driver, the environment and presumably the vehicle. Shinar (1978) presented data on the percentage of accidents from two samples that were attributed to human, vehicle and environmental causes and their combinations. One sample of 2258 accidents was investigated on-site and the other samples of 420 was investigated more thoroughly, in-depth. The results showed that human behaviour was clearly dominant cause of accident.

This paper will attempt to explore briefly the role of two important human factors, i.e., personal factors and vehicle characteristics in transport accidents and safety.

Personal Factors in Accidents

The term 'accident proneness' frequently involves with reference to personal factors in accidents. This refers to some attribute that certain people have that causes them to be accident repeaters. But it has been noted that the mere fact that individual has a high accident rate for a while is not necessarily indicative of accidents proneness. It is probable that there are some accident repeaters in driving, although such individuals do not account for a very large proportion of all accidents. Further there are statements that different types of personal factors, in certain instances, are related to driving behaviour. These can be clubbed into following categories:

Personal characteristics
Driver behaviour patterns
Driver related capabilities (experience, training, etc.)
Temporary impairments (fatigue, alcohol, drug, etc.)

Personal Characteristics

Various personal characteristics have been investigated as possible contributors to driving behaviour. These include personality factors, vision, perception, physical abilities, reaction time, intelligence, age, sex, etc. It is not possible to discuss all of these factors in details but it is probable that most of them have significant impact on driving behaviour to a limited extend and with respect to only certain individuals.

Personality

Mc Guire (1976) concluded that some highway accidents are related to people who are emotionally unstable, unhappy, unsocial, impulsive or under stress. Such attributes cause people to be less cautious, less attentive, less responsible, less caring, less knowledgeable or less capable of driving and thus increase the risks of accident.

Vision

Poor static visual acuity has been found to contribute to accident occurrence. The use of dynamic visual acuity (DVA) test that measure more specific driving related visual skills has been recommended before issuing driver's licence in addition to static visual acuity test. Shinar (1978) also urges the use of measures of visual acuity under low levels of illumination similar to those encounter in night time driving. It has been observed that night time accident involvement is related to poor visual acuity under night time levels of illumination.

Field Dependence

The visual input to driver is influenced in part by the characteristics of the drivers as well as by the nature and location of objects within their environment. Field independent people can distinguish relevant information better from irrelevant cues in their environment than those who are field dependent. It has been reported by researchers that drivers with accident records tended to be more field dependent than the accident free drivers. Although the reasons for the relationship between this type of perceptual style and driving behaviour are not yet clear but observations elsewhere suggested that field dependent drivers cannot quickly recognize developing hazards, are slower in responding to road signs, have difficulty in learning to control a skidding vehicle and failed to drive defensively at high speed traffic. Training might help field dependent individual to become more field independent.

Age and Sex of Drivers

There is good correlation between age of driver and accident. The rate is highest for teenager, decreases sharply in the 20s, with an increase in the older drivers. The role of sex is probably not an important correlate of accidents, because the differences tend to be very small when age, exposure, etc. are controlled.

Driver Behaviour Patterns

The behaviour of people during driving depends on their personal characteristics and experiences. The behaviours have been characterized as inattention, improper lookout, judgment, steering behaviour, reaction time, etc. Number of studies have been carried out to explore the relationship between such behaviours and accident occurrences.

Sensory and Perceptual Input in Driving

Sensory and perceptual input to drivers specially through the visual sense are probably responsible for recognition errors. The perceptions of individuals of what they sense serve as the basis for their judgments about things in their environment, i.e. perceptual judgments. As for example, judgments of safe distances in following other vehicles. It has been reported that such judgments are influence by road cues available. The under estimation of distances was greater on a segmented lined road than on unlined roads.

Decision Processes

A vehicle driver makes many decisions during driving and executes these in driving behaviour by controlling the steering wheel, accelerator and brake. This process of decision making is one of risk taking although the genesis of variations in risk taking on the part of different individual is not clear. There are indeed differences in the extent to which at least some people are inclined to take risks. There are variations in the same individual from time to time in the risks he or she is going to take. The risk taking tendency of people are intricately related with people's expectations that are the probability of some unhappy consequence resulting from given behaviours.

Driver Related Capabilities

Training programmes for improving the capabilities of drivers can contribute to the improvement of driving and reduction of accidents.

Temporary Impairment

The primary factors that contributed to temporary impairment of driving are fatigue, alcohol and drugs. Long travel time and associated sleep loss generally lead to fatigue that is considered a potential human factor for accident. However, the primary basis for reducing the risk from fatigue lies with the driver, who must be sensitive to the signs of the onset of fatigue and must take rest. The use of alcohol and drugs in the generation of accident are well researched and documented. Appropriate training programmes, roadside warnings on deleterious effects of alcohol and drug in driving performances and enforcement practices improve driving quality.

Vehicle Characteristics

Important vehicle design features such as displays, control processes and control mechanism, seating and space consideration, visual field and viewing angle, vibration and motion have important impact on the causation of transport accidents.

The vehicle communicates with the driver and communication is two-way because the user has to take some action on the basis of information received. The efficient operation of a vehicle depends on the ease of this two-way communication. Since the capacities of the human are limited it is necessary to design the characteristics of the vehicle so as to fit those of the user. The design of the vehicle will not achieve complete success if the driver has difficultly in reading the displays or in operating the controls and the information is not transmitted quickly and accurately. Functionally, a good display is one that allows the best combination of speed, accuracy and sensitivity when transferring the necessary information from vehicle to user. In order to meet the functional requirements of speed, accuracy and sensitivity any display must satisfy the basic criteria of detection, recognition and understanding.

Modern vehicles use a combination of qualitative, quantitative and representational types of displays to give the driver best information to avoid accidents.

Controls were developed as extensions of the human body. This extension was made to allow the generation of more power, more reach and a reduction in effort and risk to the user. Controls must be designed to meet human capabilities and limitations of the whole unit, i.e. driver and vehicle is to work at maximum efficiency. Just as a display must be detectable, readable and understandable, so, analogously, a control must be accessible, identifiable and usable.

During designing, it should be kept in mind that the driver must be able to reach controls comfortably and efficiently and it must be easy to use. Following aspects need attention during designing a vehicle controls, e.g. location in relation to the user, muscles and movement, posture and fatigue, force, speed and range and accuracy. Difficulty in using control system leads to error and accident.

People vary in their height and proportion. Two people of the same height and sex are almost certain to differ in arm or leg length, or sitting height, or hand size, or any body dimensions. By body size it is meant therefore, not just height but any body measurement which is important for the drivers cab design. It is common in ergonomics solution to aim to cater for 90 per cent of the user population and to exclude the 5 per cent furthest above and the 5 per cent furthest below the average.

Recommended distance and dimensions for the location of controls and displays for a motor car available in the international literature. Even them the lack of population statistics and compromise with motor car size and cost have been found to cause immense mismatch between driver and his workspace geometry. Areas which show maximum misfit are horizontal and lateral leg space, head clearance, hand movements, visual fields and seat dimensions. Appropriate human factors considerations in the infrastructural stages of vehicle development can reduce the cost of production, remodelling and upgradation and improvement in transport safety and reduction in injuries.

REFERENCES

McGuire, F.L., (1976), Personality factors in highway accidents. *Human Factors,* 18 (5), 433-442.

Murray, C.J.L. and Lopez, A.D. (eds), (1996), *The Global Burden of Disease.* Harvard (Boston, USA: School of Public Health).

Older, S.J. and Spicer, B.R., (1976), Traffic conflicts: A development in accident research. *Human Factors,* 18 (4), 335-350.

Shinar, D., (1978), *Psychology on the road: The human factor in traffic safety.* New York: Wiley.

INCREASING ROAD ACCIDENTS
Few Reasons and its Solutions

Raju Sarkar*

ABSTRACT

In last two decades, the economic development has shown a profound influence on lifestyles of people in India. In wake of these developments all round growth can be seen in various sectors like energy, communication, transportation, industrial, domestic, agriculture etc. These have resulted in high pace of activities in organized sectors, and even a common man is no more an exception, particularly in metropolitan areas and cities. However, this development and growth has already shown concerns about the safety and environment. The transportation, particularly the use of vehicle, is one major sector in this regard as mobility of people is essential for most economic activities. The alarming increase in the number of motor vehicles on the road has created a major social problem—the loss of lives through road accidents. The appalling human misery and the serious economic loss caused by road accidents demand the attention of the society and call for the application of the solution of the problem. Although traditional enforcement strategies are successful with the general driving population, different approaches are required for some groups, especially those who intentionally disregard the laws and become adept at avoiding apprehension, posing a high risk of injury or death to themselves or others. In the present paper, an attempt has been made to highlight possible causes and its remedial measures in road accidents.

Introduction

Every year an estimated 70,000 human lives are lost owing to over 3,00,000 road accidents resulting in an estimated Rs.3600 crores of financial loss to the Indian exchequer. The worst part being these figures do not reflect the overwhelming human misery these accidents cause. So entrenched is the culture of accidents that we realize it only when it strikes a personal note. Statistics indicate that by the year 2020 road accidents will become third leading cause for deaths worldwide and this calls for our immediate concern.

The collision of one road user with the other road user or with the fixed objects lying within the road pavement is called road accident. The overturning or running off the roadway

*Lecturer in Civil Engineering, G.B. Pant Polytechnic, New Delhi, INDIA. E-mail: sarkar_raju@yahoo.com

by any vehicle is also known as road accident. These accidents may cause damage to property, vehicles, injuries to users even death of road users. Effort should, therefore, be made to adopt adequate measures to prevent accidents on roads.

The geometric of highway should be designed to provide optimum efficiency in traffic operations with maximum safety at reasonable cost. The designer may be exposed to either planning of new highway network or improvement of existing highways to meet the requirements of the existing and the anticipated traffic.

Few Possible Reasons of Road Accidents

Single Vehicle Accident

Slippery road surface, sharp horizontal curve, absence of super elevation, surprise geometric element, absence of proper roadway delineation.

Pavement unevenness occurs due to improper compaction of the fill, use of inferior pavement materials, improper surface and subsurface drainage etc. the pavement unevenness increases fatigue which affects comfort and safety of the occupant.

Skid occurs when the wheels of a vehicle slide without revolving or when the wheels partially revolve. When the vehicle is in high speed and brakes are applied due to some obstruction, the wheels are locked partially or fully, and the vehicle moves forward due to high speed, the longitudinal skidding takes place, which may cause accident. The lateral skid is considered dangerous as the vehicle goes out of control leading to an accident.

Accidents During Wet Water

Slippery road surface, absence of super elevation at horizontal curve.

Slip occurs in a vehicle when a wheel revolves more than the corresponding longitudinal movement along the roads. When the vehicle rapidly accelerates from stationary position or from slow speed on pavement surface which is either slippery or wet or when the road surface is loose with mud, the vehicle may goes out of control leading to an accident.

Accidents During Nights

Poor lighting condition, sharp horizontal curve combined with steep grade, poor delineation of road edges, tired or drunken driver.

Night visibility very much depends upon the light reflecting characteristics of the pavement surface. Very poor visibility in nights, especially when the surface is wet may lead to a fatal accident.

Major contributing factor in road accidents is faulty road construction, bleeding infrastructure and laxity in supervision. Twenty-eight per cent of the road accidents are attributable to the present faulty road network even though hundreds of projects are pitched every year. Improper illumination of the roads especially of the highways, inadequate punitive measures, lack of proper road signs and most importantly the lack of induction of new technology in designing and construction are the main handicap.

Collision Against Fixed Objects

Trees and other fixed objects in proximity of roadway edge, narrow bridge/culvert in an otherwise wide roadway, surprise geometric element.

Sight distance is the length of road visible ahead to the driver at any instance. It is desirable to construct highways in such a way that the length of road visible ahead at every point is sufficient for safe overtaking. Restrictions to sight distance may affect the safe vehicle operation on a highway.

Aggressive or Impaired Driving

Alcohol and other drug-impaired driving continues to threaten the safety of all road users, i.e. road accidents can also be attributed to another major factor, i.e. faulty driving habits/ road rage or aggressive driving. Estimates accrue 75 per cent accidents due to human error. Aggressive road drivers frequently violate traffic rules.

Different Type of Collision and Its Solution

Rear-end Collision at Intersection

- Provision of deceleration lane right-turning pocket lane
- Provision of left infiltration lane
- Removal of sight obstructions
- Installation of advance warning signs about proper lanes for different destinations
- Increasing amber time in case of signal control

Head-on, Right Angle and Right Turn Collision at Intersection

- Removal of sight obstructions and enforcement of parking regulation
- Installation of stop lines/signs on minor arms
- Prohibition of right turns where possible
- Provision of channelising islands
- Conversion into rotary where space is available
- Installation of traffic signals where minimum warrants are met
- Creation of one-way streets

Side swiping collision at intersection

- Installation of pavement markings
- Widening of pavement
- Elimination of roadside obstructions such as caused by parked vehicles and other bottlenecks
- Installation of channelising islands

Pedestrian-Vehicular Collision at Intersection

- Installation of pedestrian cross-walk lines
- Erection of pedestrian railing barriers
- Provision of pedestrian refuge island
- Prohibition of kerb parking
- Clearing of foot-path of all encroachments
- Provision of pedestrian subways/overbridges where minimum warrants are satisfied
- Installation of signal control

Head-on Collision

- Provision of pavement lane markings, and reflector studs in case of unit highways
- Installation of no-passing restrictions at locations having poor visibility
- Installation of centre-dividing strip
- Creation of one-way streets

Collision with Fixed Object

- Application of red reflectors to fixed objects
- Banding of road wise trees
- Removal of fixed objects where possible

Vehicle Running off Roadway

- Installation of pavement edge lines
- Installation of warning reflectors, guard-rail or delineators
- Correction of super elevation at horizontal curves
- Providing skid resistant surfacing

Collision with Parked Vehicle

- Parking prohibition
- Change from angle to parallel parking
- Arranging off-street parking facilities
- Creation of one-way streets

Remedial Measures of Road Accidents

Road Geometrics - Uniform design standards recommended by the Indian Road Congress for planning and designing the alignment of roads should be adopted. While designing the alignment, kinbs in the horizontal and vertical direction should be avoided, horizontal curves provided with as large radii as feasible, and the alignment both in the vertical and horizontal directions suitably coordinated so as to provide aesthetically pleasing profile with built-in safety. Necessary sight distance as appropriate to the situation should be provided at horizontal/vertical curves and at intersections. Availability of the required sight distance should be periodically checked for removal of all obstructions.

Repair of Roads - Prompt maintenance under condition of any natural disaster should be planned in advanced and adequate equipment and personnel made available. Repairs to road in urban area should be carried out during the non-peak hours. Repair to potholes, shoulder drop-offs etc. caused during rainy season should be done soon after the said season so that minimum inconvenience is caused to traffic. The financing procedures to meet the cost of such repairs should be streamlined by the authority so that these repairs can be attended without any delay.

Visibility at Nights - For better visibility at night, rear and front of trucks and buses plying on rural highways should be painted with bright yellow. Adequate measures should be taken to ensure proper functioning of vehicle lights for enhancing visibility during night.

Location of Trees - Where it is desirable to save trees, if they are a safety hazard, such trees should be removed. Moreover trees should be painted in white colour upto a height of 1.25 m above the ground with 3000 m middle bank in black colour to warn the road users of their presence.

Motivation for Public Transport - Motivation is waning for public transport, which needs immediate revival. Parking taxes should be levied to minimize/discourage to use of private vehicles and to encourage the use of public transport for which incentives should be given. Making separate lanes for vehicles carrying two or more passengers will encourage carpooling. This will help reduce the number of vehicles plying daily on the road.

Education - Education is one of the most important factors, which will play a kingpin in the long run to drastically reverse the graph of road accidents. Through education it should be ensured that every individual plying on the road is a responsible driver. Every driver must be aware of the traffic laws, responsible and educated enough to follow them diligently. It would be a good idea to introduce traffic safety in school curriculum so that this concept of road safety is inculcated in the young minds in the formative years.

Medical care - A good number of road fatalities could be averted by timely medical attention. Efforts should be made to properly equip and operate such medical care units in all cities and on important highways. Further all hospitals should be made to attend accident cases. The medico-legal problem should be looked into and tackled.

Conclusion

The Indian road network is one of the seminal sources for the country exchequer and, therefore, a holistic approach should be adopted keeping in mind its sustainability, design, technology and machinery adopted for its construction. Roads are the artilleries of this country and, therefore, urgent course of action needs to be restored to their quality of construction to minimize the accidents and making driving safe and pleasant for its citizens.

Due to the invention of fast moving vehicles, the problem of accidents is becoming more and more complex in highway engineering. Hence it is the utmost duty of the traffic engineers to carry out systematic studies of the accident and suggest preventive and remedial measures. Further the effectiveness of traffic engineering, enforcement and educational measures may, therefore, be judged from changes in the annual accident rates, rather the total number.

Changes in driver behaviour, technology, and society itself are ongoing. In order to effectively deal with these changes, traffic safety professionals must be able to continually identify significant changes that have the potential to create significant adverse safety impacts, to define specific awareness information pertinent to the change, and establish and market a targeted campaign. A small public/private blue ribbon panel of safety experts should meet semiannually to determine if significant changes requiring attention have occurred.

Community Traffic Safety Programmes combine strategies from several traffic safety programme areas to address local highway safety problems. Communities within a county are encouraged to cooperatively develop a strategic plan, which identifies and documents the county's highway safety problems; establishes performance goals, objectives, and measures; and proposes strategies that target the problems identified. Because of the integral role local programmes play in achieving the statewide highway safety goals, increasing the number of counties participating in the programme continues to be a priority.

REFERENCES

Arora, N.L. (1992), *A Text Book of Transportation Engineering*, New India Publishing House, New Delhi.

Dhir, M.P., Bhat, A.K. (1968), "Improving the Riding Quality of Our Pavement", Proc., National Seminar on 20 years Design and Construction of Roads and Bridges, Volume I, Ministry of Transport and Shipping (Road Wing).

Khanna, S.K., Justo, C.E.G. (1990), *A Text Book on Highway Engineering*, Nem Chand & Bros., Roorkee.

Nayyar, N.K. (2003), "Engineering and Other Measures to Improve Road safety", *Souvenir,* National Seminar on Increasing Road Accidents - Causes and Remedies, Indian Institute of Petroleum, Dehradoon.

Sarkar, Raju (2003), "Metal Beam Crash Barriers to Reduce Impact of Crash", *Souvenir,* National Seminar on Increasing Road Accidents—Causes and Remedies, Indian Institute of Petroleum, Dehradoon.

Senzo Engineering (P) Limited, Mumbai (2003), Manual Senzo Road Protection System.

TRAFFIC CALMING ON HIGHWAYS
An Overview

Sudershan K. Popli*

ABSTRACT

National and State Highways accounts for 65 per cent of road fatalities in India. The pedestrians and bicyclists are mainly affected. Traffic Calming is a set of strategies/parameters to reduce negative impacts of motor vehicles and improve the safety. The main objective of traffic calming is to reduce vehicle speeds at vulnerable road sections to reduce accidents and promote safe and pleasant conditions for motorists, bicyclists, pedestrians and residents. This paper presents the traffic calming techniques that can be applied at various sections of highway. The components to be considered for implementing a traffic calming scheme is also presented. A brief has been given upon the state of practices followed in India.

Introduction

The issue of road traffic accidents has got the attention of both developed and developing countries for a very long time. This attention is not unconnected with the fact that road accidents have been estimated to be responsible for over 30 million deaths since the first road fatality was recorded. Recent study estimated that in 1999 between 750,000 and 880,000 people were killed and 23 to 34 million people were injured in road crashes worldwide with a global cost of about US$ 500 billion at 1997 prices. India accounts for 85,000 fatalities per annum (in year 2000), i.e. 10 per cent of total world's road fatalities. The national highways which constitute 2 per cent and state highways about 4 per cent of the total road network together account for over 65 per cent of the total road fallibilities. It is also seen from available accident statistics that 50 to 60 per cent of fatalities happen due to heavy commercial vehicles (trucks and buses). These commercial vehicles are involved in a higher proportion of fatalities with pedestrians and bicyclists than that of injuries. With the NHDP and other road upgradation programmes being undertaken in the States and production of modern

*Deputy General Manager (Highways), RITES Ltd., Gurgaon, (Haryana).

automobiles by vehicle manufacturers, speeds of vehicles would further increase. This will further increase the accident frequency and severity at certain locations, particularly where highway passing through habitations and near schools, markets and factories. The speeds of the vehicles are to be reduced when they pass through these settlements on rural roads. Most of the countries now recognize the use of traffic calming measures in order to reduce the problems of accidents to pedestrians and other vulnerable road users.

Background

Traffic calming has its origin in the Dutch "Woonerf" schemes of the 1970s, and since then it has been further extended and refined and applied in many countries across the World. The original "Woonerf" schemes introduced the concept of shared space between vehicles and pedestrians. Streets were reconstructed so as to have proper balance between vehicles and pedestrians/cyclists.

In the early 1980s Norway adopted traffic calming as the viable option to deal with intercity traffic speeding through its many small towns as did not have funds to construct new bypass. Inspired by Norway, Denmark also applied traffic calming measures to highways passing through the small towns. Germany experimented in late 1970s with neighbourhood traffic calming as they have learned that calming individual streets resulted in traffic diversions. A 30-kph speed limit was imposed over large areas; local streets and collector roads by various traffic calming methods. In Britain, Colin Buchanan prepared a report on traffic calming named 'Traffic in Towns' in 1963. The report author, Colin Buchanan, is considered the father of traffic calming by many Europeans. Buchanan-inspired traffic calming plans were implemented throughout Britain under the 1969 Housing Act and 1977 street design manual. Australia began its traffic calming efforts with street closures and conversions to one-way streets. By the 1980, traffic calming has been applied to number of cities in Australia. In 1989, they have developed a plan offering traffic calming as an alternative to major highway project. In U.S. the beginning for traffic Calming was made in 1971. The neighbourhood traffic calming techniques were applied to the cities of Seattle, W.A.

Objectives of Traffic Calming

Traffic Calming is a set of strategies/parameters to reduce negative impacts of motor vehicles and improve the safety of pedestrians and cyclists. The U.S. department of Transportation, Federal Highways Administration has defined the general objectives of traffic calming techniques as follows:

- To encourage citizen involvement in the traffic calming process by incorporating the preferences and requirements of the citizens.
- To reduce vehicles speeds.
- To reduce accidents and promote safe and pleasant conditions for motorists, bicyclists, pedestrians and residents.
- To reduce pollution and improve the environment and livability.
- To improve real and perceived safety for non-motorized users.

Highway Sections Needs Traffic Calming Measures

Traffic Calming measures are essentially required on a highway to keep the driver alert while driving on different sections of the highway, such as

- On Rural and Straight sections of highway
- On transition of rural to urban section
- Roads passing through the urban areas or existence of school or hospital on the highway
- At meeting point of important side roads to NH/Junctions
- Change of road cross-section from 6 lanes to 4 lanes and 4 lanes to 2 lanes
- At road-rail level crossings

Traffic Calming Techniques

The various traffic calming techniques adopted under different conditions are as follows:

Rural and Straight Sections of Highway

High speeds on rural sections of a highway are a safety concern. It causes head on collision and run-off crashes on two-lane roads. So the attempts have been made to control and reduce speeds by providing narrower lane widths, which requires drivers to slow down the vehicle. This approach is implemented either by physically creating narrower roadways or by visually decreasing the roadway widths. The visual decreasing of the roadways is generally done by painting the wider edge lines. The narrower road widths are created by putting metal crash barriers to shield the road side objects.

Narrow lane widths, Netherlands

On long straight section of the highway, motorist tends to over speed or becomes less attentive due to clear visibility for longer distances. Tree Plantation can be used an effective traffic calming measure as it narrows the field of vision/roadway widths visually.

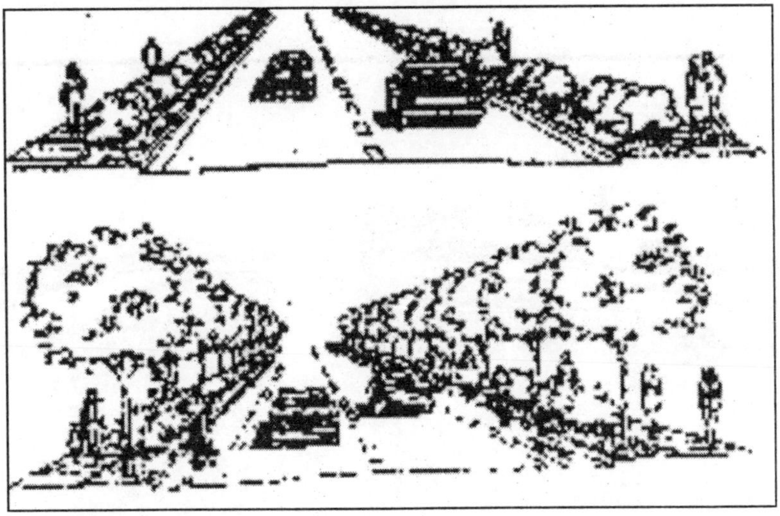

Tree Plantation on Road sides

Transition of Rural to Urban Section

This is the most vulnerable section on highways, where traffic calming techniques are to be applied as the motorists have to adopt to the urban traffic conditions ahead.

- The erection of Gateways, Gantries indicate the beginning of urban section ahead and putting road safety messages.

Gate Erection in Germany

- Extension of service roads beyond the urban limits to indicate the beginning of urban sections.
- Extension of lighting poles beyond the urban limits.
- By decreasing the road lane widths physically or visually.
- By providing different types of surfacing to alert the drivers.

Rural Bike facilities parallel to highway, Netherlands

Highways Passing Through the Urban Areas

Traffic Calming is an effective mean for controlling speeds on highways passing through the urban stretches. A variety of components are available with different uses, applications and effectiveness. These measures shall be applied depending upon the road category such as NH/SH/MDR. The various components of traffic calming are to be implemented also depends upon the settlements characteristics such as

- Length of the settlement
- Area of settlement
- Land use in the settlement and specifically along the road
- Nos. of cross roads joining the main road
- Building line from the road edge
- Settlement structure i.e. on one side of the road or both sides of the road

The principle traffic calming techniques generally used fall into five areas:

(a) Vertical deflections
(b) Horizontal deflections
(c) Road narrowing

(d) Central islands
(e) Service roads and footways

The effects of these measures may be reinforced by a range of supporting measures.

Vertical Deflections

Vertical shifts in the carriageway are the most effective and reliable of the speed reduction measures currently available. There are currently several different techniques available to achieve this:

(a) Road humps

Road humps may be of several varieties with rounded and flat topped being the most common. The latter are particularly suitable in providing crossing places for pedestrians. The width of humps may be restricted to allow the retention of existing carriageway drainage.

(b) Plateau/Speed Table

Plateau/Speed table extends to the full width of the carriageway between the kerbs and extends over a longer length of road than humps. The top surface may be of different material to the carriageway and footways. Plateau is more suitable than road humps as it causes less discomfort. The length of the plateau may be sufficient to accommodate the full wheelbase of the vehicles particularly car to reduce passenger discomfort to a minimum.

(c) Uneven Road Surface

Care is to be taken in the use of rumble strips due to the increase in the level of noise and vibration and the difficulties they create for cyclists. As a technique these are perhaps more suitable for gateways where frontage property is limited.

Speed Hump, England

Plateau/ Speed Table

Horizontal Deflections

Horizontal shifts in the carriageway are less effective than vertical ones in achieving reductions in speed; however their impact is significantly increased when used in combination with a vertical shift. Essentially all horizontal shifts may be classified as chicanes.

The speed reducing impact of chicanes is reduced if the measure has to allow for the passage of HGVs, as the wider carriageway allows car drivers to take a "racing line". The use of stone sets or similar material as a side strip can be useful in allowing the passage of large vehicles while discouraging cars.

Road Narrowing

Road narrowing may be considered as another supportive measure to vertical deflections. It cannot be considered as a speed reducing device in itself, but it can act as a reminder or encouragement to drive slowly or calmly. Narrowing the carriageway at specific locations, for example in combination with speed tables, is an effective way of combining measures to increase their effect.

Central Islands

Central islands may be provided on roads wider than two lanes and on cross roads meeting the main roads. They provide useful pedestrian refuges and prevent speeding of vehicles from cross roads to main roads.

Service Roads and Footways

Construction of service roads and footways are effective ways to provide right-of-way to pedestrians, cyclist, slow moving traffic and traffic joining from cross roads. These measures reduce the conflicts between motorised and non-motorised traffic. Service roads and footways can be provided at same or different levels from main highway. These can be separated physically by providing raised separator.

Supporting Measures

A number of supporting measures are commonly used to back up the speed reducing techniques. The use of different surface materials, the planting of trees and the use of street furniture and lighting falls into this category.

Traffic Calming on Junctions

Traffic calming on junctions plays very important role to reduce the accidents. The different traffic calming measures to be taken on junctions depends upon the location and category of roads meeting. The traffic calming measures shall be applied to village road before it joins the high speed rural road. In case of the urban section, frontage road/service road can be provided to connect various cross roads before joining the highway. The various traffic calming measures applied to junctions under different conditions are as follows:

Vertical Deflections

Raised junction areas (plateau), flat top road humps and speed tables are effective speed reducing measures at junctions. Treatment of junction corners by the use of an uneven road surface is useful for slowing down turning traffic, particularly HGVs, however, this has no effect on vehicles going straight ahead.

Changes in Alignment

This technique involves narrowing the carriageway in the vicinity of a junction so that all moves through the junction have to deviate from a straight ahead path. This measure is fairly effective in reducing the speed of straight through traffic; however, junction priorities may become confused as the boundaries of the junction are less recognisable.

Reduction of the Junction Area

The area of a junction may be reduced by building out the footways. This is carried out on junction corners at crossroads or T-junctions. In addition the footway may be built out along the straight at a T-junction. The latter technique is suitable for reducing the speed of straight ahead vehicles, while the former is only effective for turning vehicles.

Islands

The provision of islands on the approaches to a junction has only a limited effect on vehicle speeds. They will however improve the situation for pedestrians by creating a refuge, although by reducing the amount of carriageway they may hinder large vehicles when turning.

Special Junction Forms

Roundabouts and mini-roundabouts may be useful in reducing traffic speeds at junctions in semi-urban conditions.

Change of Cross-section

There are locations on the highway where the numbers of lanes reduced from 6 lanes to 4 lanes or 4 lanes to 2 lanes. The traffic calming measures are essential on these stretches. The reduction in lane width is done in stages:

- Stage I: Visually reduction in carriageway width by painting of wider strips by keeping the roadway width same.
- Stage II: Reducing the carriageway width with increasing the shoulder width in same amount such that roadway width remain same.
- Stage III: Keeping the carriageway width required and reducing the shoulder width with reduction in roadway width.

Road-Rail Level Crossing

A vehicle has to stop at rail level crossing in case of gate closure. Therefore, traffic calming measures on these locations are very essential. The provision of speed humps or speed table or pavement with different material can be provided along with proper signage as traffic calming measure.

Planning for Implementing Traffic Calming Techniques

A detailed study and planning is required to be done before implementing the traffic calming techniques to ensure the safety of road users:

- Land Use
- Road length and width
- Volume of vehicular flow
- Composition of vehicular traffic, i.e. modal split
- Identification of Black spots, road accidents history, timings and accident spots
- Major landmarks
- Encroachments
- Existing traffic signals
- Bus stop locations
- Existing traffic-calming techniques

Whenever traffic-calming techniques are planned, care should be taken to select the appropriate structures which will ensure free flow of vehicles without traffic congestion. The wrong selection leads to congestion and sometimes even responsible for accidents.

Indian State of Practice

Tree Plantation on Both Sides of Highway

This is the prevalent practice in India since ages to plant trees on both sides of the highways. These trees were planted to give shade and fruit. These trees make the road narrower visually and act as a traffic calming measure.

Installation of Community Gates

The various types of gates were installed during procession. Community gates were also installed on highways to inform the motorists about type of community staying in settlement ahead. These gates acted as traffic calming measures.

Chungee Gates by Municipalities/Gram Panchayat

The Gates were installed by municipalities/Gram Panchayat to collect the chungee for the traffic passing through the settlements. As traffic has to stop for payment, this acted as traffic calming measure.

At Start of Bridge

A special type of architectural semi arc like structure was generally constructed on left side of each motorway at start of bridge in Tamil Nadu. These structures are visible from the far-off distance on the highway and inform the driver about the presence of bridge ahead. These structures acted as a traffic calming measures.

Speed Tables in Small Settlements

When a highway cuts through the small settlements, water accumulates on the road during rains due to poor drainage. These small stretches were sometimes raised with brick paving due to shortage of funds. These small stretches acted as traffic calming measures.

Speed Humps on Rail Level Crossing

There are several rail- roads crossing at grade in India. Speed humps are generally constructed on both sides of the rail track to reduce the speed of approaching traffic.

Creation of Temporary Chicanes by Police

Police used to keep barriers before start of the settlement to check crimes. The traffic has to slow down at these barriers and acts as traffic calming measures.

Construction of Service/Frontage Road

Whenever a number of cross roads are joining the highway at short distances, these cross roads have been joined together by providing a frontage road or service road. The traffic intends to join the main road traffic has to slow down before meeting. This has also created space for movement of cyclist and pedestrians.

Roundabouts

In case of two or more highways are intersecting at grade in semi-urban areas, roundabouts are generally provided.

Conclusions

Traffic calming on highways is a relatively new concept. Traffic calming is primarily concerned with reducing the adverse impact of motor vehicles. The accidents rate and its severity can be reduced significantly by proper planning and incorporation of traffic calming measures. The highways may be made self-explaining and cross roads joining highways (main roads) shall be made self-enforcing by implementing various traffic calming measures.

REFERENCES

Arosanyin, G.T., Road Accident Costs and the Southern African Economy : An Overview, *Indian Highways,* Volume 31, no.2, 2003.

Brewer Jim, German John, Krammes Ray, Movassaghi Kam, Okamoto John, Otto Sandra, Ruff Wendell, Sillan, Seppo, Stamatiadis Nikiforos and Walter Robert Report no. FHWA-PL-01-026, Geometric Design Practices for European Roads, Federal Highway Administration, U.S Department of Transportation, 2001.

Ewing Reid, *Traffic Calming State of the Practice,* Institute of Transportation Engineers. U.S., 1998.

General Objectives of Traffic Calming, U.S. Department of Transportation, Federal Highways. Administration, 2004.

Gupta, D.P., Road Safety in India: Current Situation and Strategies Proposed, *IRC Journal,* Volume 64-3, December, 2003.

Ken Fox & Y.L. Siu, *A Review of Current Traffic Calming Techniques,* Institute of Transport Studies, University of Leeds LS2 9JT, England.

WWW.EcoCity Cleveland. Reclamation of streets.

16

SAFETY IN INDIAN TRANSPORTATION

V.D. Sharma*

ABSTRACT

The contribution of the road transportation sector in the Indian economy is enormous. The pace of growth of this sector is getting affected with the road accidents and with the result it is not giving its full to the Indian economy. As this is the dream of every citizen of India to have journey free of accidents on the Indian roads and for achieving the same the responsibility each and every stake-holder is being emphasized in this paper. In this prospective the contribution or the good maintenance of transport fleet by following the TPM practices is also discussed to a great length.

Introduction

Road transport sector may be called the backbone of the Indian economy like transportation through sea may be called the backbone of the Indian export and import. Road transport sector is holding the major share of about 70 per cent of the total cargo transported within India. Road transport industry is contributing about Rs. 40,500 crores in the Indian economy and almost 7 million vehicles are engaged in this operation. It is not surprising that it is helping about 22 million persons to earn their bread and butter through this industry. Road transportation is having definite advantage over rail transportation because of the deep penetration in the remotest nook and corner of India and cheaper fabrication of vehicle body as per requirement and more personalized service.

Transportation in the urban context assumes great economic significance, as the productive efficiency of urban areas will be maintained only if mobility requirements in the cities are fully met. However, this productive efficiency is now threatened due to the increasing number of vehicles causing congestion, and thus slower speeds on roads. An effective way of addressing this problem would be to encourage the greater use of public transport instead of personal vehicles. This requires both an increase in the carrying capacity of the public transport system and a significant improvement in the quality of public transport. Despite the high volumes of traffic, most urban centres in India do not have any rail transit

*Mother Diary, Patparganj, Delhi-110 092, E-mail: vishwa.sharma@motherdairy.com

system to cater to the requirement of public for intra-city movements. Hence, there would be substantial dependence on bus services to meet public transport needs for the next several years. This is likely to call for a major infrastructure restructuring of public transport provisioning in the urban centres.

Unplanned urbanization and increase in the number of vehicles have led to transport congestion and hence increase in road accidents. The normal drift for more and the better avenues of employment in urban centres, inadequacy of proper maintained roads, poor vehicle maintenance and improper arrangement for pedestrians have made the matter worse on the Indian roads. Transport services are largely related to the development of the infrastructure, availability of spare parts, availability of the authorize workshops at short intervals all along the roads and assurance of the quality services at these places. A large population of India lives in the rural areas where inadequacy of education and improper facilities of driving training is one of the major causes of road accident. Accidents are always a curse for any transportation as it eats away almost all your reserves. You may find these data readily available in auto journals, magazines and on web sites. Even your daily newspapers are rarely free of this information. The safety in road transportation has become a major debatable issue especially in developing countries like India. Road accidents have become one of the leading causes of death and injuries and lead to enormous cost to the society. While the economic losses can be computed, the pain and the anguish of the bereaved can neither be measured nor compensated.

I have not taken up this topic in a characteristic manner and therefore, you will not find anything under this writ up on the Government policies and what all we expect from it for a safe road transportation in India. On the contrary I have tried to emphasize on all those issues that are to be taken care of by every individual and an organization as whole being a stakeholder to ensure a safe and accident free journey on the Indian roads.

Paper Details : Road Safety Initiatives at Mother Dairy, Delhi

At Mother Dairy we take the issue of Road Transportation Safety with great concern and strongly feel that it needs to be addressed by multi-facilitated and multi-disciplinary research and technology innovation programme, incorporating infrastructure, human factors, enforcement etc. As government alone cannot find solutions, therefore, a well-coordinated effort is required by all stakeholders for better safety under road transportation. The traffic behaviour on the Delhi roads is quite complex to analyse. The main factor that can be attributed to is the prevailing mixed traffic, which comprises of different type of fast and slow moving vehicles with varied vehicular and traffic characteristic. Any analytical technique of traffic flow can only hold good on simple homogeneous vehicular traffic and we are not going to yield any results by applying it on complex Delhi traffic. To be more precise we need real life observation to analyse this unhomogeneous and complex traffic flow on Delhi roads and such procedures are quite tedious time consuming and expensive. However, at Mother Dairy we are conscious of the fact and analyzing every accident major or minor and its learnings are used for elimination of the root cause of the same. Following are some typical areas where urgent improvement is required if we really want our road transportation to be safe in reality.

- Accident data collection and its analysis
- Road safety research
- Vehicle testing/inspection
- Effective driving training and stringent testing
- Traffic education to children in schools
- Publicity and propaganda
- Effective police enforcement
- Real emergency medical services
- Effective road safety legislations

Mother Dairy is always conscious of the fact that we are also one of the stakeholder and always doing our best to ensure the road safety. Both the dimensions, i.e. Human and Equipment primarily responsible for road safety are being focused by adopting the under-mentioned ways and means.

By Keeping the Smoke Emission of Fleet within Specified Limits

Mother Dairy is being authorized by the Government of National Capital Territory of Delhi, Transport Department for issuing the PUC certificate to all our diesel vehicles, in case they are meeting the prescribed emission standards. Our pollution-checking centre is having the latest requisite testing equipments and functioning as per the recent upgraded guidelines provided to us.

Educational Qualification

Education is essential to understand the systems to read and under stand the road signs, to communicate properly with the people and the officials when required. Further education plays role in understanding the vehicle he is driving and the legal system in which he operates. In Mother Dairy the minimum qualification for a driver is Metric. We at Mother Dairy normally employing Ex-service men as drivers considering them to be better disciplined and having adequate knowledge of road signs and traffic rules. At the time of induction of drivers written, driving test and interviews are ensured.

Driving Habits

Driving habit is an important factor influencing the road accidents caused by drivers. A well-trained driver in general has better driving habits. Flawless driving, respecting the traffic signals, sign boards, pedestrians and other vehicles is essential to have accident free driving. The good habits are being inculcated in our drivers through repeated trainings.

Trainings

Education, information and training can teach drivers better and safe road behaviour and it is important that drivers under stand the meaning of road signes and the necessity of obeying the speed limits. Regular trainings are ensured in Mother Dairy for our drivers on technical as well as on traffic aspects to help in building good driving habits. We are conducting trainings on road signes and other traffic rules with the help of Delhi Traffic Police periodically. PCRA is also helping in arranging trainings at our works on our request on fuel conservation and road safety. Regular in house training on the technical aspects is ensured through our own resources. As it is our constant endeavour to have trained drivers with the Best Driving Practices for better road safety and all of this is being achieved by adopting the above practices.

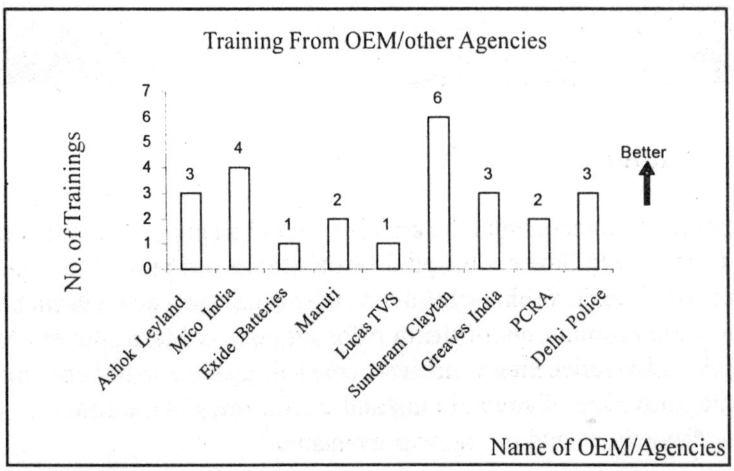

Vehicle Safety and Maintenance

Accident Free vehicular traffic can only be ensured by keeping our vehicles maintained. At Mother Dairy the maintenance team and vehicle operators are working hand in gloves. Vehicle Maintenance activities are mainly divided in two parts, i.e. Autonomous Maintenance and Planned Maintenance. Four Groups consisting of 4 persons of different discipline are reporting in round the clock shifts for the Autonomous Maintenance in coordination with the Autonomous Team of Vehicle operators identifying and eliminating all kind of running defects. For evaluating the effectiveness of Autonomous Maintenance monitoring of unreported and the reported defects is ensured. The said activities have helped us reducing the Break Downs by inducing the sense of ownership among the operators as well as Maintenance team.

Graphical Presentation of Some of the Results

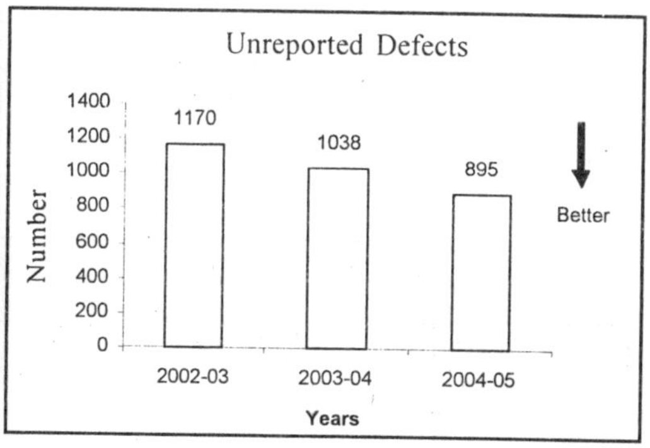

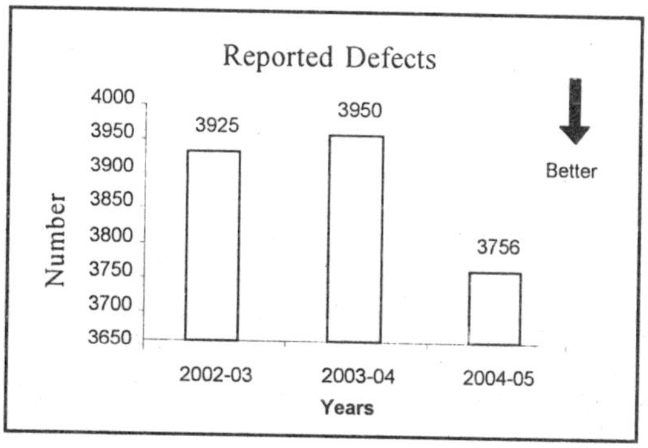

Under Planned Maintenance all the Major and specialized jobs are attended by qualified and technically competent staff. All the activities under Planned Maintenance are being carried out by adopting the laid down guidelines by the manufacturers and regular trainings through the service staff of these Original Equipment Manufacturers are ensured for updating the Knowledge and Skills. Some of the OEMs which have recently conducted training are Lucas India, Mico India, Ashok Leyland, Sundram Clayton (TVS), Greaves India, Indag Rubbers, Amron Batteries, JK Tyres, Pricol India etc.

The inputs received during these trainings helped us in improving the Maintenance system and upgrading the Maintenance Schedules in view of new developments and upgradations. On account of these some of our maintenance Schedule could be upgraded from 2000kms to 3000kms, 8000kms to 15800kms, and 16000kms to 32000kms.

This has helped us in achieving better availability of road worthy tankers, lesser break downs, better fuel and oil average, increased MTBF, reduced MTTR and thus safe and timely milk operations.

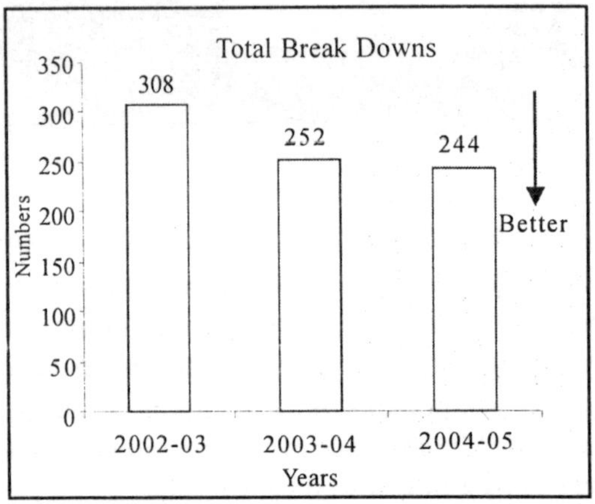

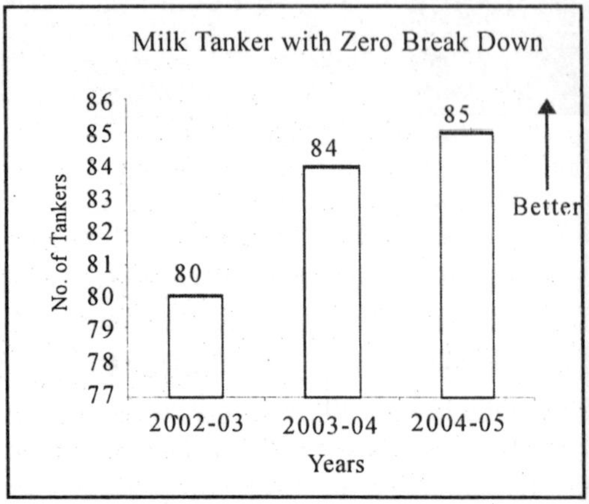

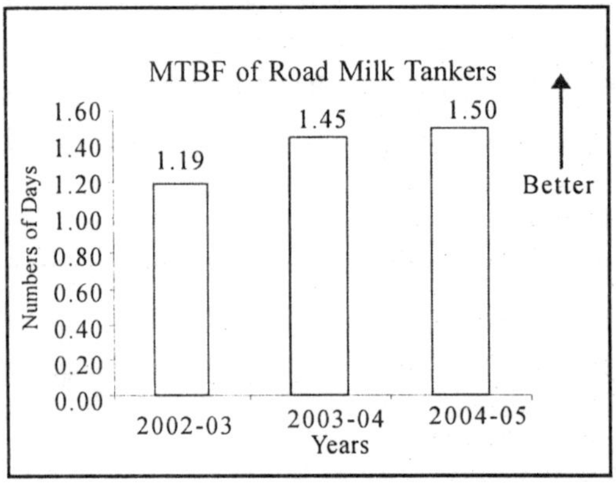

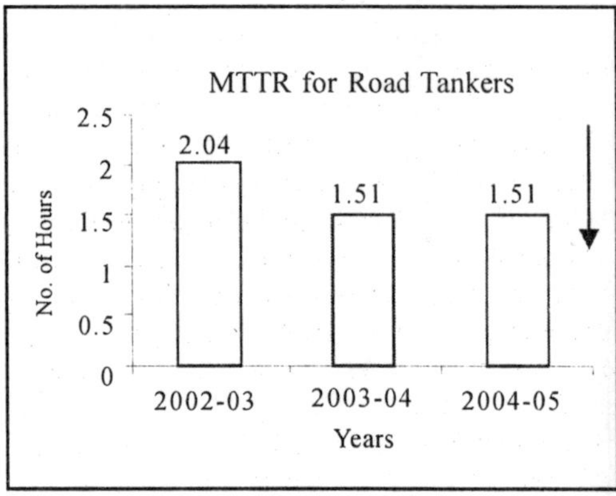

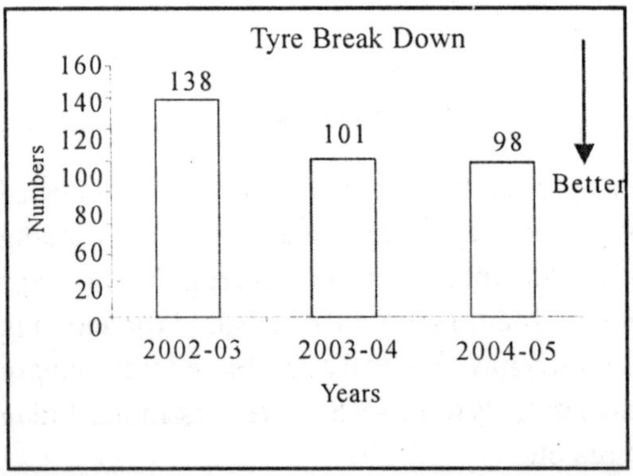

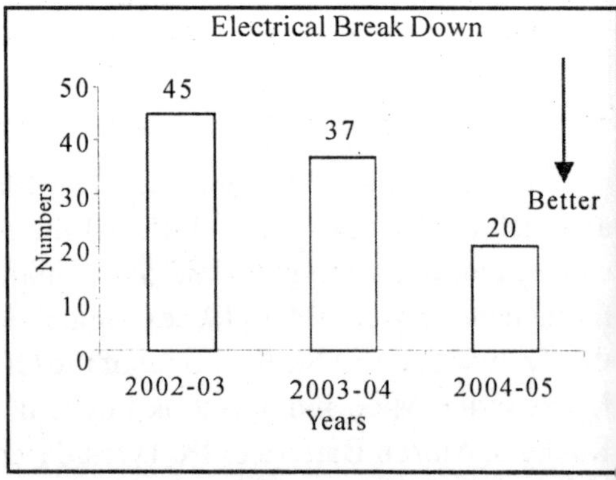

Poor up keep and ineffective maintenance is also a major factor in increasing the percentage of accidents. We in Mother Dairy are maintaining our fleet by strictly following the preventive maintenance schedules thus reducing the break down of vehicles on route and road accidents. We are discarding our commercial vehicles after 15 years of use as per as per the honourable Supreme Court guidelines.

All the above activities has helped us in improving the equipment reliability, driving skills, updating of awareness on traffic signs and rules resulting into tremendous road safety.

Design Factor

Vehicle design is an important factor influencing road accidents. For example unprotected access with in the space between the front and the rear wheels of our tankers may cause accident by trapping with in this space. To overcome this problem, necessary changes are being made in the designing of the mounting of milk barrel by providing two extra guards in a row between front and rear wheels thus obstructing the direct access to this space. Any local modification to the chassis is avoided like drilling of holes to accommodate extra accessories resulting in its failure.

Mixed Fleet

It plays a significant role in road accidents as these days variety of automobiles with almost similar specifications from different manufacturer are available in the market and all these models are having their controls at different locations thus affecting the response time of driver and may result in to accident. In Mother Dairy we are sticking to a single make of vehicle to avoid this problem.

MOTHER DAIRY
MOTHER DAIRY
B4739

Over Speeding

Over speeding and reckless driving is one of the major contributory factors for accident and moreover they also adversely affect the fuel average of the vehicle. To over come this issue we have provided Electronic speed governors in our tankers restricting the highest speed to 40 Kmph. The Honorable Supreme Court of India has also made the fitment of E.S.G. in all the commercial vehicles mandatory in the National capital region.

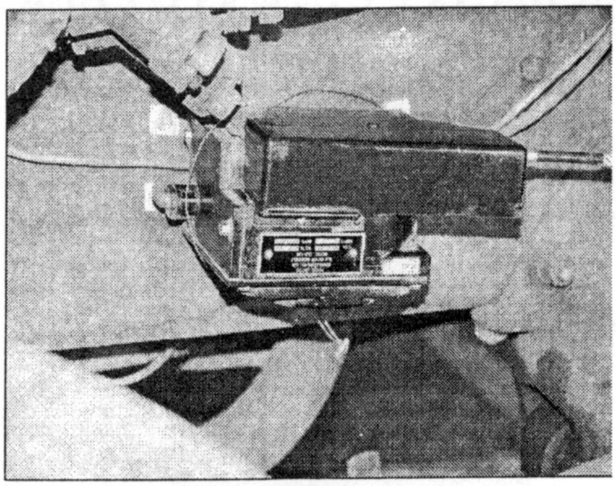

Conclusion

While concluding I would once again like to put my emphasis based on our experience and achievements that effective people's participation for improving the reliability of machines and development of safety conscious operators is the key to make Indian roads much safer for driving and free of accidents. Although Mother Dairy is having a very little share in terms of number of vehicles on Delhi roads but its importance can always be felt as its participation in catering to uninterrupted supply of milk for the citizens of Delhi all 365 days and for over last 30 years and always conscious of the road safety.

REFERENCES

B. Vittaldasa Prabhu & S. Murali on 'Dimension of road safety' in *Indian Journal of Transport Management* Vol. 28 (4), pp. 532-546.

Indian Journal of Transport Management, Volume 28, Number 4, October-Dec. 2004, pp. 475-477.

17

IS INDIAN TRANSPORT SYSTEM SAFE?

H.R. Vaish and Abhijit Rai*

Light Emitting Diode (LED) Technology—A Revolution in Lighting for Many Safety and General Applications

Incandescent bulbs, which are now considered as "heaters", generate 95 per cent heat and accidentally give out light. A renowned lighting expert, practicing and teaching in U.S.A., who has handled numerous lighting projects for organizations over 50 years, said "wholesale use of Sodium vapour lamps for street-light applications are a disaster to our society".

Worldover, lighting applications are shifting to energy efficient, long lasting and maintenance free LED lights, which not only upto 90 per cent on power costs but also greatly reduce the maintenance cost and have a half-life of 100,000 hours. In LEDs, which have properly designed power supplies and control circuitry, there is no catastrophic failure as is the case with bulbs and CFLs.

In a calculation done by the Municipal Corporation of Portland, Oregon for traffic lights the savings observed are depicted (see page 137).

Major advantages of LED vs Incandescent Lighting

- Energy savings: 80-90 %
- Life: 100 times longer than incandescent bulbs
- Maintenance Free
- Interchangeable into existing fixtures
- Extremely good quality of light—High CRI
- Best light source for Solar Power Solutions
- Environmental friendly—No Mercury
- Operates on low DC voltage

---

*Instapower Ltd., S-19, Panchshila Park, New Delhi-17

— power supplies can be designed to work on dc voltage, or
— Operates between 80VAC-300VAC

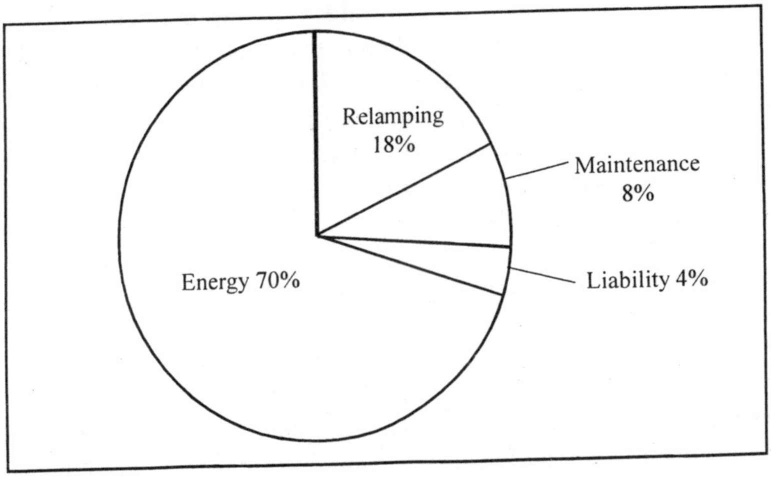

(http://www.sustainableportland.org/news_LED_case_study.pdf)

The small size and rugged construction allows design engineers to have the flexibility to design products which are aesthetically attractive and eye catching and provide special circuits and designs to give long life and conserve power.

The number of products that are already being used by industry and that will be developed using LEDs are increasing because the intensity of LEDs have exponentially increased over time. A number of products for air-field lighting are being developed such as taxiway lights, constant current light sources, helipad lights, beacon lights etc. To obtain the best life from an LED, the critical aspect in designing LED systems is a thorough knowledge of electronic circuits as these diodes are voltage and current sensitive. Even if a few milli-amps of excess current flow through a LED, there could be significant chance that it would fail.

To get the advantages, which LEDs have to offer, and thereby save energy, there is a paradigm shift required in our thinking of the design of products. Instead of going purely by thumb rules we need to understand the actual need of the user. The role of specifications based on an in-depth study of user requirements, the design keeping in view these user requirements and product specifications, so as to obtain the advantages of LEDs, becomes very very important. The understanding of optics considering the use of multiple sources of light also becomes a critical factor in product design and meeting the end user requirements.

Along with design, comes the need for having standards organizations, which oversee the quality control of the final products and test laboratories, which have manpower suitably trained to understand these requirements. Most of the government laboratories are not equipped to measure the photometric parameters for LED lights. Also, no laboratory in India has proper sources of light for calibrating colour light such as red, yellow, green, blue etc. The turn around time for measurement of even simple parameters and calibration of equipment by National Physical Laboratory is several months which is unacceptable. Since, there are large gaps in all these areas, the end users unlikely to get the expected benefits.

The attitude towards safety requires a constant vigil to ensure that accidents do not take place. There is no point crying over spilt milk when we are complacent in the first place. When a life is lost in the western world due to a lapse, there is a huge hue and cry and responsibility is fixed and significant penalties levied on the individual(s) concerned. However, in developing countries like India, there is a great deal of laxity and lack of accountability because of which even when hundreds of lives are lost, no action is taken. Committees are formed, recommendations are made, but they gather dust till the next accident takes place, after which fresh recommendations come.

A case study of a product for air safety will be used to illustrate what has been stated.

Air Safety

With the increasing number of aircrafts criss-crossing the sky, it is critical to install safety equipments not only on the air-fields but also on high rise structures like buildings, chimneys, and mobile and transmission towers. Extremely costly flying machines and precious lives are at stake and the standards of safety set up by international organizations, like International Civil Aviation Organisation (ICAO), Montreal, Canada, must be implemented in letter and spirit.

Some of the important specifications given by ICAO for aviation obstruction lights are as under :

(1) A viewing angle in the vertical plain of 6 deg. in the case of low intensity light and 3 deg. in the case of medium intensity light.
(2) A fixed light in the case of low intensity light and 20 to 60 flashes per minute from medium intensity light.
(3) 360 deg. visibility in the horizontal plain.
(4) A light intensity of minimum 10 candela for low intensity lights and 2000 ± 25% candela for medium intensity lights.

These are the specifications given by ICAO, but they have left it open to the manufacturers and the country concerned to implement these specifications in the products used for this purpose. Although the Civil Aviation Ministry or the Airports Authority of India should have involved themselves in preparing the specifications and approval of the product design for this purpose, it has been found that the products are being installed without any check by any agency. It is a disaster waiting to happen.

The status regarding implementation of the above four points in such lights is as under:

Most LED lights seen in the market have a greater angle than stipulated but this is not a case of concern. There is also no problem in regard to obtaining flashes from the LED lights. Hence the first and second points are easily achieved.

However, as regards the third point, it is worth noting that before the LEDs came in, high intensity bulbs were used for medium intensity lights. Since these bulbs radiated light in all directions, they met 360 deg. requirements. In the case of LEDs the light is in the axial direction. Many manufacturers, therefore, have tried to achieve the 360 deg. requirements by creating hexagonal or octagonal models, using vertical strips of printed circuit boards mounted with an array of a large number of LEDs. However, when measurements have been made, it has been found that there are very large gaps of darkness between the two bright areas of light. The medium intensity light being mounted on high-rise structures is to be visible from a long distance of several kms. These large dark gaps, visible even at one-metre distance, would obviously create a dark zone of several kms. in width at a distance.

The figure 17.1 below illustrates the problem:

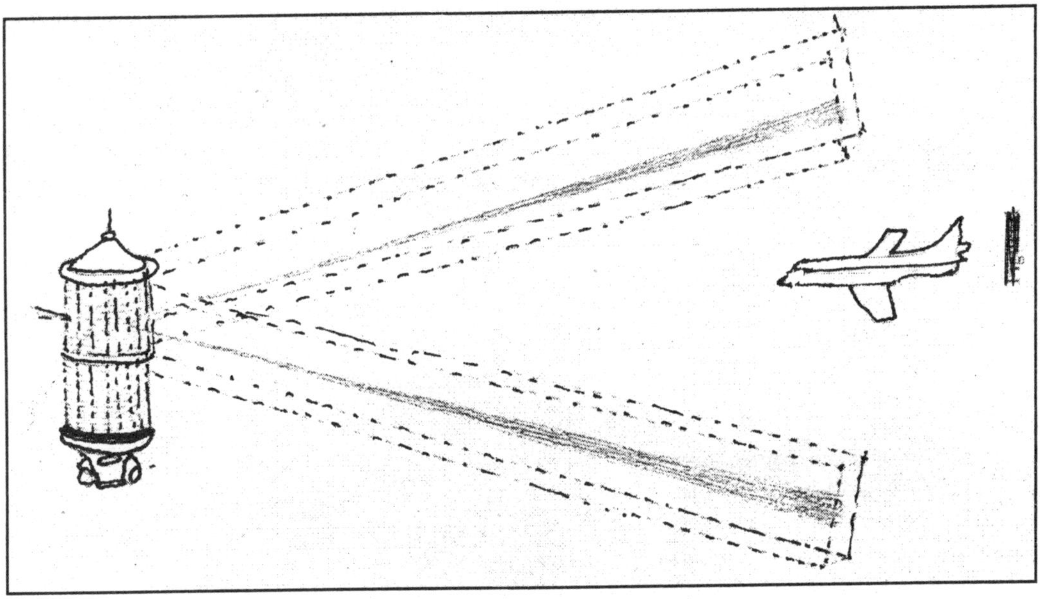

As regards the fourth point, the intensity of light required has been stipulated as 2000 ± 25 per cent candela. The obvious implication is that at all times when the light is working, ICAO wants 1600 to 2500 candela. It is well known that the intensity of any light source, including LEDs, deteriorate with time and a light source which is 1600 candela at the time of purchase, will fail to meet the standard within a few months thereafter.

The measurement of intensity of a light source and the calibration of such measurement equipment is not so easy in India and the average user is unable to do anything in this regard. Some manufacturers have in a very simplistic manner claimed that having put 200 LEDs of 20 candela each on a platter, they are providing 4000 candela of light, which would take care of deterioration in time also. Unfortunately, when a measurement is done, these multiple sources of light do not add up to show that the light source is providing 4000 candela. In fact, it may be only about 500 candela. However, given below are the test results obtained by two manufacturers of the same product, from the same laboratory which is India's most respected:

A. METHODOLOGY OF TEST

THE EFFECTIVE I IS CALCULATED FROM INTEGRATING THE INSTANTANEOUS LUMINOUR INTENSITY, I, BASED ON BLONDEL RAY EQUATION.

INSTANTANEOUS INTENSITY IS MEASURED FROM THE OSCILLOSCOPE TRACE OF PHOTODETECTOR SIGNALS.

RESULT

OPERATING INPUT VOLTAGE 230 VOLT AC

INSTATENOUS LUMINOUS INTENSITY VARIATIONS OF FLASHING LIGHT ARE GIVEN BELOW.

EFFECTIVE INTENSITY FOR NIGHT MODE IS 4000 Cd.

COMBIND RELATIVE UNCERTAINITY IN MEASUREMENT OF EFFECTIVE INTENSITY IS +2.5% AT K=2

B. ELECTRICAL PARAMETERS

	VOLTATE	CURRENT	WATTS
	12	6.624	79.48

Flashing Frequency

On Time = 1.0 Sec. Off Time = 1.0 Sec. (Variable)

OPTICAL PARAMETERS

I effective = · ?1.dt
0.2+(t2-t1)

MEASUREMENT OF EFFECTIVE INTENSITY

Intensity value	Effective Intensity cd
1826	1611 Peak.

It may be noted that the report nowhere states that the product tested is a medium intensity aviation warning light and that it is tested as per ICAO specifications.

Not being satisfied with the above implementation, to achieve the specifications of ICAO, the author experimented and developed a product using just 12 super-bright LEDs with secondary optics, to achieve all the four requirements of ICAO listed above. LEDs have been focused such that the illumination as measured, is more than 2400 candela, while maintaining the minimum 3 deg. angle.

The figure is given below:

This light source is revolved at an r.p.m., which is adjustable from 20 to 60, so as to provide not only the 360 deg. visibility, but also 20 to 60 flashes per minute, as stipulated. A patent has been filed for this product, which is now being manufactured by M/s Instapower Ltd.

Extracts of the test report obtained by the author are as under:

Nomenclature			Led Based Medium Intensity Aviation Obstruction Light
Applicable Standard Specification IS/IEC/JSS/other			ICAO (ANNEX-14)
Parameter	National Value Requirement	Measured Value Observations	Remarks.
Illumination Measurement at 18VDC for 1 metre distance.	2000+/- 25%	Satisfactory 2480 Lux	--
Vertical Beam Spread at 18 VDC for 1 metre distance.	3 Deg. (mim)	Satisfactory 4.58 Deg.	---

It is worth noting that the test laboratory has mentioned lux measured at 1m. It is well known that equipment, which measures lux. At 1m, lux is equal to candela but the test laboratory was not knowing this and not willing to certify this even when informed.

Instapower is using the LEDs from a company owned by M/s Philips, which is the most reputed manufacturer of such LEDs in the world. There are some other manufacturers where the intensity of the LEDs reduce even within a month and, hence, quality control over each aspect of the product is very critical to ensure that the customer gets the benefits. Needless to say there are large numbers of products in the market from manufacturers who do not even have the measuring equipment required for testing their products. It is worth noting that the test house certificate very clearly brings out that the light meets the ICAO Specifications.

What ICAO has not talked about is the energy consumption of the product, the warnings to be given in case of its failure and synchronization of lights. The author has taken care to minimize the energy consumption which is about 12 watts and likely to become lower, as against other products which take more than 50 watts and still do not meet the specifications as narrated above. It may be mentioned that many of these lights are installed on structures where it is difficult to provide power and hence, they operate on solar power. Over 12 hours of daily operation with autonomy for a few days depending on the solar insulation of the area, this difference in power consumption accounts for a saving of lacs of rupees in the cost of solar panels and batteries. Moreover, the product has been designed to run directly from a single 12-volt battery. Not only is there a major saving in the capital investment, but also in the recurring maintenance cost necessary for replacement of products.

In case of any failure of an LED, a micro controller ensures that the intensity of the other LEDs is increased to compensate for the loss of light from the failed LED. Beyond a certain level of failure, an alarm has been provided to give an audio and visual signal at the base station. With the advancement of the technology it is now possible to also give wireless signal, for example, by using cell phones. In many applications multiple lights are installed on a structure and the user may specify synchronization of this light. This has also been incorporated in the product.

It is thus a foolproof product design which will provide years of trouble-free service, as contemplated by ICAO, but not mentioned in their brief specifications.

Similar aspects are involved in LED products designed for road and rail safety.

Rail Safety

Instapower, after dedicating four man years in designing, has designed a state-of-the-art LED Train Signalling System. The author has applied for national and international patents on this design. This system is foolproof and useful for the Railways worldwide. The train signalling system using LEDs is much more reliable than the conventional systems and has numerous features, including early warning and system failure alarms. With the computerization and technology now available, it is possible to time the failure and rectification of such signals and fix responsibility for lapses.

Road Safety

Vehicular traffic and speeds are increasing exponentially. It therefore becomes imperative to have the vehicle brake and the sidelights in good working condition at all times.

The filament bulbs fail from jerks and considering Indian road conditions, most of the buses, trucks, trailers and heavy vehicles do not have working taillights. In many cases even the headlights don't work. It is, therefore, important that LED products be used for preventing accidents and casualties. A study done by General Motors has found that LEDs light up one million times faster than incandescent bulbs thereby giving an extra breaking distance of 17.5 ft. at a speed of 60 miles per hour. This greatly helps in reducing accidents on roads and highways. The author is working on a new concept of traffic signalling, which can increase not only safety but make the signals unambiguous, reduce down times and installation expenses. Synchronization of signals and wireless messaging and control can provide many features to manage the complexities of traffic on the Indian roads.

Traffic lights connected with solar panels make the traffic light self-reliant and eliminates the need of power. Similarly, the road navigation signages and markers have become critical to provide information and demarcate the obstacles. The LED technology when clubbed with solar provides an ideal arrangement for road navigation. In case of any blackout or bad weather they prove to be a boon for the drivers.

Social Economic Reasons to use LED Lights

LED Lights do not have poisonous and non bio-degradable such as mercury or tungsten, which are a cause of health hazard to the society on the whole. Light from the LEDs are clean and pollution free. By using LED lights in a big way in India the power shortages in the country would tremendously reduce as also the need to construct new thermal power plants. USA has already adopted this approach and it would be wise for us to do so, too.

In conclusion, the author would like to urge the Indian engineers, sociologists, decision makers and all those who can contribute for the cause of safety to:

1. Consider safety as a major issue like it is in the Western world.
2. Provide a safer environment for our society and family.
3. Use LEDs for the various advantages enumerated above.
4. Develop institutions like R.D.S.O. of Railways, which can set safety standards, certify products which meet those standards, and test the products on an ongoing basis.

For more information on Instapower product range, please visit us at www.instapowerindia.com or send us an email at sales@instapower.com

Section–II

Safety in Energy Sector

18

OCCUPATIONAL SAFETY AND HEALTH AT INDIAN OIL

S. Kaul*

ABSTRACT

Indian Oil handles large quantities of crude oil and petroleum products and is conscious of the fact that safe operations are necessary for higher productivity and well-being of all concerned. This paper probes how Indian Oil takes lead not only complying with applicable legislation but also by developing sound and systematic safety, health and environment management systems. It also takes stock of its Corporate Safety Health and Environment policy which provides the basic framework to continuously strive for higher standards of performance.

Introduction

The success of any business depends in large part on the men and women who work for it. Ensuring their safety and health on the job makes good business sense. It also is the right thing to do. To be successful in injury prevention, an organization must construct a solid foundation from which to build. The foundation is built by the organization's clear vision of the future and the specific means by which it will get there through the achievement of the mission. As a principle of "how we do business within this organization", a healthy corporate culture is based on positive and respectful values, principles, and beliefs towards safety, health and the conservation of environment. In successful organizations this culture is established throughout all business strategies reflecting an integrated approach and philosophy. By establishing and integrating these fundamental themes into a common managed system, organizations are then able to build meaningful standards system and procedures/practices resulting in the reduction/elimination of injury causes, losses to the environment, property, process, equipment, materials, as well as personal injuries and adverse health effects.

It is a matter of fact that safety, health and environment systems require a continual and systematically managed effort in order to achieve sustainable success. The key SH&E organizational success components included (but not limited) the following:

*Indian Oil Corporation Limited, Corporate Office: SCOPE Complex, Core-2, 7, Institutional Area, Lodhi Road, New Delhi-110 003. E-mail: www.indianoilcorp.com

- Successful organizations have a common set of fundamental values, beliefs and principles that are integrated into all organizational business strategis.
- There exists a belief that all occupational injuries, illnesses and losses are preventable.
- Legislative requirements reflect minimum standards only.
- Ensuring safety, health of workers, clients, community and protection of environment are fundamental to organizational culture.
- Each and every employee is responsible for the successful management of S H&E.
- S H&E requires a continuous improvement process and integrated managed system.
- Organizations are required to possess the internal capacity to successfully sustain a system for continuous improvement and integration of S H&E.
- Full commitment and active participation of all workplace stakeholders are required for the continued success and development of effective workplace programmes. Stakeholders include employees, collectives, management, clients, suppliers, community, government etc.
- Leadership, commitment and participation are required by top management and board of directors.

Occupational Safety and Health

Occupational safety and health refers to the physiological and socio-psychological conditions of an organisation's workforce, resulting from the work environment. It encompasses varied responses to number of compelling influences, the most basic of which is a sense of social and humanitarian responsibility.

Today, health and safety management is a complex activity requiring the expertise of specialists from many disciplines, such as industrial hygiene, occupational medicine, ecology, psychology, and safety engineering, to name only a few. Moreover, concerns in health and safety management now reach beyond physical conditions in the workplace to embrace a regard for worker's mental and emotional well-being, and a commitment to protecting the surrounding community from pollution and exposure to toxic substances.

The Growing Importance of Occupational Safety and Health (OH&S)

Presently a number of factors are focusing attention on OH&S that may take it in new directions. The working environment has to be considered in the management of day-to-day operations because it affects workers both as employees and as members of society as a whole. Safe, health and environmentally sound working conditions must be a priority for socially responsible employers. They are part of an organization's public image, contribute to positive public relations, and should be reflected in a commitment to employee safety'and health that extends beyond economic benefits to long term consequences for workers and their families.

It is obvious that helping employees to stay healthy besides preventing unnecessary hardships to employees also benefits the organization. The old adage "Prevention is better than Cure" is applicable more in an industrial environment than anywhere else.

Some of the major developments predicted by the International Labour Organization over the next decade include: increased awareness and support of OH&S on the part of government, workers and employers, significant improvement in the control of chemical hazards and programmes of injury prevention.

When organizations reduce the rate and severity of occupational accidents and improve the quality of work life for their employees, the following benefits normally occur:

- Productivity is higher because fewer man days are lost;
- Employees become more efficient due to increased involvement with their jobs;
- Medical and insurance costs are reduced due to fewer claims;
- Worker compensation rate and direct payments are also lowered;
- The workforce is more flexible and adaptable a result of increased participation and a feeling of ownership;
- The organization is more attractive as place to work; and
- There are fewer fatalities.

Organizations are spending more time, energy and resources in promoting health and safety in the work place not only because of the alarming costs of inadequate occupational heath and safety, but also because safe conditions are legally required, and constitute sound management practice.

Integrated Safety. Health and Environment Management system at Indian Oil

The explosive and flammability hazards of crude oil and products are intrinsic. Indian Oil handles large quantities of crude oil and petroleum products and is conscious of the fact that safe operations are necessary for higher productivity and well being of all concerned. Safety, therefore, is an integral part of our operations and a very important factor in business strategy. The vision of Indian Oil is to be "A major, diversified, transnational, integrated energy company, with national leadership and a strong environment conscience, playmg a national role in oil security and public distribution". One of our missions is "To help enrich the quality of life of the community and preserve, ecological balance and heritage through a strong environment conscience".

Indian Oil has taken a lead not only by complying with applicable legislation but also by developing sound and systematic safety, health and environment management systems. Our corporate Safety, Health and Environment Policy provides the basic framework to continuously strive for higher standards of performance. In order to translate the elements of the Policy into action, guidelines on seventeen identified elements have been developed for formulating and standardizing the sound and systematic safety, health and environment management systems for fulfilling the corporate Vision and Mission.

Elements of SH&E Management System

SH&E Leadership and Commitment

The corporation has visible management leadership, commitment and accountability at all levels. Senior Management establishes policy, provides guidelines, sets expectations and ensures availability of resources. Managers at operating facilities establish the scope, set priorities and speed up implementation of SH&E Managernent System. Employees at all levels are accountable for Safety, Health and Environmental performance. SH&E performance parameters have been established at the corporate level and monitored on a quarterly basis. Two tier safety committees, namely Management Safety Committee and Shop-floor Safety Committee have been formed at locations. Safety Awareness Survey is carried out to assess the awareness level and identify the areas of concern and necessary inputs are given. Developing a green canopy by tree plantation in and around our installations is one of our sustainable efforts.

Employee Participation

Employees' participation in management is culturally ingrained in Indian Oil. All concerned employees participate in shop-floor safety committee meetings for better safety performance. The suggestions are reviewed at departmental level and the accepted suggestions are implemented. Their involvement as partners in day-to-day activities has made implementation of S, H & E Management System more effective. An Employees Suggesting Scheme is in place throughout the corporation where suggestions related to Safety, Occupational Health and Environment are also given and are reviewed and acted upon.

Process Safety Information

Process Safety information is essential to the safety and reliability of operations. At Indian Oil, Process Safety Management Systems consist of comprehensive procedures, and practices designed to ensure that barriers to major accidents are in place, in use, and effective. We have always kept pace with developments in the area of Process Safety. Information is accurate, comprehensive and accessible.

The major initiatives in improving process safety are:

- Proper attention to ensure appropriate lines of defence in design, their quality by way of Quality Assurance Plan and operation within the design intent.
- Implementation of the state-of-the-art technology, e.g. DCS for process optimization and enhancing safety by way of redundancy.
- Practice of HAZOP (hazard Operability) studies and Quantitative Risk Analysis to identify Process Safety Hazards and remedial actions for new facilities/extension of plants. Compliance with statutory requirements and applicable standards/codes.
- Institutionalization of ISO 9000 Quality Systems across the organization to standardize procedures and practices.

For Process Safety management, three lines of defence are very important. As a first line of defence the process is operated and maintained as designed and the chemicals are contained. The first line comprising of three layers is essentially provided to avoid the release of chemicals. Therefore, attention is paid to different layers of this line in design engineering stage.

The first line is backed by a second line of defence with two layers, one prevents unwanted release and other controls or mitigates the consequences in case there is an unwanted release.

The third line of defence is the "Emergency Preparedness (Disaster Management) Plan", both onsite and offsite. The third line of defence is required only when other two lines of defence fail.

Risk Analysis and Management

Systematic reviews are carried out to evaluate risks to heip prevent incidents from happening. Qualified personnel from both within and outside each operating facility conduct regular assessments of potential hazards to employees, facilities, the public and environment. HAZOP studies and quantitative risk analysis are carried out for new facilities and augmentation of the existing ones. Major risk contributors are identified and remedial actions taken before implementation of the project/facility. Based on the findings of Risk Analysis studies, we have resited many plants and also installed mounded bullets for storage of LPG. At some places Metallurgy upgradation has been done especially with respect to Hydrogen sulphide service.

Reliability of Critical Systems and Devices

Systems have been put in place to ensure that safety critical and environmental critical systems and devices are identified, tested and undergo maintenance for reliability. Each critical system or device is periodically tested. Systems are in place, which control the defeat of critical systems and devices by temporary disarming or deactivation. Through pre-commissioning check up, reliability of safety system and devices of a new facility is ensured. Single and double redundancy has been provided based on the criticality of the system.

Facilities Design and Construction

All construction projects, from small improvements to major expansions and grassroots units are evaluated early in the design stage for their safety, health and environmental impact. For each project, thorough reviews are done to ensure that:

- Measures are taken to prevent operating incidents,
- Emergency procedures are in place,
- Personnel are trained,
- All regulatory requirements are met, and
- Risks have been assessed and are being managed.

Operation and Maintenance Procedure

Operation and Maintenance procedures are assessed and modified to improve safety and environmental performance. Operation, maintenance and inspection manuals are updated to include changes in the respective procedures. Units and locations have structured operation, maintenance and inspection systems, reliable safety and control equipment and qualified people who continuously upgrade their knowledge on safety procedures and practices. To maintain safe and uninterrupted operation, preventive and predictive maintenance are carried out both for stationary as well as rotary equipment in plants and offsite facilities. 100 per cent Positive Material Identification (PMI) is done for all metallurgy.

Occupational Health and Hygiene

All refinery locations have well equipped Occupational Health Centres. Doctors and paramedics are specially trained to monitor the health of employees working in hazardous areas. At locations where these facilities are not available, employees' health check up is carried out through retainer doctors. All Occupational Health Centres are certified under Occupational Health and Safety Management System (OHSMS)

The employees are regularly exposed to training programmes in work-related hazards and ways and means of protection from such hazards.

The health care personnel regularly interact with shop floor managers and staff in the shop floor meetings on health and safety.

Various media of Communication like House Journals, Posters, Video Films, etc. are extensively used for creating awareness.

Environmental Management

Pollution prevention plans, developmental activities for Environment Management are established, communicated and their execution stewarded in line with the Environmental Management System of each Refinery, Marketing and Pipelines facilities to support continual improvement. The latest statutory requirements and changes in legislation are communicated to all the divisions for compliance and status monitored. Treated wastes from our facilities are monitored for compliance with MINAS and or limits set by State Pollution Control Boards. Recycling and reuse of treated effluent is maximized. Sludge inventory at all locations is monitored and action plans drawn for liquidation of the same.

All Refineries and major Marketing and Pipeline Installations are certified under ISO-14001 for Environmental Management Systems. Green belts have been developed in refinery units and Marketing and Pipeline locations. Ecological Parks have been developed in some refinery units. Maintenance of the scientifically designed Green Belts is done to ensure maximum survival of the planted trees and also fresh tree plantation is carried out.

Personal Safety

The implementation of accident prevention measures is monitored at various levels to ensure the safety of all personnel. Employees are safeguarded by equipping them individually with conventional and specialized personal protective equipment (PPE). Input on proper use of PPE's is imparted to personnel working in the plant and project, through training and demonstration.

Training

Systems are in place to ensure that employees are qualified and trained to meet specific job requirements. In addition, ongoing and periodic refresher training is imparted to mitigate potential safety, health and environmental hazards. Special workshops at corporate level are organized with participation of delegates from all the divisions. Assistance of external faculty is also requisitioned.

Contractors and Business Associates

Contractors and temporary workforce are important links to safe operations. The contractors are educated to perform in a manner that is compatible with Indian Oil's policies, procedures and practices, especially with regard to safety, health and environment. Contractors' performance requirements are defined and communicated and they include responsibility for providing trained personnel and maintaining a system for self-monitoring and stewardship. A full chapter on S, H &E management system is incorporated in the tender document itself and compliance of the provisions therein is strictly monitored. Thrust is maintained on use of appropriate PPE by the contractor workmen with adequate supervision and effort is made to ensure adherence to safety systems and procedures by all contractors. Accident related Penalty clause has been firmed up and is incorporated in General conditions of Contracts (GCC) of all contract jobs.

Work Permit System

Work Permit System is an element of safe work practices. Requirements of OISD Standard 105 on "Work Permit System" are complied with. For performing any work in the plant/ facility by any person other than the operating personnel of that area, duly authorized written permit is obtained before commencement of the work. Specific clearance is given for man entry into confined spaces. Separate clearance is given for, blinding/deblinding operations.

Management of Change

Set procedures have been developed to manage change to process chemicals, technology equipment and facilities. Changes in operations, procedures, regulations, site standards, facilities or personnel are tested for safety, health and environmental impact. The review process covers such areas as authority for approval of changes, analysis of safety, health and environmental implications and standards, documentation and communication and training requirements.

Emergency Planning and Response

Good preparation can significantly reduce the impact of an accident. Indian Oil maintains a comprehensive emergency planning and preparedness system to ensure that in the event of an accident, all necessary actions are taken to protect the public, the environment and company personnel and assets. Onsite and offsite emergency response drills are conducted at units and installations. The District Civic authorities and Mutual Aid Scheme members are involved. The areas of improvement are identified and remedial action taken. The Disaster Preparedness Plans are updated. All units and installations also maintain regular contact with local communities to keep them apprised of contingency plans and other matters of important concern.

Incident Investigation and Analysis

All Incidents are investigated. Procedures are in place to provide timely reporting and for investigation of incidents including identification of root causes and contributory factors. Recommended actions are taken to reduce the risk of similar incidences.

Safety Audit

Safety systems are audited for continuous improvement. ESA and SSC are conducted by OISD, MoP&NG. The status of implementation of the recommendations OISD safety audits and checks are placed before the board. Internal safety audits are also carried by the Divisions. A Safety Index system has been developed and audits are carried out as per this system. The status of implementation of recommendations is monitored at the divisional level.

Conclusion

The Safety and Health of employees in their occupational environment is becoming increasingly important. Employers are becoming more aware of the benefits of having a healthy workforce. The Government thr7ough statutory regulations is making it all the more necessary for employers to be concerned with employee safety and health

The image formed of an organization regarding its concern for occupational safety and health is also a positive factor in attracting new employees. To sum up, effective programmes for the improvements made on the Safety, Health and Environment front can enhance significantly both employee health and effectiveness of fhe organization.

19

ELECTRICAL SAFETY ON INDIAN RAILWAYS

R.S. Grover*

ABSTRACT

Indian Railways are the third largest railway system in the World under a single management and is the principal mode of transportation for bulk freight and long distance passenger traffic. It handles daily 1.7 million tonnes freight and 15 million passengers across the country. Therefore, safety is *sine qua non* for train operation on this gigantic railway system covering 63,000 Rkms. With emphasis on prevention of accidents on Indian Railways, avoidance of Electrical accidents is one critical area, which needs to be addressed with due seriousness and importance. Electricity if not handled with due care, may lead to heavy damages and casualty. Indian Railways have formulated legislations and laid down rules/regulations to ensure public safety due to electrical equipments in locomotives, coaches and railway stations/installations. These regulations are derived from Electricity Act and accordingly appropriate Electricity Rules are made. There also exists a suitable link between Electricity Act/Rules with Indian Railway Act. Railway have special requirements in train operations, certain amount of discretion has been provided in the IE Act as well as IE Rules, which in turn are expected to be taken care of in IR Act. Additional Member Electrical at apex level, i.e. Ministry of Railways and Chief Electrical Engineers of Zonal Railways have also been assigned role of Electrical Inspector—a statutory authority under the IE Act.

This paper is an effort to bring out obligations and responsibilities stipulated in these Regulations and to assess level of compliance at present, and also to identify areas where implementation needs to be streamlined and suggest modalities for strengthening the institution of *Electrical Inspector* in accordance with the Electricity Act/Rules as part of corporate responsibility.

Introduction

Indian Railways is the principal mode of transport of the country and have completed 150 years of service to the nation. It has played a vital role in the overall development of the country and national integration. During 2004-05, Indian Railways had carried 604 million tonnes of freight and 5513 million passengers. Therefore, safety in train operation is one of the important mission areas of Indian Railways. Accordingly steps have been taken to reduce train accidents and tangible results have been achieved during last 2 decades. One of important area of concern is electrical safety in train operation and maintenance of rolling stock and other fixed assets.

*Advisor Electrical, Rail Vikas Nigam Ltd., New Delhi, Former Addl. Member Electrical, Rly. Board., Ministry of Railways

Indian Railways—Use of Electricity

The Indian Railways use various types of carriages and vehicles (referred to as Rolling Stock) that have electrical equipment for traction and controls, lighting, ventilation, air-conditioning and other passenger amenities. In addition electrical power is required for various other fixed installations for convenience of passengers at railway station, besides maintenance of various assets. Some of the important applications are as under:

- Passenger carriages are provided with electrical equipments for generation and distribution of electricity at various voltages, e.g 110 V AC and DC, 230 V AC, 415 V AC, 750 V AC etc., for lighting, ventilation, air-conditioning, heating and various types of pantry car equipment. Further, the development of equipment using high voltage and power-electronics system of capacity up to 500 kW is in progress.
- In Electrical/Diesel locomotives and Electrical Multiple Units and Diesel Multiple Units, high, medium and low voltage systems for traction and controls.
- 25 kV AC and 1500 V DC system traction Over Head Equipment as well as 750 V DC third rail system for feeding power to locomotives and electrical rolling stock.
- Railway traction substations are fed at 132/220kV for traction and 66/33/11kV for general services. Supply of power supply for lighting and ventilation in Railway stations, workshops, maintenance yards, goods sheds, administrative and ancillary offices, pumping and sewage installations, signalling and telecommunication equipments, computer installations for operations, commercial and other purposes, construction equipments etc. With technological improvements and need for automation, electrically operated equipments are increasing use in various applications. Supply is either taken from the local electricity supply authorities, or OHE and supplemented with other sources such as diesel generating sets, uninterrupted power supply systems etc.

Electrical Safety Regulations

Provision in the Act

Basic provision in Indian Electricity Act for ensuring electrical safety is Section 53 which empowers the Central Electricity Authority to specify suitable measures for:

"Protecting the public from dangers arising of electricity and specifying action to be taken in relation to any electric line or electrical plant, or any electrical appliance under the control of a consumer for the purpose of eliminating or reducing the risk of personal injury or damage to property or interference with its use."

Chief Electrical Inspector and Electrical Inspector are authorities who are to be appointed by Appropriate Government under Section 162 of Act to ensure enforcement of provisions of Act and Rules.

In respect of Railways, the Chief Electrical Engineers are also the Electrical Inspectors to the Government, as decided by the Appropriate Government. In addition to this the Additional Member Electrical of Ministry of Railways is also designated as the Chief Electrical Inspector for Ministry of Railways.

Applicability and Extent of Compliance

With dual role of Chief Electrical Engineer and Electrical Inspector, enforcement of various provisions in Act and Rule are by and large followed and in-built with regard to electrical equipments and works directly under technical and administrative control of Chief Electrical Engineer. In respect of fixed installations, there exists a practice of concerned officers and staff making a formal application with all related documents to Electrical Inspector for commissioning of new and modified electrical works.

Strict compliance of the following sections of the Act and the Rule are not being observed by the Railways:

(a) Section 45 of Rules titled Precautions to be adopted by consumers, owners, occupants, electrical contractors, electrical workmen and suppliers stipulates:
"No electrical installation work should be done by a person not holding a certificate of competency and by a person not holding a permit issued or recognized by the State Government.
No electrical installation work which has been carried out in contravention of sub-rule (1) shall either be energized or connected to the works of any supplier".
No exemption has been officially notified by the Railways under this section and there are several areas where electrical works are sometimes executed by persons not having the requisite permit and are also not supervised by competent personnel.

(b) Section 5 of Rules titled "Entry and Inspection" authorizes the Electrical Inspector or any other officer appointed to assist him to enter into any place, carriage or vessel where there is an electrical equipment or apparatus for generation, transmission, transformation, conversion, distribution or use of energy takes place and carry out tests therein and check observance of provisions of the Act.
No set practice exists in the Railways presently to cover all such areas indicated above under the purview of the Electrical Inspector or his assistant.

(c) Section 47(A) of the Rules titled "Installation and testing of generating units" stipulates:

Any consumer or occupier installs a generating plant, he shall give a thirty days' notice of his intention to commission the plant to the supplier as well as the Inspector and shall not commission his generating plant of a capacity exceeding 10 KW without the approval in writing of the Inspector.
Strict compliance of this provision is not being followed at times, especially with regard to Rolling Stock applications.

(d) Section 161 of the Act titled Notice of Accidents and Inquiries stipulates prompt reporting of electrical accidents by person in charge of equipment to Electrical Inspector.

Strict compliance of this provision is also not being followed, again especially with regard to Rolling Stock applications.

Special Provisions for Railways

A specific case is regarding the exemption of Railways Rolling stock from the applicability of Chapter IX (Electric Traction) of the Rules as detailed in Section 94:

Section 94: Additional rules for electric traction: The rules in this Chapter apply only where energy is used for purposes of traction

The items covered under this chapter pertain to the conductors used for electric traction and for the Railways; these items are in turn covered elaborately in the AC Traction Manual.

Another case is the exemption of Railways from issue of advance notice to prior to transmission or use of electricity in public places, as detailed in Section 54 of the Act:

Section 54: Control of Transmission and Use of Electricity: no person other than the Central Transmission Utility or a State Transmission Utility, or a licensee shall transmit or use electricity at a rate exceeding two hundred and fifty watts and one hundred volts in any street, or in any place in which one hundred or more persons are ordinarily likely to be assembled; or which is a factory within the meaning of the Factories Act, 1948 or a mine within the meaning of the Mines Act of 1952, or

In this case also, with regard to fixed installations Railways have taken adequate precautions under the AC Traction Manual and other laid down procedures. But with regard to Rolling Stock, adequate steps have to be taken by the Railways.

Role of Electrical Inspector in Electrical safety of Rolling Stock

Railways have laid down a detailed procedure for opening of new lines, introduction of new rolling stock etc., *vide* Gazette notification dated 21st July 2000. However, the role of Electrical Inspector is defined clearly only in case of approval of electric traction equipment.

As regards, use of new types of locomotive or rolling stock, an application is required to be made to the Government through Commissioner along with six items. One of six items is a Safety certificate, to be jointly signed by Chief Engineer, Chief Mechanical Engineer and Chief Electrical Engineer (for Electric stock). Chief Signal and Telecommunication Engineer and Chief Operations Manager are also required to sign certificate for certain specified cases. However, definition of Electric stock is absent in the document. Also since the certificate is to be issued by the Chief Electrical Engineer as different from the Electrical Inspector, it is presumed that the maintenance aspects and interference with OHE etc., only are taken care of and not the fundamental issue of Electrical safety. There is an urgent need to include a separate certification from the Electrical Inspector clearing the electrical safety of rolling stock additionally, while making the application to the Commissioner.

Penalties

Another important aspect in the Act, as provided under Section 148 is the applicability of penalty for non-conformity to the provisions, even for works pertaining to the Appropriate Government, as detailed below:

Section 148: Penalty where works belong to Government: The provisions of this Act shall, so far as they are applicable, be deemed to apply also when the acts made punishable there under are committed in the case of electricity supplied by or of works belonging to the Appropriate Government.

This makes it all the more imperative that a proper machinery be installed for proper monitoring and implementation of all the applicable provisions in the Act.

Responsibility of Railways

From the foregoing, it is evident that there is a definite responsibility on Indian Railways not only as a public service provider involved in generation, transmission, distribution and utilization of electricity, but also as Appropriate Government, to institute adequate machineries and lay down rules, regulations and procedures for strict compliance of various provisions of Indian Electricity Act and Indian Electricity Rules truly in letter and spirit.

In this context, some of the observations made by the Hon'ble High Court of Delhi while delivering the judgment on the Association of victims of Uphaar Tragedy and Others vs. Union of India and Others are very pertinent and relevant.

The Government and its agencies would also be liable for not having ensured strict compliance with Rules and Regulations which have been created to ensure safety.

Safety is the first aspect which has to be looked into by all concerned including the Board and even if there is any apprehension of the safety standards having not been followed, the transformer ought not have been made functional.

This litigation has highlighted the malaise of indifference, sloth and indeed "chalta hai" attitude, which affected the statutory authorities required to ensure and guarantee public safety. It is very clear that the tragic loss of innocent lives could have been avoided/ reduced if rules for public safety were given the importance, which the legislature intended, and the executive disregarded.

Functions of Electrical Inspector to the Government on the Railways

Presently Chief Electrical Engineers of Zonal Railways are expected to discharge functions of Electrical Inspector in Railways. The duality of role both as a maintainer and as an enforcer of rules creates a situation of complicity for items under his technical and administrative control. A similar situation existed in State Governments but now they have established an independent institution of Electrical Inspectorate to deal with issue of electrical safety exclusively.

The present system of implementing institution of Electrical Inspector on Railways is becoming increasingly inadequate and ineffective on account of large number of electrically operated equipments getting introduced into system as already explained and number of agencies involved, many of them technically and professionally not competent to operate and maintain such assets safely. With volume and complexity of electrical assets increasing manifold and need for exercising adequate control over various agencies involved with such assets, there is a pressing need for augmenting, if not separating institution of Electrical Inspector on Indian Railways, before system goes out of control leading to public trauma and tragedy of a magnitude witnessed as in case of Uphaar Cinema fire disaster.

Operative Mechanism

Various options are available for organizational structure of Institution of Electrical Inspector on the Indian Railways:

- Maintain existing system with necessary augmentation.
- Modify existing system to extent of creating a separate and exclusive identity of Electrical Inspector under the Chief Electrical Engineer.
- Differentiate the functions of Chief Electrical Engineer and Electrical Inspector by setting up a separate entity.

In addition to structure of organization, several additional systems, regulations, procedures and instructions need to be introduced to take care of various safety aspects and statutory obligations brought out above, which have so far either not been addressed or implemented adequately.

Conclusion

Railways are a gigantic organization who are involved in transportation of men and material, therefore, safety is of paramount importance. In order to get out of the complicity arising out of the dual roles and also loss of objectivity arising out of common administrative control of Electrical Inspector and Chief Electrical Inspector, it would always be preferable to have a separate entity who will be directly under the Ministry of Railways.

SAFETY MANAGEMENT IN INDIAN COAL MINES

S.K. Puri*

ABSTRACT

This paper deals with the safety aspects in Coal mining industry in India. Taking stock of mining hazards, it looks at the problems of stability of underground opening and underground environment. Further studying the frequency rates of mine accidents and fatality rates, it describes the measures taken for improvement of safety.

India is the third largest coal producing country. The production of coal has increased substantially after its nationalization in 1972 and 1973. In 1955-56 the Coal Production in India was barely 38.4 Million Tonnes (Mt), which increased to 375 Mt during 2003-04, while that of Coal India was 323 Mt. Coal is the most important and abundant fossil fuel in India. It accounts for 55 per cent of the country's energy need. Commercial primary energy consumption in India has grown by about 700 per cent in the last four decades. The current per capita commercial primary energy consumption in India is about 350 kgoe/year which is well below that of developed countries. Driven by the rising population, expanding economy and a quest for improved quality of life, energy usage in India is expected to rise around 450 kgoe/year in 2010. Considering the limited reserve potentiality of petroleum and natural gas, eco-conservation restriction on hydel project and geo-political constraints of nuclear power, coal will continue to occupy centre-stage of India's energy scenario.

With hard coal reserves around 246 billion tonnes, of which 92 billion tonnes are proven, Indian coal offers a unique eco-friendly fuel source to domestic energy market for the next century and beyond. Hard coal deposit spread over 27 major coalfields, are mainly confined to eastern and south central parts of the the country. The lignite reserves stand at a level around 36 billion tonnes, of which 90 per cent occur in the southern State of Tamil Nadu.

Safety in Coal Mining Industry

Mining is said to be most hazardous peacetime operation. All out efforts are being taken by Coal Mining Industry in India to reduce the hazards to the minimum. The Coal Mining is

*B-7/77/1, Safdarjung Enclave, New Delhi-110 029. E-mail: surinderk_puri@yahoo.co.in

broadly categorized as Underground Mining and Opencast or Strip-mining. Of the two methods of Mining the Underground mining is relatively more dangerous. There are hazards inherent in underground mining that individually and collectively are not found in any other industry. Mining industry as a whole is characterized by high accident rate and high fatality rates, and high incidence of occupational diseases. The main reason is that by its very nature, mining is a continuous fight against nature. Fresh ground is exposed every moment and everyday miner works in an area where none has been there earlier than him. Unlike in the factory the working conditions in an underground mine change everyday and quite often they change drastically and suddenly. The risk in underground mining are primarily because of the following problems are:

The Problem of Stability of Underground Openings

Making an underground excavation involves the withdrawal of natural support from the ground around the excavation which therefore becomes unstable. To overcome the problem artificial supports are erected by the mines management, falls of roof and sides do take place in underground mines causing accidents to mine workers and damage to mine property. Due to increase in depth and size of underground mines workings with the passage of time the problem is becoming acute day by day.

The Problem of Underground Environment

The environment of the mine is far from natural. Ventilation is artificial, the intake air may be charged with noxious and inflammable gases, noise and dust are inevitable byproducts of all mining operations. Besides there is total darkness except in artificially lit areas. Thus there are night like darkness even during day hours minus moon and stars. Temperature and Humidity conditions vary from place to place and the conditions may be very tough in deep mines. At a few underground mines there may be constraints of height, and the miners may not be able to stand erect. Keeping the body in bent posture creates its own problem.

Additionally there are serious risks of fires, explosions, inundation release of noxious gases and affliction with occupational diseases. Roadways are not always level, quite often they have steep gradients and dripping water may make them highly slippery. Moreover the machines have to be worked and maintained in the workings which may be narrow, of low height and totally dark.

The problems are no doubt gigantic but over the years much progress has been made in making mine conditions relatively better. Change in methods of working, equipments, appliances, and instruments, and better understanding of problems and awareness of dangers, effective preventive measures have helped in reducing accidents in Undergound Mines. However, new Technologies have brought in new hazards, for example use of larger size and more powerful machines, higher voltages and higher speeds produce more noise, more dust, reduced clearances, greater risk of overloading, electrical sparks and shocks and narrower margins of error permissible in human judgement. These necessitate frequent reviews, better training, stringent monitoring and measurement of legal compliance and greater inputs in research, development and design.

Safety

Safety is an abstract term rather difficult to define. The progress of the mankind over the ages through Stone Age, Copper Age to Iron Age has been in pursuit of safety. Men could conquer the hazards of wild nature, namely extreme hostile weather and danger from wild animals. As he learnt agriculture, use of fertilizers and chemicals became a necessity. These brought in new hazards. These hazards were not only hazardous to human health but have also started affecting the environment as well. The growth of civilization brought in its trail hazards of his own making chemical, nuclear, vehicular etc. Hazard has been defined under Occupational Health and Safety Management System (OHSAS -18001) as source or situation with a potential for harm in terms of injury, ill health, damage to property, damage to workplace environment or combination of these. Classical definition of Hazard is potential of an accident. Mine safety connotes the condition of being free from the occupational hazards of mining.

Mining hazards are of three types:

- Those that cause (or could cause) bodily injury.
- Those that are injurious to health.
- Those that can damage the property.

While an accident is immediately and easily recognized, the health hazards are not always apparent. The occupational diseases develop over a long period of time and the worker may not be aware of the hazard until some loss or impairment of the body function becomes apparent. International experience proves that more people are killed or disabled by mine accidents, yet the accidents receive spotlight of concern while the silent insidious killers go unnoticed.

What is an Accident ?

Any mishap in which one or more persons are killed or injured or cause damage to property constitutes an accident. It may be defined as an unexpected event causing loss of life or bodily injury or damage to company property. The Mines Act does not define an accident. Section 23 of Mines Act is titled a notice to be given of accidents but makes no attempt to define an accident. It states that, whenever there occurs in or about a mine : an accident causing loss of life or serious bodily injury, or an explosion, spontaneous heating, outbreak of fire, inrush of water or gas premature collapse of any part of the workings over winding of cages etc. a notice has to be given in prescribed form to the prescribed authorities. Even if these occurrences do not cause of life or bodily injury, they are deemed to be accidents because of the important aspect is the unexpected occurrence and not the effect it produces. It might just be a matter of chance that nobody was present at the place when the event occurred. In the interest of safety it would be necessary to find out the cause and circumstances leading the occurrence so the suitable measure can be taken to prevent such a mishap in future.

In the year 2002 CIL achieved a major breakthrough as regards reduction of fatal accidents and fatalities from a range of 80 to 95 accidents and 100 plus fatalities to 62 accidents and 69 fatalities. In 2003 it was possible to further reduce the number of fatal accidents and fatalities over those achieved in 2002 to 56 fatal accidents and 60 fatalities.

On an average 250 persons got killed in mine accidents every year upto 1999-2000. Of these about 175 got killed in Coal mines and 75 in metal mines. The number of persons receiving serious bodily injury in mines was about 11000 per year. Of these 800 are injured in coal mines and about 300 in metal mines. Detailed statistics of fatal and serious accidents are given in the following tables.

Measurement of Safety Performance

So far it has not been possible to quantify the safety status of a mine or a factory. Although it is true that absence of accidents is not safety, yet accident statistics are the only means available for judging safety standards. The yardsticks of accident frequency rates and severity rates are universally accepted for comparison of safety performance.

Frequency Rates of Mine Accidents

In order to evaluate the standard of safety in Mines or to compare the safety status of one group of mines with another or to make international comparisons, it has become a common practice to express accident rates in relation to the exposure to risk, that is howmany persons are killed or injured against the number of persons and duration for which they are exposed to occupational risks. The usual practice is to calculate the death rates and injury rates based on :

(a) One million man hours worked,
(b) One lakh shifts worked, or
(c) One thousand persons employed.

In India the figures of man-hours worked are not easily available, and therefore, the frequency rates are calculated on the basis of one lakh manshifts worked or 1000 persons employed. Safety Statistics for Coal India Ltd., at ten-year intervals since formation of the company in 1975, till 2003 are given below in Table 20.1. Company-wise accident statistics for Coal India Ltd. for 2003 are given in Table 20.2.

Table 20.1: Safety Statistics for CIL since formation at 10 year intervals till 2003.

Year	*Fatal Accidents*		*Serious Accidents*		*Fatality Rate*		*Serious Injury Rate*	
	Accidents	*Fatalities*	*Accidents*	*Injuries*	*Per MT*	*Per 3 lakh manshifts*	*Per MT*	*Per 3 lakh manshifts*
1975	177	233	1456	1515	2.62	0.52	17.03	3.41
1985	136	152	507	524	1.15	0.31	3.97	1.07
1995*	107	115	575	612	0.50	0.27	2.65	1.14
2000	80	100	510	540	0.37	0.25	1.99	1.36
2003	56	60	339	354	0.20	0.17	1.18	1.00

* Figures for 1995 are excluding disaster. Figures for 2003 are subject to reconciliation with DGMS.

Table 20.2: Company-wise accident statistics for Coal India Ltd. for 2003

Company	*Fatal Accidents*	*Fatalities*	*Serious Accidents*	*Serious Injuries*	*Fatality Rate per M.T.*	*Fatality Rate Per 3 lakh manshifts*	*Serious Injury Rate Per M.T.*	*Serious in Rate Per 3 lakh Manshifts*
ECL	12	12	144	151	0.43	0.14	5.47	1.77
BCCL	12	13	33	36	0.56	0.21	1.54	0.57
CCL	5	5	14	15	0.14	0.11	0.41	0.32
NCL	3	3	14	14	0.07	0.20	0.30	0.93
WCL	7	10	46	47	0.26	0.17	1.22	0.82
SECL	10	10	77	80	0.14	0.15	1.15	1.18
MCL	7	7	11	11	0.12	0.44	0.19	0.69
NEC	0	0	0	0	0.00	0.00	0.00	0.00
CIL	56	60	339	354	0.20	0.17	1.18	1.00

Table 20.3: The figures for the death rates and serious injury rates per thousand persons employed in the last few years for coal and metal mines.

Year	*Death Rate*		*Serious Injury Rates*	
	Coal Mines	*Metal Mines*	*Coal Mines*	*Metal Mines*
1994	0.46	0.42	1.48	1.25
1995	0.43	0.35	1.58	1.38
1996	0.29	0.41	1.43	1.45
1997	0.33	0.43	1.43	1.49
1998	0.30	0.38	1.14	1.55
1999	0.29	0.45	0.86	1.14

Table 20.4: Fatality Rate per Million Tonne of Coal Produced

Year	*Australia*	*France*	*India*	*CIL*	*Japan*	*USA*	*Germany*
1995	0.02	0.12	0.50	0.50	0.32	0.05	0.26
1995*	0.02	0.12	0.77	0.83	0.32	0.10	0.26
1996	NA	0.37	0.48	0.44	0.00	0.04	0.25
1997	NA	0.15	0.54	0.43	0.47	0.03	0.19
1998	0.02	0.15	0.47	0.40	0.47	0.03	0.19
1999	0.02	0.00	0.45	0.40	N.A.	0.03	NA
2000	N.A.	0.29	0.46	0.37	N.A.	N.A.	N.A.
2001	N.A.	N.A.	N.A.	0.37	N.A.	N.A.	N.A.
2002	0	N.A.	0.28	0.24	N.A.	0.03	N.A.
2003	N.A.	N.A.	N.A.	0.20	N.A.	N.A.	N.A.

Fatality Rates of India Compared to some Developed Countries

Comparison of fatality rates with some developed countries indicates that though fatality rates per million tonne of production for India and CIL is not favourable, fatality rates per 1000 persons employed and per 3 lakh manshifts deployed is comparable with some developed countries like USA and Japan.

Table 20.5: Fatality Rate per 1000 Persons Employed

Year	*Australia*	*France*	*India*	*CIL*	*Japan*	*USA*	*Germany*
1995	0.15	0.07	0.28	0.26	0.41	0.36	0.15
1995*	0.15	0.07	0.43	0.46	0.41	0.36	0.15
1996	NA	0.23	0.26	0.27	0.00	0.31	0.14
1997	NA	0.08	0.33	0.28	0.65	0.24	0.11
1998	0.18	0.08	0.30	0.24	0.06	0.25	0.11
1999	0.29	0.00	0.29	0.27	NA	0.30	NA
2000	N.A.	0.13	0.30	0.25	N.A.	N.A.	N.A
2001	N.A.	N.A.	N.A.	N.A.	N.A.	N.A.	N.A.
2002	0	N.A.	0.21	0.19	N.A.	0.32	N.A.
2003	N.A.	N.A.	N.A.		N.A.	N.A.	N.A.

Table 20.6: Fatality Rate per 3 Lakh Manshifts

Year	*Australia*	*France*	*India*	*CIL*	*Japan*	*USA*	*Germany*
1995	0.17	0.16	0.27	0.27	0.42	0.48	0.33
1995*	0.17	0.16	0.42	0.45	0.42	0.48	0.33
1996	0.39	0.51	0.28	0.24	0.00	0.41	0.32
1997	0.24	0.20	0.31	0.25	0.63	0.31	0.25
1998	0.39	NA	0.29	0.24	NA	NA	NA
1999	0.31	0.00	0.27	0.25	NA	0.40	NA
2000	N.A.	0.29	0.46	0.37	N.A.	N.A.	N.A
2001	N.A.	N.A.	N.A.	0.27	N.A.	N.A.	N.A.
2002	N.A.	N.A.	N.A.	0.19	N.A.	N.A.	N.A.
2003	N.A.	N.A.	N.A.	0.19	N.A.	N.A.	N.A.

Safety Scenario in Open-Cast Mining

Nearly 75 per cent of total Coal production is being obtained by opencast mining. The opencast workings being free from many hazards of underground mining like roof falls, explosions, suffocation by gases, inundations are inherently safer than underground mining. Unfortunately the accident experience in Indian open cast mines shows a different trend.

Measures taken for Improvement of Safety

The following major safety measures were taken in the year 2003 for further enhancement of safety in the coal mines of CIL:

1. Introduction for Risk Assessment as a tool for improvement of Safety-Risk Assessment has been conducted in some mines in each coal producing company of CIL
2. Conducting of Safety Audits of mines by external experts for assessment of threats and implementation of the recommendations thereof were continued.
3. Thrust was maintained on preparedness for emergencies through:

   (a) Reviewing Emergency Action plans for each and every working districts in underground mines.
   (b) Demarcating escape routes on plans as well as below ground.
   (c) Conducting of mock rehearsals and monitoring failure points for further improvement.
4. Towards avoidance of accidents due to inundation:
   (a) Taking the required preventive measures against danger of inundation after assessment of danger from underground and surface sources of water in each mine.
   (b) Conducting of check surveys by company's surveyors as well as cross-checking the same in some cases by external agencies like ISM & CMPDIL.
   (c) Efforts were taken for identification of water bodies below ground without physically proving the barrier by drilling through.
      (i) Electrical Resistivity survey by CMRI in ECL.
      (ii) Seismic method for identification of water body in MCL by NIRM.
      (iii) Ground Penetrating Radar by NIRM.
   (d) Establishment of centralised emergency pumping arrangement for use during emergencies.
5. Thrust on measures for reduction of roof and side fall accidents in below ground mines were continued by:
   (a) Drawing up of Support Plans based on RMR and implementation of the same.
   (b) Greater use of Roof Bolting/ Stitching methods of roof support by quick setting cement capsules.
   (c) Introduction of Resin grouted Roof Bolts in watery seams.
   (d) Training of Support Personnel and Supervisors.
6. Measures for reduction of accidents in opencast mines were pursued through:
   (a) Implementation of various Codes of Practices, Traffic Rules checking of contractor's vehicles/equipment by company's engineers.
   (b) Training of Heavy Earth Moving Machinery operators and contractor's workers.
   (c) Enhancement of safety awareness through publicity and propaganda, safety drives, safety weeks, etc.
7. A drive was conducted for examination of CHPs/bunkers and taking corrective action wherever necessary for prevention of accidents.
8. Thrust on training and retraining of workmen, supervisors to increase safety awareness of the workmen. CIL has taken up a programme of preparation of Video films for imparting standardized training in all its training centres and 33 out of 58 planned films have been completed.

Measures to Improve Safety in Mines

Mines safety legislation in India has evolved as one of the most comprehensive corpus encompassing practically every aspect of mining operations. The statutory protection given to the Indian miner in matters of health and safety compares well with the protection provided in the most countries. Probably no Indian citizen today enjoys greater protection by law than the Indian miner. In order that critical equipments and materials used in mines have adequate safety features such items are required to be approved by Director General of Mines Safety and no such equipment can be used unless it is of approved type. Such items include Flame safety lamps, permitted explosives, cap lamps, exploders, methanometers, CO detector, self-rescuers, friction and hydraulic roof supports, cage suspension gear, winding rope, electrical cable, flameproof and intrinsically safe equipment etc. Every case of fatal and serious accident is enquired by DGMS Dy Director with a view to remove its shortcoming to absorb technical advancements and to incorporate new developments in the safety management. The institutions of Workmen's Inspectors and Safety Committee at every mine ensure that the workmen themselves are involved in the safety management. An excellent system of inspection by the officers of DGMS is in force since the enforcement of Mines Act in 1901. Promotional measures taken are, celebration of Safety Week, the institution of National Safety Awards (Mines). The other measures taken to improve safety in Mines are self-regulation through Internal Safety Organisation. Every company to have a well defined safety policy and safety objectives.

The further steps suggested to bring safety management to world class standard is having a Safety Management System and its certification through third party certification by adopting OHSAS 18001:99 (Occupational Health and Safety Management).

During the last decade, the idea of systematic occupational health and safety (OHS) management system has found favour with regulators, employers, and other parties both in developed and developing countries. ISO 9001:2000 and ISO 14001:96 management systems do not directly address corporate responsibility to OHS issues in the organization, so a need was felt for developing and implementing cost effective occupational health and safety management system. The demand for a certification scheme for occupational health and safety has been intensified with the publication of the British Standards Institution (BS8800:96, guide to Occupational Health and Safety Management Systems OHSMS) and the American Industrial Hygiene Association (AIHA) guidance document on ISO 9000-based OHSMS (AIHA1996). An association of national standard bodies, certification and specialist consultants has developed the Standard for Certification of Occupational Health and Safety management systems. The Standard Occupational Health and Safety Assessment Series Specification, OHSAS 18001, officially is a specification against which certification is awarded. Already some mines in India have reaped benefits by establishing, implementing and maintaining the Occupational Health and safety Management systems in their organizations. Mines of Coal India and other public sector mines like SCCL, Neyveli Lignite can take similar steps for improving their safety performance. In conclusion the effort should be directed to merge the OHS MS in the management system of the organization.

OCCUPATIONAL HEALTH MANAGEMENT IN INDIAN OIL

M. Ahmad*

ABSTRACT

This paper deals with the occupational health management in Indian Oil. Going deep into Indian Oil's objectives of protecting the health of the employees as well as community around and detecting the occupational disease early, it discusses in brief the four steps followed in occupational health management: anticipation of hazards under different heads, recognition of hazards; monitoring of hazards, and control of hazards, both engineering and medical.

Indian Oil firmly believes that the human resource is an asset to the organization and key element of the business. It is actually the man behind the machine who is running the show. Keeping them healthy is in the interest of business and sustainable development of the organization. Every citizen of our society has a right to healthy and safe work environment that enables him or her to live a socially and economically productive life. Indian Oil has integrated occupational safety and health protection with health promotion activities into a coordinated system that address both workplace and employees' health.

In spite of our good intentions, the cost of occupational diseases and injuries is enormous, and estimated global economic losses resulting from them are to the tune of 4 per cent world's Gross National Product.

According to WHO and ILO, the Global burden of occupational diseases and injuries may increase considerably in the first half of 21st century. Contributing factors include the ever-greater transfer of industrial process to developing countries and improved reporting of occupational illness and injuries worldwide.

It is estimated that 160 million new cases of work-related diseases and injuries occur in the world annually, including respiratory and cardio-vascular diseases, cancers, hearing loss, muscular skeletal and reproductive disorders, mental and neurological illness. Each year the work related injuries and diseases kill an estimated 1.1 million working people worldwide, which obviously affect productivity and sustainable development of an organization.

*DGM (HS), S, H&E Department, Corporate Office, Indian Oil Corporation Limited, Lodhi Road, New Delhi-110 003

Occupational Health care contributes by improving the increased availability of workforce by anticipating, identification, monitoring and control of hazards, monitoring health of the employee and initiate corrective action if indicated. The initiative supports the view that illness and injury should be prevented when possible, controlled when necessary and treated when appropriate.

Indian Oil in its Mission Statement has committed: "To help enrich the quality of life of the community and preserve ecological balance and heritage through a strong environment conscience." Safety, Heath and Environment policy has been formulated taking into account the Mission statement.

Our Objective is:

1. Protect the health of the employees and community around,
2. Detect the Occupational disease early, when the condition is reversible.

We Aim at:

1. Anticipating the hazard
2. Identification and monitoring of the hazard
3. Protection of employee, population and environment
4. Promotion of health of employees
5. Treatment
6. Rehabilitation

The following 4 steps are followed in occupational health management

1. Anticipation
2. Recognition
3. Monitoring and
4. Control of hazards

For **Anticipation** we list hazards under different heads

- *Physical:* heat, light, electricity, radiation, noise, vibration, high altitude.
- *Chemical:* organic and inorganic gases, liquids, solids, acids, alkalies, salts, and minerals.
- *Biological:* plants/animals, allergy, viral, bacterial.
- *Mechanical:* injuries—acute and repeated, deformity, and loss of any member of human body.
- *Psychological:* boredom, monotony, competition, rivalry and stress.

We also make use of MSDS extensively for this purpose.

For **Recognition** of hazards, We

- use designated format,
- list hazards with its location,

- identify employees exposed to the hazard.

Monitoring of hazards are being done in Two different ways

- Work environment monitoring
- Personal monitoring of hazard

Any abnormal finding during work environment monitoring is discussed with concerned HOD. In case of any abnormal finding during Personal monitoring health is monitored more carefully, in relation to that hazard. The point to remember is 2 or more hazards with similar toxicity have additive effect at lesser concentration.

For **Control of hazards** two ways are employed : (i) Engineering control and (ii) Medical control.

Engineering control is exercised by

- Survey and Identification of hazardous area.
- Monitoring of levels of hazardous substances.
- Substitution of toxic substances—TEL, phenol.
- Isolation and enclosed processes.
- Ventilation—natural, forced, exhaust.
- Wet—procedure.
- Use of PPEs.
- Frequent supervision for strict adherence to safety measures.
- Treatment of waste products and proper disposal

Medical control is exercised by

- Pre-employment Medical examination for selecting the right candidate for the job, for which a guideline has been issued.
- Pre-placement Medical examination for placement of employees from Non-battery area to battery area.
- Periodic Medical examination for employees working in Hazardous area.
- Well person screening for detecting lifestyle diseases/disease prevalent in the community, so that preventive action could be initiated.
- All medical records are kept confidential.
- Suitable medical counseling is provided on results of findings of medical examination.

Indian Oil have 7 fully equipped occupational health centres under well-trained manpower at all refinery locations for carrying out these activities. At locations other than the Refineries, retainer doctors carry out the medical examinations as per the guidelines.

Information and Education being an integral part of Occupational Health management, We provide class room training in First Aid, Occupational Heath awareness. We also participate in Shop floor safety meetings. Health awareness programme is conducted based on disease prevalence rate in the community. Open talks on lifestyle changes like

Diabetes	Hypertension
Obesity	Alcoholism
Smoking	Cancer awareness

are regularly arranged.

Rehabilitation: In case of need employees are rehabilitated in other suitable jobs.
System improvement: Occupational Health professionals participate in ISO-14001, OHSMS, and ISRS certifications.

22

OCCUPATIONAL HEALTH DISEASE, HISTORY AND LEGISLATION

Mathura Prasad*

ABSTRACT

The occupational health diseases are the health problem caused by or manifest at work. Some health problem not caused by the job may be aggravated by it. Some substances are particularly liable to damage certain organs, e.g. hydrocarbon solvents may affect the liver, cadmium may damage the lungs or kidneys and mercury may affect the brain. According to ILO/WHO the promotion and maintenance of the highest degree of physical, mental and social well being of workers in all occupations, the prevention of disease amongst worker caused by their working conditions. The protection of workers in their employment from risk to health, the placing and maintenance of the worker in an occupational environment adapted to his physiological and psychological equipment and summerised the adoption of work to men and each men to his job. This aim could be achieved by suitable medical services in an Industrial establishment. In India Factories Act, 1948, have provisions for the establishments carrying 'hazardous process' to provide and maintained in good order an occupational health centre with the services and facilities prescribed in state factories rules.

History

In ancient times, many hazards associated with work were recognized, but very little was done to prevent occupational disease. By second century AD some miners were using bladders to protect themselves from dust inhalation, probably this was the first example of Personnel Protective Equipment (PPE) used by Industrial workers. A mining engineer and doctor *Agricola* published 12 books on metal mining in year 1556 in central Europe. In one of the book it was advised to use of loose veils worn over the face to protect the miner against dust and ventilating machines to purify the air

In 1700 a book, *Bernardino Ramazzini* on trade disease was published by an Italian physician. Dr. Ramazzini who was the first person to advise that, during diagnosing illness of patients, their occupation must be specially asked.

*General Manager (R&R and Safety), NTPC

In 1832 a doctor of city Leeds (UK) wrote a book on *Occupational Diseases.*

In UK first time in 1844, factory inspectors appointed Certifying Surgeons and by 1855 they were required to investigate industrial accidents.

In 1978, the Royal College of Physicians of London established a Faculty of Occupational Medicine as an academic centre for the subject. The Faculty has established criteria for the training and examination of specialists in the field. In USA the National Institute for Occupational Safety and Health (NIOSH) determines standards of occupational health and safety at work and organizes training and research facilities. In India Factories Act, 1948 and amended in 1987 notified 29 diseases under its third schedule.

The environmental factors such as the presence of contaminants in the air, physical factors like heat, humidity, noise and radiations etc. has got effect on state of health of worker. Workers health can also be adversely affected by substance like metals, pesticides, gas fumes, asbestos, and fibers. The workers exposed to all these toxic have risk of suffering from their harmful effects. It is, therefore, of vital importance that the employees who are required to work in a industry are protected and kept in a healthy state.

The Main Area Rout of Entry of Toxic Agent in Human Body

- Inhalation
- Absorpsion through skin
- Injection
- Metabolison and Extritedation

Functions of Occupational Health Department

Functions of Occupational Health Department are:

- Health Assessments
- Advisory services.

Health Assessments

- Pre-employment and other medical examinations, i.e. employees returning from sickness or those changing jobs.
- Examination of the employees exposed to specific occupational hazards.
- Emergency treatment of illness or injury occurring at work.

Advisory Services

- The study and prevention of occupational disease.
- Advice on problems of medical legislation and codes of practice.
- Advice on medical aspects of new processes and plant.
- The study of sickness absence.
- Advice on the reduction or prevention of common non-occupational diseases such as alcoholism and the effects of smoking.

- Advice to employees prior to retirement.
- The preparation of contingency plans for major disasters at the place of work.

In India Director General Factory Advisory Services organize three months full-time Occupational Health Course for doctors at Central Labour Institute Mumbai.

Risk to Health at Work

The main hazards are of three kinds—physical, chemical and biological, although occupational psychological factors may also cause illness.

Physical Hazards

Noise, vibration, light, heat, cold, ultra-violet and infrared rays, ionizing radiations.

Chemical Hazards

These are liable to occur as a result of exposure to any of a wide range of chemicals. Their effects may arise atonce or a considerable period of time may elapse before signs and symptoms of disease are noticed. By this time the effects are often permanent.

Biological Hazards

These may occur in workers using bacteria, viruses or plants or in animal handlers and workers dealing with meat and other foods. Diseases produced range from infective hepatitis in hospital workers (virus infection) to ringworm in farm labourers (fungus infection).

Stress

This may be caused by work or may present problems in the time spent at work. Work related stresses may be due to difficulties in coping with the amount of work (quantitative stress) or the nature of the job (qualitative stress).

Method of Control

The control method can be at three different points, these are:

- Engineering Control at source
- Environmental control
- Medical control

Engineering Control at Source

Engineering control can be achieved by following techniques:

I. Substitution
II. Change of Process
III. Enclosure of Process
IV. Wet Methods
V. Local Exhaust Ventilation
VI. Adequate Maintenance Programmes

Environmental Control

Environmental Control can be achieved by following techniques:

I. House Keeping
II. General Exhaust Ventilation (Roof Fan)
III. Dilution Ventilation
IV. Increasing Distance Between Source and Receiver (Semi-automatic or remote control)
V. Continuous Area Monitoring (Pre-set alarm)
VI. Adequate Maintenance Programme

Medical Control

Medical control can be achieved by following techniques:

I. Rotation of Worker
II. Medical Examination
III. Pre-employment Medical Examination
IV. Periodic Medical Examination
V. Health Survey
VI. PPE

The variation in fatal occupational injuries in the age group of young workers in US in the year 1992-2002 as given by NIOSH (National Institute of Safety and Health):

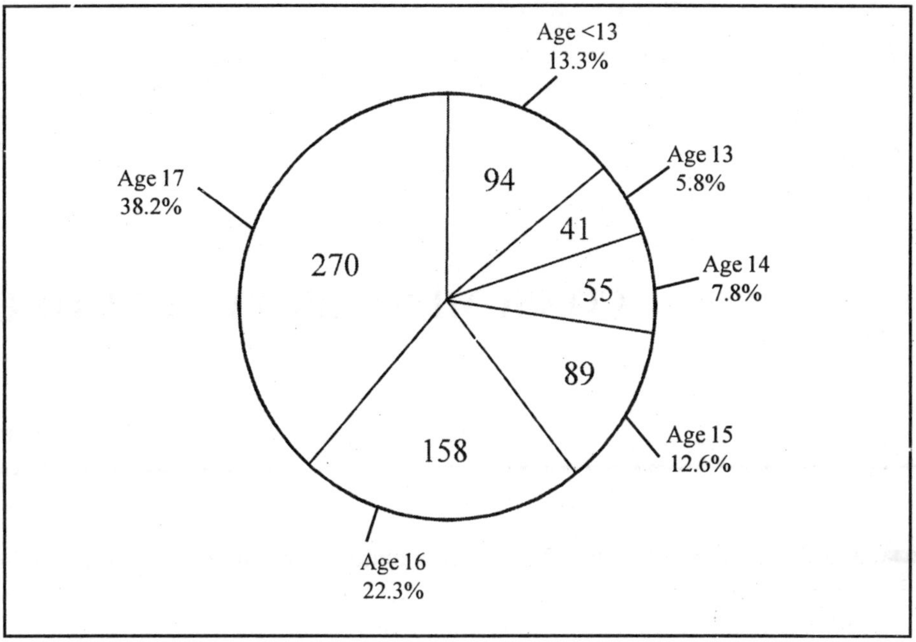

Conclusion

In India Factories Act, 1948 and its amendment in1997 has various provisions for mitigation of occupational heath hazards at workplace. Chapter III of Factories Act is especially for health. Section 91A of Factory Act has provisions for safety and occupational health surveys. Central Labour Institute, Mumbai has Occupational Health Centre to provide training for doctors.

Since companies productivity is dependent on workers productivity and workers productivity is directly related to their health ,it becomes utmost important for companies to maintain good environmental conditions at workplace. Companies should also ensure compulsory health checkup of employees working in hazardous atmosphere on yearly basis.

23

CONSUMER SAFETY-ROLE OF HIS

Neeta Sharma*

Industrial and Personal Safety has always been a matter of great concern and remained a major issue for governments, employees and workers. In view of vast growth in industry, both in large and small scale sectors in recent, years, great emphasis on safety is being given by various sectors to minimize accidents and other losses.

BIS is also developing standards which one dealing solely with safety aspects from the point of view of unacceptable risks of harm i.e. not becoming a source of hazard for a product, process or personnel. These can also be termed as 'Safety Standards' covering various types of hazards arising from electricity, industrial gases, chemicals, fire, construction, food etc. The work of development of standards in these areas are covered by various technical departments like Civil Engineering, Chemical, Electrical Engineering, Mechanical Engineering, Petrochemicals, Textiles, Food and Agriculture of the Bureau.

A number of standards one made in each department which are based on latest practices being followed in the country besides utilizing the information of other developed countries as applicable to Indian scenario.

Consumer at large one generally not aware of the hazards associated with various item of daily use nor is it easy to make them aware of such hazards. The need to have inbuilt safety in such items becomes imperative. Common examples being gas stoves, toys used by children, cosmetics, fire extinguishers, electrical items etc. The list is endless. Role of BIS is, therefore, enhanced to see that all safety aspects are covered in the standards, so that 'Item' conforming to Indian Standards would automatically be 'Safe' for its intended use.

At an International level, Committee on Consumer Policy (COPOLCO) has been established by ISO, in 1978, to-address the need for closer iinks between the standards world and that of its ultimate customers and beneficiaries—consumers.

Scope of the committee is—To promote adequate representation of consumer interests within technical bodies that develop standards of specific interest for consumers and to, ensure that the work progresses satisfactorily from the consumer point of view.

---

*Director (Standard Promotion & Consumers Affairs), Bureau of Indian Standards

Through COPOLCO, ISO Undertakes to

— Study how the consumers can benefit from standardization.
— Promote consumers input into the development of standards, both nationally and internationally.
— Encourage the exchange of experience on standards work of consumer interest.
— Channel consolidated views from consumers, both on current projects and on proposals for nowwork in areas of interest to them.

And aims to:

— Identify issues and work areas of priority interest both to consumer and International Standardization.
— Provides ISO with appropriate feedback.
— Proposes new areas of standardization activity.
— Advices ISO on policy directives in relation to consumer issues; and
— Focuses on activities to promote the benefits of standardization among consumers.

It has published many documents and has made tremendous contributions on various guidelines like

— ISO/IEC Guide 71—Guidelines for Standards Developments to Address the Needs of Older Persons and Persons with Disabilities
— ISO/IEC Guide 51—Safety Aspects - Guidelines for their Inclusion in Standards
— ISO/IEC Guide 74— Graphical Symbols—Technical Guidelines for the Consideration of Consumers' Needs (Under'Development) .

BIS has been 'P' member of ISO COPOLCO for several years and has now formulated NATIONAL MIRROR COMMITTEE OF COPOLCO to address need for links between standards and consumers. BIS also become 'P' member of the following Working Groups under COPOLCO:

1. Product Safety
2. Generic Guidelines for Standardization of Services
3. Consumer Participation in the Global Market
4. Enhancement of Consumer Participation in Standardization
5. Revision of ISO/IEC Guide 37
6. Priorities for Consumer point of view

Out of several areas of work in hand at COPOLCO level, National Mirror Committee has proposed to initiate work on Child Safety, Protection against dumping of unsafe products, Complaint Handling, CSR, Services, customer/organizational disputes etc. to name a few.

Though lot of work has been done at national /intemational level, in the area of consumer safety and continuous affects are ON, merely by formulating guidelines/standards consumer is not benefited. The real benefit would come when the consumer behaves intelligently and becomes aware of his own RIGHTS and RESPONSIBILIIES. This is possible only when the society as a whole becomes responsible and each individual acts as a wise consumer, by not only using, but also responding on the standards. This will not only help in bringing out the best possible standard addressing all safety aspects of consumers, but also uplift the morale of the standards malcing bodies who are putting lot of efforts, brain bank, finance etc. in the formation of standards for safety for consumers.

I, therefore, wish that every consumer becomes aware of his Rights and Responsibilities.

24

FOOD SAFETY THROUGH ISO 22000

Madhulika Prakash*

Introduction

Food and agriculture is the largest sector of the national economy. The food system comprises of production of food, its post-harvest handling, distribution and consumption. It is complex, multidisciplinary covering varying scales of operations, i.e. small, medium and large. Before reaching the consumer, food passes through many hands and is handled and processed in several ways. To retain its quality and nutritive value, it is essential to ensure the integrity and the safety of food throughout the food chain. Any measures taken to ensure the above will also add to its value as a commodity and create better demand, both domestic and foreign.

With globalization and the coming in of the SPS and TBT Agreement under WTO, there is an international requirement towards implementation of science-based risk assessment and protection measures using International Codex standards as a benchmark. This has led to many industrialized and less industrialized countries to review, reassess and update their food control systems ensuring food safety.

For a well developed and efficient food system ensuring food safety and quality, the principles and practices to govern the same are based on laws and government regulations, as well as the requirements and expectations of customers and consumers. In addition to the basic need for food quality and food safety activities, operations in the food industry have been influenced by numerous factors since the 1970s, including:

- Consumer expectations relating to various aspects of food (e.g., nutrition, convenience, additives)
- Incidents relating to food safety
- Changes in government regulatory processes
- Traceability in food production and processing—technological changes
- Foods derived from biotechnology

*Head (Food & Agri.), Bureau of Indian Standards, New Delhi-110002

— Irradiated foods
— Organic foods
— Economic factors
— Issues relating to international trade
— Food security concerns related to bioterrorism

As a result, in the last decade, many countries have seen substantial changes in food industry operations, including significant developments in food quality and food safety activities.

In its continual efforts to address the food quality and food safety requirements of government, customers and consumers and in the face of the challenges mentioned earlier, the food industry has embraced generic quality systems and quality programmes similar to those used by other industries. At the same time, it has adopted safety programmes that have been developed specifically for use within the food industry. Consequently, food quality and food safety requirements are addressed through the use of systems and programmes that include quality management, quality assurance, quality control, the hazard analysis critical control point (HACCP) system, and good manufacturing practices (GMPs). Within a particular food company, the food quality and food safety activities are likely to be covered by some combination of these programmes or systems.

Scope of Food Quality and Food Safety

In addressing food quality and food safety, it is important to keep in mind that the term "food" covers any unprocessed, semi-processed, or processed item that is intended to be used as food or drink. This includes any ingredient incorporated into a food or drink, and any substance that comes into direct contact with a food during processing, preparation, or treatment. Therefore, food quality and food safety principles and practices are applied to foods from farm produce and livestock production; manufactured and processed food products for consumers; and all raw materials, ingredients, processing aids, food-contact packaging materials, and food-contact surfaces that are used in the preparation of food and beverage products.

The scope of food quality and food safety covers foods already in the marketplace and new or modified foods. When new or modified foods are developed for the marketplace, quality and safety must be considered at the conception, design, and development stages.

Responsibility for Food Quality and Food Safety

The overall responsibility for food quality and food safety is shared by all segments of the food system, including the various food industry sectors, government regulatory agencies, and consumers in general. The food industry has both the legal and moral responsibility for providing customers and consumers with foods that meet all established quality and safety requirements. Within a food company, overall responsibility for the implementation and effective use of these programmes and systems rests with senior management.

Governments worldwide have enacted food laws and regulations designed to ensure that foods are fit for human consumption. Such laws protect consumers from harm resulting from unsafe foods and from deception resulting from misrepresentation or fraud relating to certain established food quality characteristics. Governments have also established various agencies that enforce these food laws and regulations; this legal framework is intended to provide consumers with confidence in the safety and quality of foods.

Within the food supply chain, customers who purchase raw materials, ingredients and food contact packaging materials for manufacture of consumer foods, must ensure that these materials are safe and fit for use. When making purchases, consumers need to be vigilant in their assessment of foods for safety and quality. In particular, customers and consumers must pay attention to the instructions for handling, storage, preparation, and use of foods.

Distinction between Food Quality and Food Safety

While the terms food quality and food safety are often used interchangeably, it is important for the food industry professional to distinguish between them. Food quality is the extent to which all the established requirements relating to the characteristics of a food are met. Food safety is the extent to which those requirements relating specifically to characteristics or properties that have the potential to be harmful to health or to cause illness or injury are met. Some food quality characteristics (e.g. counts of total bacteria, coliform bacteria) can be used as indicators of food safety, although they are not considered specifically as food safety characteristics. This distinction between food quality and food safety needs to be made, primarily because of the much greater importance that must be attached to protecting consumers from food borne illnesses or injuries. A food that does not conform to the food safety requirements automatically does not conform to the food quality requirements. On the other hand, a food can conform to the food safety requirements, but not conform to the other quality requirements.

Food Safety as Part of Food Quality

In the food industry, food safety principles and practices have always been integrated into activities identified within quality assurance or quality control programmes, or within quality management systems; therefore, these programmes and systems can address both food quality and food safety simultaneously. The more recent use of HACCP systems in some food companies has resulted in a well-defined set of activities that are specifically devoted to food safety. The principles and practices of the HACCP system are similar to those of quality systems and, therefore, the specific activities required by the HACCP system can be integrated within quality systems. A food company that does not operate with the HACCP system must continue to incorporate food safety activities and GMPs within its existing quality programme or quality system.

As a result, a number of countries have developed national standards for the supply of safe food and individual companies and groupings in the food sector have developed their own standards or programmes for auditing their suppliers. The plethora of more than 20 different such schemes worldwide generates risks of uneven levels of food safety, confusion over requirements, and increased cost and complication for suppliers that find themselves obliged to conform to multiple programmes.

Food Safety Management Through ISO 22000

Integrating quality and safety, ISO has recently brought out ISO 22000 Food safety management systems—Requirements for any organization in the food chain, backed by international consensus, it harmonizes the requirements for systematically managing safety in food supply chains and offers a unique solution for good practice on a worldwide basis. In addition, food safety management systems that conform to ISO 22000 can be certified—which answers the growing demand in the food sector for the certification of suppliers—although the standard can be implemented without certification of conformity, solely for the benefits it provides.

Developed with the participation of food sector experts, ISO 22000 incorporates the principles of HACCP, and covers the requirements of key standards developed by various global food retailer syndicates, in a single document.

"Public sector participation the development of the ISO 22000 family is also significant," ISO Secretary-General Alan Bryden commented, "notably that of the FAO/WHO's Codex Alimentarius Commission, which is responsible for the well-known HACCP (Hazard Analysis and Critical Control Point) system for food hygiene. Thanks to the strong partnership between ISO and Codex, ISO 22000 will facilitate the implementation of HACCP and the food hygiene principles developed by this pre-eminent body in this field."

Another benefit of ISO 22000 is that it extends the successful management system approach of the ISO 9001:2000 quality management system standard which is widely implented in all sectors but does not itself specifically address food safety. The development of ISO 22000 was based on the assumption that the most effective food safety systems are designed, operated and continually improved within the framework of a structured management system, and incorporated into the overall management activities of the organization.

While ISO 22000 can be implemented on its own, it is designed to be fully compatible with ISO 9001:2000 and companies already certified to ISO 9001 will find it easy to extend this to certification to ISO 22000. To help users to do so, ISO 22000 includes a table showing the correspondence of its requirements with those of ISO 9001:2000.

ISO 22000 is applicable to all organizations, regardless of size, which are involved in any aspect of the food chain and want to implement systems that consistently provide safe products. The means of meeting any requirements of this International Standard can be accomplished through the use of internal and/or external resources.

This International Standard specifies requirements to enable an organization:

(a) to plan, implement, operate, maintain and update a food safety management system aimed at providing products that, according to their intended use, are safe for the consumer,
(b) to demonstrate compliance with applicable statutory and regulatory food safety requirements,
(c) to evaluate and assess customer requirements and demonstrate conformity with those mutually agreed customer requirements that relate to food safety, in order to enhance customer satisfaction,

(d) to effectively communicate food safety issues to their suppliers, customers and relevant interested parties in the food chain,
(e) to ensure that the organization conforms to its stated food safety policy,
(f) to demonstrate such conformity to relevant interested parties, and
(g) to seek certification or registration of its food safety management system by an external organization, or make a self-assessment or self-declaration of conformity to this International Standard.

All requirements of this International Standard are generic and are intended to be applicable to all organizations in the food chain regardless of size and complexity. This includes organizations directly or indirectly involved in one or more steps of the food chain. Organizations that are directly involved include, but are not limited to, feed producers, harvesters, farmers, producers of ingredients, food manufacturers, retailers, food services, catering services, organizations providing cleaning and sanitation services, transportation, storage and distribution services. Other organizations that are indirectly involved include, but are not limited to, suppliers of equipment, cleaning and sanitizing agents, packaging material, and other food contact materials.

This International Standard allows an organization, such as a small and/or less developed organization (e.g. a small farm, a small packer-distributor, a small retail or food service outlet), to implement an externally developed combination of control measures.

25

SAFETY CULTURE

M.S.R. Sarma*

ABSTRACT

It is directed especially to the senior management of all organizations whose activities affect Plant Safety. Safety Culture is that assembly of characteristics and attitudes in organizations and individuals which establishes that, as an overriding priority, Plant safety issues receive the attention warranted by their significance. The definition relates Safety Culture to personal attitudes and habits of thought and to the style of organizations. Such matters are generally intangible. Nevertheless such qualities lead to tangible manifestations. Safety Culture requires all duties important to safety to be carried out correctly, with alertness, due thought and full knowledge, sound judgment and a proper sense of accountability. Safety Culture has two major components: .the framework determined by organizational policy and by managerial action, and the response of individuals in working within and benefiting by the framework. Success depends, however, on commitment and competence provided both in the policy and managerial context and by individuals themselves. "But this is what we do already is a negative road block". More details on the tangible characteristics of a satisfactory safety culture in different kinds of organizations.

Introduction

The phrase "Safety Culture" was coined by INSAG (International Nuclear Safety Advisory Group of International Atomic Energy Agency), while reviewing the accident at Chernobyl 1 - 4 of erstwhile USSR.

Any problems arising at a Plant originate in some way in human error. Beyond adherence to defined procedures, they must act in accordance with 'Safety Culture'. Particular attention to operating organizations because the link between human performance and plant safety is closest there. Of course, safety depends critically on those who designed, constructed and commissioned it.

It is also the experience the world over that Safety Culture by itself cannot be dealt with in isolation. It has to be part of, or one component of the total culture of the organization. Hence it is proposed to deal with "Culture" and its manifestations in individuals and organizations first before we embark on the main topic "Safety Culture".

*Exeutive Director (Retd.), Atomic Energy Regulatory Board

Culture

This term has been freely used and misused. There is hardly any clarity of concept between tradition and culture. If I keep on doing what my grandfather did, does it form culture? Can a policy the corporate head set up long ago continue for ever. Does it contribute to the organizational growth and health if the same practices are continued in a violently changing social and business environment? It is necessary to conceive of the definition which will stand alone. Thus, Culture is defined as "the integral sum of values which individuals, groups of individuals and the society at large conceive and practise".

Values

If the values are only conveyed and not practiced they become redundant and are without any intrinsic value. If Culture is a value based concept it is to be understood how values enter human mind and how they translate to inspire and motivate the rest of the people in society or an organization around the individual.

The Mahatma lived as an example of an integrated personality with perfect unison in concept and practice. He lived the values he conceived and taught all his life. It is a natural order in human beings that only practice enlivens society and the surroundings. Verbosity fails to inspire. If we look back into history only such individuals who lived their values could create order and evolution in society. Similarly, if the head of an organization preaches simple needs like economy, orderliness etc., but appropriates all the perquisites for his personal comforts, he is bound to attract criticism from others. The result is chaos, failure of purpose and decadence.

There are some eternal values. Speak the truth—Do not hurt others etc. However, they are written and hardly understood or, practiced. How do the values enter human mind? Some are inherent and circumstantial. The same family contains a scholar and an idiot. Similarly any organization also is a mixture of gentlemen and rogues. Yet values can be acquired. Values belong to the thought process. While an individual hardly has any control on his own thoughts, a different individual certainly cannot, even guess the thoughts or the reasons for these thoughts in the former.

The acquisition of values by observation and voluntary absorption becomes stable. This essentially sets the values in the thoughts of the person and makes him cultured to that extent. Culture is an integral concept. You do not expect a person to be cultured in office, namely, the work environment and become a boor when he goes to his house or *vice versa*. When they are absorbed voluntarily they stay, and become inherent values.

Can one try to implant values by methods of Indoctrination tried by the foul Nazi Regime and subsequently by other chauvinist political concepts? It is well known that if they take a hold at all, they are only temporary. At the most critical juncture, the implanted and partially accepted values fail and the inherent nature of the individual manifests.

There can be no action without precursor thoughts. It is erroneous to say that a fellow acted thoughtlessly. The only word should have been the fellow acted without discrimination. Untoward thoughts have to be dissipated by the individual who has the right values—the one who is cultured. He is the one who is wise. Else he is otherwise. There is no distinction between an animal and such an individual who lets his indiscriminate thoughts produce indiscriminate actions. Even societal laws just do not permit such licentious behaviours.

There is neither reliability nor repetability nor responsibility for human behavior. For a given input no two human being behave alike. The reaction to an action is totally beyond the control of the performer.

Discipline is nothing but regulated action. Regulated action due to inherent values through the process of discriminated thoughts constitute culture but, discipline is concerned only with actions. What goes on ,in any individual's mind does not disturb anything around till it is put into action. But when required to be disciplined, the majority rebel on the imposition. This is nothing but eliminating the feed back into the mind. Calmness always results in efficiency of thinking and dexterity in action. One has to sincerely cultivate, train and absorb this value in his own mind which performing such a balanced action. Herein lies the justification of the intellect of man. Otherwise he is not performing has actions for which he is really intended.

The whole story has been put in half a verse by the great saint Adi Sankara thus *chittasya suddhaye karma natu vastoopalabdhaye*. Human actions, human life, human discipline, human values are meant for purification of the-mind and not for material benefits alone; Let us not get confused: Nowhere it is said that one should not strive to obtain material comforts: The world is created only for the benefit of the beings living in it. But the attitude with which the material comforts are enjoyed changes the face of the world. There is a tremendous misgiving of the difference between comfort and happiness. The fellow on the pavement is miserable bee cause he does not know where his next meal comes from. His very existence is threatened. The group with the Five Star comforts are also miserable because they are constantly chasing something. The difference is one fellow is uncomfortably miserable while the Five-Star group is comfortably miserable: Misery is the common factor: Do you call that culture? Let us understand that culture produces calmness of the mind. This results in efficiency and extreme capability in actions. The real manifestation of culture is neutralizing the feed back cycle so that pristine purity of the thought process stays alive, effulgent and efficient.

Behaviour of personnel in the organization during normal day to day situations reflects the culture at the work place. When one walks around and sees people sitting and whiling away time the least to expect is that when a supervisor orr engineer is walking around such people get up and go to their work location. The supervisor or engineer is expected to take note of it. If it is not done either way that means it is an organizational habit. When people are sitting around unmindful of just who is going by, it shows that the management does not really pay any attention to what the workmen are doing and *vice-versa.* There is no work discipline. That is the culture, of course, a bad example.

Safety Culture

The INSAG, constituted, by the IAEA to advise the Director General on safety issues, have coined this phrase during the review of the Chernobyl accident. INSAG after the review of

Chernobyl accident commented that the safety culture in that plant was certainly not up to the mark.

To illustrate: the operator was blamed. He was there and he was caught in a sort of a situation which was beyond his control. An experiment was required to be conducted. The procedure to conduct the experiment was not reviewed by safety bodies. The operator was told that he should conduct and complete the experiment. In that anxiety he bypassed various safeties: the experiment already included some trips bypassed for fear that in case the unit trips during the experiment the results of the experiment will not be available and it will have to be conducted only at the next annual maintenance shutdown. Therefore, various trips and safeties were further jumpered and eventually the plant landed in a very unsafe configuration. The fall out of this accident is reflection on the entire nuclear industry. The significance of safety culture in this situation is that such a thing cannot be done just by any individual operator or shift supervisor or anyone at that level. The corollary one can draw is that if the experiment was successful may be no one would have raised an everybody. From this it can be inferred that the safety culture was more or less non-existent in that organization. The objective was only to complete the assigned task and nothing else including safety was of any consideration. This is how the phrase "Safety Culture" was coined at the time of review of Chernobyl accident and it has come into vogue the world over. Pursuing this phrase, the International Atomic Energy Agency brought out a booklet on "Safety Culture" INSAG-4 which attempts to answer the questions like; what are the elements of safety culture? How to monitor them? What are its attributes? etc.

Difinition and Character of Safety Culture

"Safety Culture is that assembly of characteristics and attitudes in organizations and individuals which establishes that as an overriding priority, Plant Safety issues receive the attention warranted by their significance."

Safety Culture refers to the "personal dedication and accountability of all individuals engaged in any activity which has a bearing on safety of Plants". "An all pervading safety thinking" which allows "an inherently questioning attitude, the prevention of complacency, a commitment to excellence and the fostering of both personal accountability and corporate self-regulation in safety matters".

Good practices in themselves, are not sufficient if applied mechanically.

Universal Features of Safety Culture

Individual awareness
Knowledge and competence
Commitment
Motivation
Supervision
Responsibility
Safety Culture has two general components

First: Necessary framework within an organization;

Responsibility of the Management.

Second : Attitude of staff at all levels in responding to and benefiting from the framework.

Requirements at Policy Level

The highest level affecting Plant safety is the legislative level, at which the national basis for safety culture is set.

Legislation is backed by the necessary advisory and regulatory bodies which have sufficient staff, funding and powers, to perform their duties and the freedom to do so without undue interference.

In this way, national climates are fostered in which attention to safety is a manner of everyday concern.

International Exchanges

Within an organization, similar considerations apply. Policies promoted at a high level create the working environment and conditions individual's behaviour.

Statements of Safety Policy

An organization pursuing activities with a bearing on Plant safety makes its responsibilities well known and understood in a safety policy statement. This statement is provided as guidance to staff, and to declare the organization's objectives and the public commitment of corporate management to Plant safety.

Plant safety has the utmost priority, overriding if necessary the demands of production or project schedules.

A regulatory body has a weighty influence on the safety of Plants within its purview and an effective safety culture pervades, its own organization and its staff.

Organizations responsible for design, manufacture, construction and research influence greatly the safety of Plants. Their primary responsibility is for quality of the product. Basis is the directive establishing policy and practices to achieve quality.

Management Structures

Implementation of these safety policies requires that accountability in safety matters is clear. Strong lines of authority are established for those matters bearing on Plant safety. The formal responsibility for plant safety lies with the operating organization and the delegated authority with the plant manager.

Large organizations with significant impact on Plant safety, provide independent internal management units with responsibility for the surveillance of Plant safety activities. In operating organizations, these units have the role of scrutinizing safety practices at the plant. They report at a senior management level.

Supporting organizations adopt similar methods to achieve product quality, involving audit and review practices.

Resources

Adequate resources are devoted to safety. Sufficient experienced staff are available, supplemented as necessary by consultants or contractors; have available to them the necessary equipment, facilities and supporting technical infrastructure.

Self Regulation

As a matter of policy, all organizations arrange for regular review of those of their practices that contribute to Plant safety.

Commitment

This commitment is publicly asserted demonstrates also an organization's willingness to be open in safety matters.

Regular review of the processes that bear on plant safety, taking direct interest in the more significant questions of plant safety. Plant safety is an important agenda item at meetings of boards of operating organizations.

Requirements on Managers

The practices moulding the environment and fostering attitudes conducive to safety. Responsibility of managers is to institute such practices in accordance with their organization's safety policy and objectives.

Definition of Responsibilities

Discharge of individual responsibilities is facilitated by unique and clear lines of authority. Managers ensure that individuals understand not only their own responsibilities but also those of their immediate colleagues and of their management unit, and how these responsibilities complement those of other groups. Operating organizations to assure themselves, by means of third parties if necessary, that other organizations whose activities contribute to the technical basis of plant safety discharge their responsibilities satisfactorily.

Definition and Control of Working Practicer

Managers ensure that work on matters related to plant safety is carried out in a rigorous manner. The necessary basis is generally a hierarchy of up-to-date documents ranging from policy directives to detailed working procedures. Managers institute systems for supervision and control and insist upon orderliness and good housekeeping.

Qualifications and Training

Managers ensure that their staff are fully competent for their duties. The assessment of technical competence is an integral part of training programmes. Judgment of fitness for duties includes physical and psychological considerations.

Instruction instills more than technical skills or familiarity with detailed procedures to be followed rigorously. Sufficient to ensure that individuals understand the significance of their duties and the consequences of mistakes arising from misconceptions or lack of diligence or lack of comprehension of the risks involved.

Rewards and Sanctions

Managers encourage and praise and seek to provide tangible rewards for particularly commendable attitudes in safety matters. Incentives are therefore not based on production levels alone but are also related to safety performance. Individuals are encouraged to identify, report and correct imperfections in their own work in order to help others as well as themselves to avert future problems. When necessary they are assisted to improve their subsequent performance.

For repeated deficiency or gross negligence, managers accept their responsibility for taking disciplinary measures, since safety may otherwise be prejudiced. There is, however, a delicate balance—sanctions are not applied in such a way as to encourage the concealment of errors.

Audit Review and Comparison

Managerial responsibilities include their implementation of a range of monitoring practices which go beyond the implementation of quality, assurance measures and include, for example, regular reviews of training programmes, staff appointment procedures, working practices, document control and quality assurance systems.

In design, manufacturing and operating organizations, they include scrutiny of the means by which design or engineering changes are controlled. In the plant operations context, they include scrutiny of changes to operating parameters, maintenance requirements, modifications to plant, plant configuration control and any non-routine operation of the plant. It is good practice to augment such processes by calling on experts from functions other than those concerned or from outside the organization.

Managers make arrangements to benefit from all sources of relevant experience.

Commitment

The practices identified, structure the environment in which people work. Fostered by demands for orderly work, by clarity of understanding of duties, by rewards and any necessary sanctions and by the invitation of external scrutiny. It is the task of managers to ensure that-their staff respond to and benefit from this established framework of practices and, by attitude and example, to ensure that their staff are continuously motivated towards high levels of personal performance in their duties.

Response of Individuals

The response of all those who strive for excellence in matters affecting plant safety is characterized by

A QUESTIONING ATTITUDE
PLUS
A REGOROUS AND PRUDENT APPROACH
PLUS
COMMUNICATION

The result will be a major contribution to:

Safety

His or her Questioning Attitude raises issues such as

— Do I have the necessary knowledge to proceed?
— Do I need any assistance?
— What can go wrong?
— What do I do if a fault occurs?

Rigorous and Prudent Approach. This involves

— Being alert for the unexpected;
— Seeking help if necessary;
— Forgoing shortcuts

Communicative Approach is Essential to Safety. This involves:

— obtaining useful information from others;
— transmitting information to others;
— reporting on and documenting results of work both routine and unusual.

A questioning attitude, a rigorous and prudent approach, and necessary communication are all aspects of an effective safety culture in individuals. The product contributes to a high level of safety and generates a personal pride in dealing with important tasks in a professional manner.

Tangible Evidence

Assembly of commendable attributes of any organization or individual—starting point for self-examination by organizations. Questions which can be used to aid judgment of the effectiveness of safety culture.

Government and its Organization

The following aspects demonstrate government commitment:

— Legislation and Government policies.
— Governments assign the responsibilities, conflict of interest is minimized, safety matters are addressed on their merits, without interference or undue pressure.
— Governments provide strong support for regulatory agencies, adequate powers, sufficient funds, guarantees without undue interference.
— Governments promote and contribute to the international exchange of safety related information.
— Relation with operating organizations that are open and co-operative and yet have the formality and separateness as appropriate. Controversial topics are dealt with in an open fashion.
— Standards are adopted that call for appropriate levels of safety while recognizing the inevitable residual risk. By this means a consistent and realistic approach to safety is achieved.
— Regulatory requirements are clear but not so prescriptive as to set undue constraints.
— Improvements in safety result from a well judged combination of innovation and reliance on proven techniques.

Operating Organization

Corporate Policy Level

Safety culture flows down from action by the senior management of an organization. Corporate policy level since it is there that attitudes, decisions and methods of operation demonstrate the real priority given to safety matters. Reference is made in the statement to the vital importance of safety such that concern for safety may on occasion override production objectives. Establishment of a management structure, assignment of responsibilities and allocation of resources are all primary responsibilities at corporate policy level. Senior management initiate regular reviews of the safety performance of the organization. As specific examples:

— Evaluation of the attitudes of individuals to safety is part of the process of selection and promotion of personnel.

Plant Level

At the plant itself, safely is an immediate concern, and an effective Safety Culture is an essential feature of day to day activities. Three different aspects are considered:

—the environment created by the local management;
—the attitude of individuals;
—actual safety experience at the plant

The Working Environment

Safety responsibilities and detailed practices at all levels at the plant are defined. Particular care is taken in the treatment of special activities, a systematic independent examination is required. Reviews of documentation and records are carried out; training is founded on a basic understanding of the safety questions involved, includes consideration of the possible consequences of such errors, and deals specially with how they may be avoided, or corrected if committed. Specific example:

— Simulator retraining.
— Training sessions with mock-ups or video recordings.
— The results of safety analysis are consulted regularly to support decisions as specific issues arise.

Plant safety is kept constantly under scrutiny through plant inspections and audits, visits by senior officers, and internal discussions and seminars. Findings are evaluated and acted upon. Satisfactory facilities must be provided.

The work—loads of individuals

Relationship between the plant management and the regulatory authority is open and based on a common concern for plant safety, but with a mutual understanding of the different accountabilities.

Individual Attitudes

The attitudes of individuals may be examined in exchanges with staff members at various levels.

— Are procedures strictly followed even when quicker methods are available?
— Do staff members stop and think when facing an unforeseen situation?
— Is a good safety attitude respected by management and within peer groups of the staff?
— Do staff take the initiative in suggesting safety improvements?

Managers attitudes are demonstrated. Discussion with staff concerned about delay in restarting the plant for reasons of safety makes clear the commitment to safety as a primary objective.

The presence of managers at the work site provides opportunities to emphasize directly the importance assigned to safety.

Plant Safety Experience

All significant events that have occurred on the site are analysed in close cooperation with the staff concerned. The thoroughness of the review and the strength of the corrective responses are important safety culture indicator.

Supporting Organization

Design organizations keep up to date with developments in Plant safety technology and safety analysis techniques by active participation in national and international activities.

Concluding Remarks

"As an overriding priority, plant safety issues receive the attention warranted by their significance".

Our pioneer Dr. Bhabha in one of his directives in the initial days, when handling radioactive material was started, said that "Radioactive material and sources of radiation should be handled in a manner which not only ensures that no harm can come to workers or anyone else, also in an exemplary manner so as to set up a standard which other organizations in the country may be asked to emulate". That was the directive he gave which shows how much of importance he gave to safety. This was the "Safety Culture" with, which the Department started its activities.

To reiterate the same principle in the nuclear business a comment was made by C.K. Hinton of the UKAEA in the' year 1957. He says "All our engineering technology has advanced, not on the basis of their success but on the basis of their failures. The bridges that have collapsed under the load have added more to our knowledge of bridge design than projects that have been successful. The boilers that have blown up added more to our knowledge than the boilers which have been free from failures. But Atomic Energy must forego this luxury of progressing on the basis of knowledge gained by failures". That has been the aim of the Department and that has been the guiding principle in all the activities: "Safety First".

Acknowledgements

The initial part of the write up on "Culture" was taken from an article on "Work Culture" by Dr. L.G.K. Murthy, Director H & SG (Retd.) of NPC. The later part on "Safety Culture" is abridged from INSAG - 4 of IAEA.

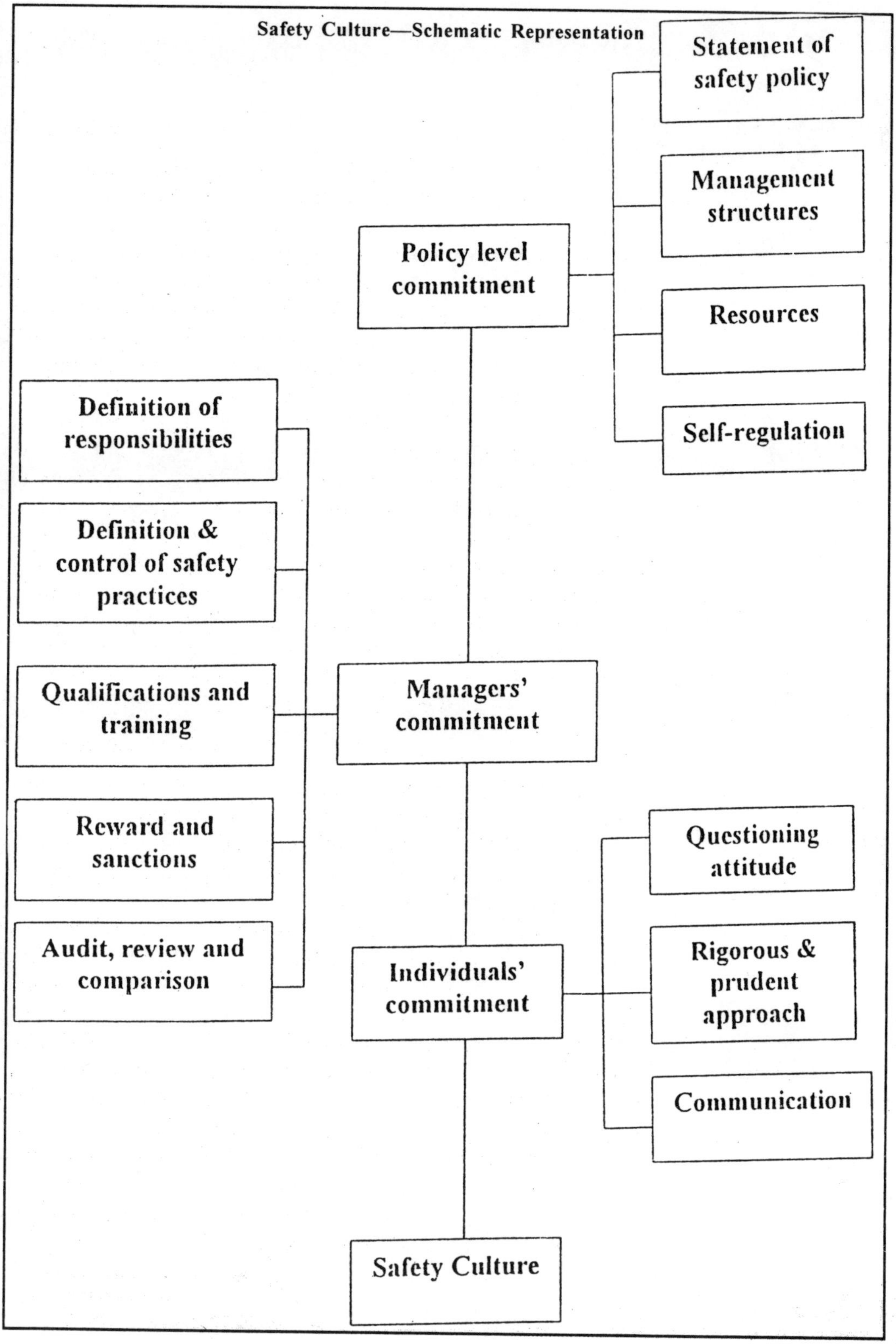
Safety Culture—Schematic Representation
Statement of safety policy
Management structures
Policy level commitment
Resources
Self-regulation
Definition of responsibilities
Definition & control of safety practices
Qualifications and training
Managers' commitment
Reward and sanctions
Audit, review and comparison
Questioning attitude
Rigorous & prudent approach
Individuals' commitment
Communication
Safety Culture

Section–III
Safety in Industry

26

EMISSIONS AND DISCHARGES IN INTEGRATED STEEL PLANTS, HEALTH HAZARDS AND ROLE OF ENGINEERS

Suresh Chandra Bhatia*

ABSTRACT

The Iron and Steel Industry is a prodigious consumer of water and hence a monstrous water polluter. Similarly it is a notorious emitter of obnoxious wastes into the air. This industry creates tremendous environmental problems. Much attempts have been made all over the world to study the actual problems of pollution caused by the various processes involved in the industry—right from preparation of iron ore to steel making to finished steel products. In a situation as at present, where the delicate balance of nature is violently upset by our explosive demands for technological transformation of the natural resources to urban complexes, industrial plants, agricultural outputs and consumer goods, preservation of the environment is the business of everybody. To relieve the great convulsions of anxiety and ingrained hostility that now rock the society, what we need is a new coalition, a relentless international effort, not only to combat the negative forces—man-made perils of pollution, noise and wastes—that are destroying our environment, but to shape the positive forces that are vital to restore its beauty, majesty, salubriousness and wholesomeness and thereby, to halt the gradual erosion of the quality of life, the upkeep of which is the ultimate goal of all our endeavours. This paper attempts to highlight the Health Hazards posed by the Environmental Pollution in the Indian Iron and Steel Plants and the Role of Engineers for minimizing the ill effects on the society.

Introduction

Dust and smoke emissions and other sources like smog, which lead to atmospheric pollution, affect the human, animal and plant environments in many ways, besides deteriorating buildings and damaging machinery. Release of toxic discharges, harm the respiratory and digestive systems and the skin and eyes mainly. The exhaust gases and vapours reduce visibility and penetration of sunlight. Plants are more susceptible to deleterious effects of sulphur oxides, dust falls and humidity changes. However, studies have still not fully identified the ill effects of air pollution on animal life.

*Chief Engineer (Design)/ SAIL (Retd.), President, Engg. Sciences, 89th Indian Science Congress-2002 284, Sector - I, Chiranjiv Vihar, GHAZIABAD - 201 002

Contaminations

The discharges and contaminations, which pollute the air and water in the steel industry may by gases, liquids, solids or a combination of these. They give rise to psychological, toxic and allergic reactions. Dust produced by many resources in the industry is irritating, harmful and toxic and gets absorbed in the bloodstream in many cases. Inorganic and organic—metallic gases and fumes, which may be irritants and asphyxiants, all cause damage and inflammation of the skin, eyes and respiratory track of prolonged exposure and some of them may lead to fatigue and diseases.

Increasing population concentrations also aggravate the pollution problems. Water pollution by industrial effluents represents a serious threat to man, fish, wildlife and vegetation. Such pollution by liquid wastes also causes serious difficulties in the operation and maintenance of water supply and storage and the sewer installations.

Pollution Sources

The sources and agents of pollution most commonly requiring serious attention from the health, safety and regulation considerations in the Integrated Iron and Steel Plants are : dust, smoke, fumes, steam, liquid oxygen, electricity, blast furnace and coke oven gases. Carbon dioxide and monoxide, cynaide concentrations, hydrogen sulphide, sulphur dioxide, ammonia, phenol, chlorine, nitrogen oxides, toxic and chemical substances, suspended mineral particles, polluted waters, dissolved and emulsified oils, high temperatures, noise, vibrations, molten metals, radio-active matter and coloured process wastes.

Characteristics of Health Hazards

The generally reported characteristics and their effects on health, resulting from the processes, are briefly summarised below :

Sulphur Dioxide (SO_2)

Extremely irritating and poisonous gas. Causes destruction of tissues of respiratory system and oedema of lungs, headache, dizziness, arterial dilation, fall in Blood Pressure, corrosion of teeth, etc.

Carbon Dioxide (CO_2)

Prolonged working in an atmosphere containing excess carbon dioxide causes giddiness and also leads to unconsciousness and other effects as above.

Nitrogen Oxides

Irritating gases cause breathing and lung troubles, as also other effects, explained above.

Carbon Monoxide (CO)

Poisonous and asphyxiating. Becomes explosive, when in excess, in the presence of air. Exposure and inhalation cause headache and unconsciousness.

Hydrogen Sulphide (H_2S)

Irritating and nauseating. Affects eyes, lungs and breathing.

Ammonis (NH_3)

Pungent and asphyxiating. Causes eye pain, giddiness, nausea and oedema of lungs.

Benzol

Poisonous vapours. Inhalation causes excitement. Affects blood and nervous system.

Carbon Disulphide (CS_2)

Inhalation of fumes causes headache, nausea, giddiness, slight intoxication and irregular breathing.

Phenolic Acid

Causes giddiness, heaviness in the head, weakness, breathing difficulty, hoarseness, nausea, insomnia and respiratory irritation.

Pyridine

Very poisonous, irritating and toxic fumes. Causes nervous and digestion imbalances.

Sulphuric Acid (H_2SO_4)

Irritating and poisonous. SO_3 fumes form acidic fog with steam. Irritates respiratory organs and affects lungs.

Naphthalene

Inhalation of fumes causes headache, nausea, retching and cough and irritates mucous membrane, eyes and respiratory organs.

High Temp and Molten Metals

High temperature are harmful to body tissues and eyes causing glare and blindness. Molten metal splashes causing serious burn injuries. Explosion of molten metal or slag dumped in wet locations cause heavy damage to equipment and human life.

Electricity

Electric shocks and electrocution are very common accidents in a steel plant, mostly due to inadequate knowledge of electrical safety, lack of proper care and improper handling. Also cause fire when grease, oil, cotton waste or other inflammable substances are thrown around.

Steam

High temperatures, pressures and hazardous operations are involved.

Oxygen

Very low temperature liquid oxygen is very hazardous and promotes combustion and fire. It has killing effect on the skin and deadens the body tissues.

Chemical and Radioactive Substances

Special care and safety precautions are necessary in dealing with these chemicals in protecting the workers from the effects of radioactive radiations. Water in human body gets ionised by the radiation and effects lead to deformation of limbs.

Running Machinery and Material Handling

Hazards due to these are very common and adopting adequate and proper safeguards are crucial.

Dust, Noise and Vibrations

Dust is a terrible nuisance, irritating and injurious and affects the respiratory system and lungs. Prolonged inhalation leads to asthma and other ailments. Silica dust is very dangerous and causes silicosis, etc. Prolonged inhalation of carbon in the form of soot, get absorbed and the silica present is slightly soluble in the body fluid, forms acid product causing fibrous growth around lungs. Noise and vibrations in many units cause great uneasiness and affects normal hearing.

Industrial Safety

In view of the aforesaid, the accident prevention, specially in an industry like the one we are now dealing with, is a complex subject involving knowledge of many disciplines—engineering, technology, medical science, psychology, etc. and those who have to put it into practice, have to look for answers to specific problems by establishing channels of communication and taking assistance from working personnel on the one hand and specialists on the other. This naturally needs agencies, which can pool collective consciousness and experience in accident prevention and provide them for the benefit of the industry as a whole. In India, there are a few voluntary organizations devoted to the field of industrial safety, like the

Safety First Association of India and some set up by the Government like The National Safety Council, helping the industry in reducing the toll of accidents by propagating the message of safety and creating safety consciousness among the workers and supervisory staff. They are also propagating training programmes and taking aids from the relevant Indian Standards on protective devices, safety measures in the industry, etc.

Global Coordination

World Health Organization (WHO) has laid down guidelines to achieve global coordination in order to help mankind avert the dangers and destructive forces, which threaten the environment and establish conscious, purposeful and cooperative strategies and techniques to combat pollution through qualitative, rather than quantitative development. In this background there is great urgency on at least two counts:

1. For an emphatic recognition of the guidelines.
2. For intensified, high priority efforts by professional and engineering to promote and implement practices based on the guidelines for environmental preservation and control of pollution.

Guidelines

The guidelines laid down by WHO, lay stress on the following aspects:

1. Identification and determination of environmental commitments.
2. Determination of the levels of pollutants in environmental samples, including ecosystems, thus tracing the pathways of pollutants in the environment, contributing to the understanding of their roles and also assessing the magnitudes of their efforts.
3. Linking of pollution to specific origins and identifying the sources and their relative contributions.
4. Studying and understanding of such processes as dispersion, washout and deposition of contaminants.
5. Determination of the operational efficiencies of the pollution control devices installed and control of water management operations.
6. Selection of outfall locations and designs.

Role of Engineers

The engineers of various disciplines and the Professional Engineering Institutions, have a definite Role to play in Environmental Pollution Control and Management. They can contribute to the world plan for abatement of pollution in a variety of ways. They can extend technical guidance, through their specialised engineering disciplines, to the governmental agencies engaged at the national and regional levels in matters of pollution control, in the following ways:

1. In initiating and coordinating research and development activities in relation to the health, comfort, nuisance, engineering and economic aspects of the pollution problems.
2. By collaborating in the formulation and publication of standards and codes of practices for emissions of smoke and dust and effluent discharges from industrial processes, which are based on sound engineering principles and economic possibilities.
3. By encouraging, sponsoring and assisting in the preparation, presentation and publication of technical papers and periodicals on various aspects of air, noise and water pollution.
4. By instituting a clearing house and data bank for dissemination and interchange of knowledge and information on pollution control trends, techniques, products, equipment and projects.
5. By discussing and specifying, by means of symposia and seminars, the economic and public health requirements for the control and prevention of nuisance or troublesome emissions and effluents.
6. By conducting exploratory studies of the chemical, biological and bacteriological characteristics of the groundwater, rivers and water courses and in undertaking air and water pollution sampling and surveys, as also the compilation of analytical data, which would aid measurement of the levels of different pollutions and their effects on the natural community and industrial environments.
7. By encouraging study groups on : efficient uses of fuels of all types, handling of materials, operational practices and aids in Iron and Steel Industry and other related aspects, such as characterisation and beneficial utilisation of wastes, analytical air monitoring, fabrication of indigenous testing equipment, improved safety and protective devices, etc.
8. By stimulating and co-ordinating research for determining the permissible upper limits of emissions and discharges of air and water pollutants in relation to topography, meteorology, location, etc. and for forecasting the water demands for the foreseeable industrial expansion and population growth.
9. In developing the precise, approximate and visual devices for detecting, recording and measuring contaminations.
10. By arranging training of environmental engineers through continuing education and refresher courses at regular intervals.
11. By imparting knowledge and information on occupational hazards resulting from pollution and offering instructions on industrial safety and protection measures and devices in collaboration with national safety councils.

The above is only a partial list of what the engineers and professional institutions could really undertake. There are many other areas, which include quality criteria and controlled disposal of process waste waters, re-use of treated industrial effluents, reclamation, segregation and stockpiling of selected solid wastes for possible future use, administration of pollution control laws, application of systems and treatment techniques such as re-cycling plants, reverse osmosis units, etc. use of nuclear energy for storage of concentrated wastes, etc. on which constant and devoted attention by the engineering profession as a whole should be focused in the context of its obligation to the society and the immensely broad opportunities and challenges that are being thrown open before it by the pollution crisis.

Conclusion

Paradoxically, industry does pollute the environment but it is essential to produce the necessities of life. Industry and pollution are thus inevitably interlinked. A great deal of research is therefore essential to provide us with a better understanding of the overall resolution of this dilemma and what is necessary is close cooperation between technologists and ecologists, economists and social and physical scientists to discover and apply a whole new range of technologies, not only to advance the technological development of the nation but help them combat the ecological dangers consequent to pollution, preserve the environment in its natural state and protect the health and spirit of man to the best possible extent.

REFERENCES

Bhatia, S.C. (1985). *Re-use & management of solid wastes from iron & Steel industries.* Kolkata: Institution of Engineers.

Bhattacharya, A. & Subramanyam, B.R. (1975). *Pollution from iron & steel industries in Indian context.* Kolkata: Institution of Engineers.

Environmental Management. New Delhi: SAIL documents.

Guidelines for Health Hazards. New York: WHO documents.

Pollution Management. Kolkata: IEI documents.

VULNERABILITY OF INFORMAL SECTOR IN NATURAL HAZARDS

Pawan Kumar*

ABSTRACT

Informal sector population is compelled to use unsuitable and unsafe terrain prone to natural hazard due to non-availability to affordable land. The use of non-engineering construction techniques, combustible building material, etc. in unplanned and highly dense informal settlements increases vulnerability. The non-availability of data on number of residents, morbidity, mortality, etc. gives invisible status of this segment of society creates further increments in their vulnerability. Reduction in vulnerability of informal sector requires multi-dimensional approach at different levels of planning having long-term and short-term measures. Informal sector is an asset for the society and hence it is the duty of local authorities to develop a comprehensive development plan to protect them and build up the capacity to prevent and mitigate the impacts of natural hazards.

Introduction

Natural events such as tsunami, cyclones, flood, earthquakes are considered as hazards or disasters only when they result in loss of life and property, affecting overall fabric of the society. Natural disaster is defined as disruption of human ecology, which can't be absorbed by the adjustment capacity of affected community within its own resources. It is one that causes maximum and continuous human suffering in informal sector because they are least equipped to cope with social, economic, cultural and political disruption due to natural disasters.

Pre-disaster activities can be subdivided into disaster prevention, disaster mitigation and disaster preparedness. Disaster prevention is event focused and the objective is to prevent the disaster from occurring at all. Disaster mitigation accepts the fact that some natural events may occur but tries to reduce the losses. Disaster preparedness assumes that disaster will occur but focuses on wide range of measures to save the life and limit the damages. In developing country, a large segment of urban residents are excluded from enjoying the benefits of urban life. They are often forced to live in informal settlements/

*Town & Country Planning Organization, (Ministry of Urban Development, Govt. of India), E-Block, Vikas Bhawan, I.P.Estate, New Delhi-110002. E-Mail: pawan612@sify.com

slums, which are neither recognized nor serviced by city authorities. The poor and socially disadvantaged groups of informal sector affected by flood, drought, desertification and other types of natural disasters are more vulnerable and insufficient equipped to mitigate them.

Vulnerability of Informal Sector

Informal sector is unorganized sector, which has easy entry, reliance on indigenous resources, highly dense residential pockets, concentration of more labour and service class population etc. Large-scale urban concentration is more vulnerable because it is characterized by explosion of population and inadequate infrastructure. However, the urban local bodies are unable to provide safe land, basic infrastructure and services to them and hence they are living in densely populated informal settlements, which are always at high vulnerability.

The magnitude, complexity and frequency of natural disaster are increasing due to increased vulnerability of the population. Similarly, informal sector is the worst affected by the natural hazards because they are compelled to use unsuitable and unsafe terrain due to non-availability of affordable land. Now-a-days cities are characterized by shortage of land but excess of informal population, which forced to use unsafe and risked areas, causes more vulnerability in natural hazards.

Land is the platform for all activities but the land unsuitable for construction, close to hazards industrial areas, coastal areas, flood plain, wetland, land beneath unstable soil and slope itself increased vulnerability of informal settlements in natural hazards.

The informal settlements are unplanned, highly dense and overpopulated. The dwelling units and other structures are constructed as quickly as possible with minimum cost. The combustible materials and non-engineered techniques used in construction are just to protect the occupants from rain, sun and wind and hence cause the vulnerability. In informal sector, data on number of resident's morbidity, mortality, migration etc. are almost not available. An invisible status of this group of people creates further increment in their vulnerability.

Environmental degradation, unhygienic conditions, dilapidated stock and lack of awareness among urban poor lead to more vulnerable and disaster prone society. The older age population, senior citizens, disabled persons, women, children etc. are the most vulnerable segment of the informal settlement in natural hazards, which cause maximum human losses if disaster strikes.

Informal sector affected by natural disaster brings typical disastrous effects on the society. A large group of affected population shifts from one place to other which causes additional demands on infrastructure. If urban local government is unable to provide, it leads a society vulnerable to crimes, violence, unsocial actions etc.

Hazard Assessment and Vulnerability Analysis

Hazard assessment is a tool to assess the intensity, frequency and location of natural hazards. It requires historical records available (written/memory), as well as regional and local physical characteristics of areas and their scientific analysis. Hence location criteria of the site are important. Generally, location and intensity are much easier to predict than frequency. A detailed hazard analysis provides strong base for preparation of risk maps.

Vulnerability analysis is a method of risk identification and its evaluation. Vulnerability = Natural hazard risk × Damage probability. Vulnerability plan of the area is the first pre-requisite to implement pre-disaster measures. Vulnerability analysis helps to take rational decision on how best the impacts of potentially disastrous natural events can be mitigated through proper control on land development. Infact, hazard assessment and vulnerability analysis suggest preventive measures and preparedness action plan.

Reduction in Vulnerability of Informal Sector: A Multi-dimensional Approach

Planning is one of the most efficient tools to deal with disaster. It can be applied in the physical planning as well as advocacy planning. An accurate and meticulous planning not only produces an effective plan but also provides a strong base for successful disaster management overall.

Regional Landuse Policy

It is one of the possible strategies to mitigate disaster. Regional land use policy is dealt with land development with economical, social and environmental goals. Hence disaster management policy should be formulated as a part of regional policy, which guides location, and spatial distribution of development within a region. It also regulates activities of Private sector through location restrictions and minimum standards on specific type of land use. Hence an approach towards land development and land use control helps to take rational decision to reduce vulnerability.

Zoning Regulations

Macro zoning establishes land use planning zones at regional level and hence is efficient tool to control overall location of human activity patterns. It has special importance in reduction of vulnerability. Similarly, micro zoning in urban settlements details out the density at particular sites and establish a detailed land use pattern within various risk zones. In fact it relates risk assessment to land use planning and suggests provision of wide roads, open space, adequate infrastructures, safe location of human concentration, which reduce vulnerability of informal settlements.

Sub-Division Regulations

It provides public control over the development of land through approved plans, provision of public facilities, infrastructures, etc. It also ensures safe development patterns of hazardous land where concentration of informal settlements are more. It prohibits sub-division of areas most vulnerable to natural hazards and eliminates development.

Urban Renewal and Extension Programme

These may be a part of development plan of any city or town. The same may be used to plan and zone for new land use in order to limit or prevent potential disasters. Before that, it is necessary to assess the potential risks of new land use area. Hence, it is appropriate step to shift /establish the informal settlements on less vulnerable site.

The local authorities at grassroots level can reduce vulnerability of informal sector in natural hazards. It suggests the following measures:

— *Planning Strategy:* Architects, Engineers, Town Planners, Policy makers, Decision Makers, etc. are able to encourage the development of informal settlements away from natural hazard risk zones by creating favourable environment at different levels of planning for investment in less vulnerable regions and areas.

— *Economic Incentive:* Government are often able to extend economic incentives such as provision of safe land, loans, grants, favourable credits and taxation, technical assistance, etc. to the local people and community based organization in order to encourage development away from hazardous areas. It helps to reduce vulnerability of informal sectors.

— *Disaster Resistance Economic Activities:* Economic opportunity is one of the reasons in the formation of informal settlements. Some economic activities are relatively unaffected by certain type of disasters. Hence efforts may be to identify and encourage the development of the enterprises that are less vulnerable to the hazards.

— *Economic Diversification:* It is important in those areas /regions where economy is based on just one or two occupations. Areas prone to particular type of natural hazards can diversify their economy selecting suitable occupation, which can sustain the communities.

— *Safe Structures:* The duty of local authority is to advise and encourage homeowners to use disaster resistant construction techniques rather than simply enforce the building codes.

— *Open Space Protection:* Open space is and around the informal settlements may be protected. The same may be used as sites for emergency shelters, rescue place, camps during calamity.

Conclusion

Informal sector is an asset for the society and it is the duty of local authorities to reduce vulnerability of the same in natural hazards at grassroots level effectively. Dense and overpopulated informal settlements in hazard prone areas should be banned and replaced by change in land use other than residential. The areas may be changed into permanents zones for agriculture or recreational use. Similarly, areas more vulnerable to natural hazards may be preserved as parks, wild life sanctuary, etc.

The urban local authorities should create awareness among informal residents through print media, electronic media, local media, etc. The role of NGO, CBO, NSS, Red Cross, etc. should be recognized and promoted to take active participation. At the same time, Government should express the political commitment to reduce vulnerability through declarations, legislation, policy decision, etc. The implementation of vulnerability reduction measures may be a component of the development plan.

The special responsibility should be assigned to "Disaster Mitigation Cell" under Town Planning and Building Engineering Department to promote disaster resistance construction technique and methods through public participation in informal settlements. Local Government should encourage the programmes that promote community-based approach to disaster mitigation and vulnerability reduction. Community based disaster preparedness utilizes the resources within the community and neighbouring areas, which build capacity of the community to prevent and mitigate the impacts of natural hazards.

REFERENCES

Carter, W.N. (1992), *Disaster Management—A Disaster Manager's Hand Book,* Asia Development Bank, Manila.

Kumar, Pawan (1998), Community Participation in Natural Disaster Management, paper presented in National Conference on "Disaster and Technology", Dept. of Architecture, MIT, Manipal.

Kumar, Pawan (1999), Disaster Mitigation and Rehabilitation, Paper presented in South Asian Countries Conference on "Challenges to Civil Engineers and Architects in 21st Century", Nepal Engineering College, Kathmandu.

The Institution of Civil Engineers (1995), *Mega Cities—Reduction Vulnerability to Natural Disasters,* Thomas Telford, London.

UNCHS (1996), *The Habitat II Agenda,* Nairobi, Kenya.

INDUSTRIAL SAFETY
Proactive Measures—Emergency Preparedness

M. Ramachandra Nayak*

ABSTRACT

The main theme of Industrial Safety proactive measures towards Emergency Preparedness is the Life Saving measures taken at Bharat Electronics, Bangalore complex keeping in view that BEL is considered as Hazardous Process Industry as we are holding 10 Tonnes of LPG (Liquefied petroleum gas) and several types of imported toxic gases including the in-house manufacturing of Oxygen, Hydrogen, and Nitrogen. Also we are using about 6 tonnes of various chemicals including cyanides for Electro-plating activities inside the factory premises annually.

Introduction

Bharat Electronics fulfils the needs of Defence equipments for Army, Navy and Air Force. As per Factories Act, 1948, Section 2 (cb) BEL is categorized as hazardous process industry. As per Section 41 (B) a detailed "ON-SITE EMERGENCY PLAN" and detailed control measures are to be established. Also the employees of the Factory including Trainees, Contract workers and visitors to be safeguarded in case of FIRE, EXPLOSION, and TOXIC GAS release disastrous eventualities. BEL has established in-built safety measures like Sirens, Hooters, Emergency phones, Ambulances, Automatic water sprinklers, Fire and Smoke detectors, Carbon dioxide flooding system, Emergency escape routes etc. to face any emergency situations.

Paper Details

The theme of this presentation is life saving of employees in case of Fire, Toxic gas release and Chemical mishaps. This paper throws light on Human Safety measures taken at Industrial scenario. Several processes are being carrying out in BEL including IC and Silicon devices

*Additional General Manager, Management,Service Division, Bharat Electronics Limited, Bangalore

manufacturing, Electro-plating process like, Gold plating, Silver plating, Nickel plating, Zinc plating, Tin plating etc. Also several gases like Oxygen, Hydrogen and Nitrogen are being manufactured in BEL utilising the in-house facility.

In the present Industrial Scenario Human Safety stands first. Bharat Electronics is a pioneer in Electronic Industry which fulfils the defence needs of our nation. Our products are from tiny component to a huge Radar to fulfil the needs of Army, Navy and Air Force.

Our industry is coming under Hazardous Process Industry in which we are using more than 200 chemicals and 10 Tonnes of LPG for the manufacturing process and canteens. In addition to this our Industry is using several imported gases, which are toxic, pyroforic and poisonous in nature, these gases are used in manufacturing process of IC's and Semiconductors at IC Department, Gas plant and Silicon devices department. Apart from these Gases, we manufacture Oxygen, Hydrogen and Nitrogen in our gas plant department.

Keeping in view all the facts mentioned above, we have to upkeep a systematic Emergency Plan to rescue the employees from Fire, Toxic Release and Explosion Disastrous events, which are common Emergencies at industrial scenario.

As per Factories Act, 1948 Section 41(B) we have an "ON-SITE EMERGENCY PLAN" which is a Statutory need to safeguard the employees at crisis/dangerous situation.

We have formed a network of Emergency declarer and Emergency Officers including Incident Controllers spread all over the factory from various Departments/Divisions. One of the senior General Manager is an Emergency Declarer and two other Additional General Managers are Alternate Emergency Declarers.

Several Emergency systems are kept on line 24 hours to mitigate the fire situations / other Emergency situation inside the organisation.

Communication Facilities

(a) Sirens, hooters
(b) Digital phones
(c) Wailing siren
(d) Mobile walkie-talkie
(e) Public address system
(f) Electronic attendance system

Siren and Hooters plays a vital role in case of Emergency situations, which are located at various divisions/departments including A/C rooms to cover nuke and corner of the factory.

Digital Cordless Telephones are distributed among Emergency Officers and Incident Controllers to obtain the communication towards Emergency situation occurred in a particular division/department.

Wailing Siren is a coded siren which is maintained 5 seconds on and 5 seconds off for 3 times at high pitch which will bring the awareness to employees who are at work place to identify there is some danger at that moment.

If the situation is brought under control, a long siren for 2 to 3 minutes will be blown at high pitch. This is a clear indication that the situation is quite normal and employees need not feel panicky at the work place.

Mobile Radio Trunk System or Walkie Talkie is under the control of chief security officer and distributed to vigilance staff, control room staff and Sr. Fire Officer for online mobile communication system while the staffs are on walk through surveys inside the complex.

Public Address System is well established at nuke and corners of the factory to communicate the message to all the employees inside the organisation.

Electronic Attendance System is established by Information systems division at important places like security control room, chief security officer's cabin, safety engineering department, all SBU heads and senior executive's cabin for the purpose of head count during emergency situations, under any circumstances at anytime, online attendance is available to assess the situation by Emergency declarer and his team towards On-Site Emergency plan.

Fire Fighting Facilities

Fire department under the Sr. Fire Officer has a 24 member dedicated staff who are engaged round the clock to mitigate the Emergency situations.

— Towards this 2 dedicated multipurpose fire tenders are there which will land at accident spot with in few minutes any place inside the factory.

— There are 316 fire hydrants always kept at 7kg's pressure located in vulnerable areas, which is capable of supplying water to high-rise buildings also for 4 hours to mitigate the fire situation.

— As per the statutory requirement and Tariff advisory committee advice, we have established automatic water sprinklers at various fire vulnerable areas, i.e. central gas cylinder stores, inflammable stores, gas plant (i.e., Hydrogen storage and Hydrogen truck manifold area) and LPG yard. As soon as the temperature reaches 60° C to 70° C, the quartz bulb kept at air pressure will burst open and thereby due to pressure reduction in the circuit the wet raiser valve will burst open suddenly supplying sprinkled water with in 55 seconds to mitigate the fire/gas leakage situation.

— Fire emergency staff has been supplied with fire proximity suits to enter the fire premises without any fear to control the gas line valve or to fight the fire manually using fire hydrant tender water hoses, which are located at 10 prominent places inside the factory premises.

— There are two pairs of self-contained breathing apparatus which is always kept on- line with pure air to fight the Fire/Gas leakage/Explosion situations inside the factory. One is kept at fire department and the other one is at Semiconductors/IC department.

— Fire department has 2 ladders of sophisticated variety to reach the high rise building for rescuing the employees who have gone helter shelter during Fire and other emergency situations. While bringing down the victims down the building safely they use standard nylon ropes.

— Bangalore Complex has established Auto dialing system from fire vulnerable areas to Fire department Security control room, Safety department, Chief security officers cabin, Emergency control room and Emergency declarer's cabin. In case of fire the accident message will automatically send through phone to Fire department Security control room, Safety department, Emergency control room and Emergency declarer's office sequentially at 20 seconds time gap.

LPG Bullets-Kicated at an isolated place inside the Factory Premises.

- In information systems division carbon dioxide flooding system is installed in magnetic tape and document storage section. In case of fire the quartz bulb linked to online CO_2 system will burst open and CO_2 will spread spontaneously, thereby mitigating the fire.
- In fire vulnerable areas like CADDS, MR, LPE assembly etc., smoke detectors are fixed on the roofs wherein even a smoke will sense by the detector and sends a prompt alarm and visual red light signal to fire station and helps the fire staff to reach the accident spot before the fire starts.
- Personal protective equipments like helmets, goggles, safety belts etc., are supplied by the divisions to the workers to mitigate the emergency situations and to avoid accidents during working at heights, excavation works and other civil works during modernization of buildings as well as new construction projects.

Discipline Order Service Facilities

- Dedicated teams of well-trained home guards are there in the complex who will render their service at crisis and dangerous situations along with other emergency officers and team members.
- Volunteers are also nominated by divisional heads towards on-site emergency/ local emergency mock drills and recently about 80 members have been trained and it is a continuous process to train up these groups suitably to face Fire situation, Rescue operation, medical services etc., towards improvement of emergency plan in our organisation.

— During emergency situations the security team will co-ordinate with other emergency officers to mitigate the dangerous situation as well as control the mob in the event of fast walk of employees to a safe assembly point to bring them from the dangerous spot and save life.
— Towards Emergency needs, Fork lifts and Montex machines will also be used to rescue the employees from 1st floor and other high-rise buildings.
— There are vans and jeeps under the control of Chief Security Officer. These can be used for emergency situation inside the organisation.

Medical Rescue (First AID) Facilities

— A senor medical officer and his team are well trained in participating emergency systems and mitigating the seriousness of accidents and help to transport the victims to BE main hospital and other notified hospitals.
— There are 2 Ambulance vans with in built wireless communication, oxygen cylinder and stretcher facilities to transport the victims. 2 drivers are always in ready mode to transport the patients as well as accident victims to hospital.
— In all the divisions/departments stretchers and first aid boxes are kept for the usage of emergency situations.
— BEL Main hospital has 60 bed facility to treat the patients as well as accident victims at crisis and doctors in specialized field are always rendering their service towards health and safety aspects of employees.

During Mock drill-Automatic Sprinkler is Maitigating the Situation

Water Supply Facilities

- There are 2 dedicated water sumps in which potable water up to capacity 800KL (at IC backyard), 1250KL (Near Inflammable stores) is stored, which are connected to 288 fire hydrants all over the factory and automatic water sprinklers at fire prone areas. The water is sufficient to fight the fire situation or gas leakage up to 4 hours.
- The water lines are always kept under pressure by powerful electrical or diesel engine driven pumps. Also one high capacity jockey pump is also kept online with the system as a stand by in case of power failures etc.

Transport Facilities

- There are 3 ambulance vans, 4 jeeps, 80 buses, 25 vans and 10 cars in the factory premises. Which we can use to transport the victims from dangerous zone to notified hospitals or to transport employees to their original pick up stops to reach home safely.

Emergency Control Room Facilities

- There are six telephones online with emergency systems to receive the message through all the phones simultaneously.
- A board with names, telephones and addresses of all emergency officers is displayed for ready reference to contact them at crisis.
- A digital wind velocity meter is installed in emergency control room to estimate the wind velocity and direction in case of emergency to vacate the employees safely from dangerous zone to safe zone in the opposite direction of the wind.
- Attendance board is situated in emergency control room for display of daily attendance towards ready reference of head count in case of emergencies.
- Emergency escape route map is also displayed at emergency control room for the ready reference of emergency declarer and his team.

Mutual AID Facilities

- In case the Fire/Toxic gas/LPG leakage becomes catastrophic, the mutual help from Karnataka state fire force, HMT fire force and Air Force fire station Jalahalli etc., can be taken to mitigate the situation successfully.

Notified Hospitals

- Since BE main hospital has 60 bed capacity we can get the medical aid from MSR hospital, St. Martha's hospital, Victoria hospital, St. John Medical College hospital and Bowring hospital to safeguard our employees and to give them suitable medical facility apart from life saving.

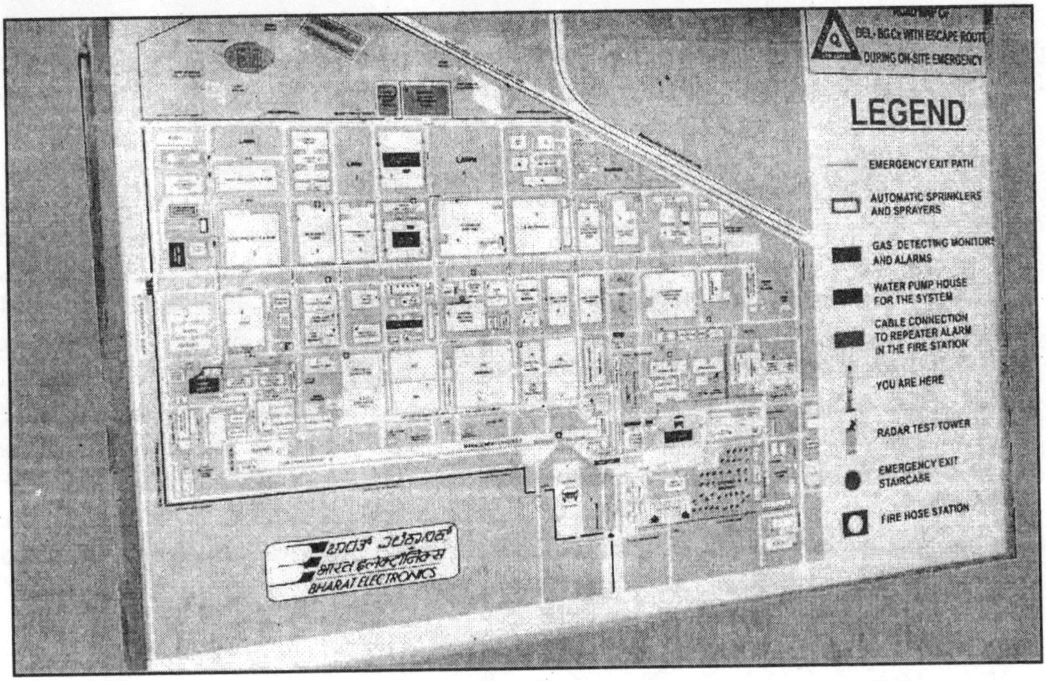

Emergency Escape route map

Other Emergency Improvements

- Safety engineering department is maintaining emergency check list towards the performance of emergency systems once in a month and ask the concerned to set right the brake down systems online to working condition.
- Emergency route map is designed considering important factors like, Wind velocity, Wind direction, Hazardous storage area, Hazardous chemical quantity, Density of persons, Width of the road, exit roads etc. using computer so that most optimum solution was arrived. The Emergency routes are clearly marked with fluorescent type ink in the map for easy identification even at power failure situations during nights. These maps are displayed at prominent places inside the factory premises. Apart from this every corner of the road is provided with a guide post to identify Emergency route.
- There are about 36 safety health and environment committees well organized in the factory under the chairpersonship of a senior executive of the division/ department, wherein this team will discuss all the emergency needs and safety improvements once in a month in the meetings and solve the problems at shop level itself.
- There is a safety council apex body which meets once in 3 months under the chairmanship of a senior General Manager and senior executives with negotiating trade union nominated persons and discuss the overall emergency and safety aspects of the organisation and solve the problems immediately.

— All the emergency officers and incident controllers will meet once in a month under the chairmanship of a General Manager and discuss the emergency systems functioning style and any improvements to be made for up-keeping emergency systems inside the factory.

— Training is a ongoing process at training centre in which faculties from various safety health and environmental aspects will give lecture classes for the benefit of employees, trainees and contract workers and train up the staff batch by batch in the organisation. Also on-site emergency plan cards will be issued to the participants to have an idea how to behave calmly and to over come the situation at the time of emergency situations inside the factory.

— Towards improvement of "ON-SITE EMERGENCY PLAN" inside the organisation each and every departments are conducting local emergency mock drills to educate their staff to learn about the department/factory emergency escape routes, dangerous process in which they are involved, dangerous chemicals they are handling at their work places and first aid facilities in their work places including handling of fire fighting equipments/fire extinguishers, to rescue the physically handicapped and other trapped people inside at crisis etc.

Conclusion

BEL has taken all the required Safety measures to combat the worst situation of disastrous situation like FIRE, TOXIC GAS RELEASE and CHEMICAL MISHAPS, with the available well built infrastructure.

REFERENCES

Factories Act, 1948.
Onsite Emergency Plan of BEL.
British Safety Council Chronicles.
Other Industries Emergency Plan.

METHODOLOGY FOR ASSESSMENT OF IMPACT OF INDUSTRIAL DISCHARGES ON STREAM WATER QUALITY

Prateek Sharma,* Vijay K. Minocha** and P.R. Bose***

ABSTRACT

With the rapid industrial development the potential polluters become larger, as well as numerous the limited assimilative water resource capacity will play an intrinsic part in technical solutions. Natural water-bodies, in general, have an assimilative capacity for most of the wastes discharged into it. While the pollution control agency wants the pollution to be within an acceptable limit, the discharger prefers to use the assimilative capacity of the river system to minimize the waster treatment cost. The aspiration of the pollution-controlling agency is to ensure the required quality in the natural water bodies by imposing certain effluent standards on the dischargers. The problem lies in optimal allocation of waste load discharge from a series of dischargers in a river basin.

A methodology is proposed for the determination of optimal waste allocation and wastewater treatment alternatives that minimizes the overall cost of wastewater treatment. The aim of this paper is to demonstrate how to optimize the overall treatment cost for achieving target DO levels in a stream.

Introduction

The most common problem arises in the planning context is the amount of pollutant removal required at a number of discharge points along a water body that must be determined in order to maintain an acceptable level of water quality in an optimal manner. This is important considering the vast amount of money that is spent on treatment facilities for various sources of pollution and also considering the limited assimilative capacity of many water bodies. Water quality management for a river network is a multiple objective optimization problem with conflicting goals of those who are responsible for maintaining the water quality of the river system and those who make use of the assimilative capacity of the river system by discharging waste to the water body.

*Lecturer, School of Environment Management, GGS Indraprastha University, Delhi, 110006
**Asst. Professor, Civil Engineering Department. Delhi College of Engineering, Delhi, 110042
***Professor, Civil Engineering Department, Delhi College of Engineering, Delhi, 110042

The aspiration of the pollution control agency is to ensure that pollution is within an acceptable limit by imposing certain water quality and effluent standards. On the other hand, dischargers prefer to use the assimilative capacity of the river system to minimize the waste treatment cost.

A number of water quality management models have been developed in the past for the allocation of assimilative capacity of the river system. Model results help in setting the amount of waste that can be disposed into the river from various points and non-point sources without violating the water quality standards. The presented methodology can be used to manage the river water quality considering the disposal of waste from various sources by allocating the removal to be done at each source.

Literature Review

The wastewater generation has been invariably related to water demand, therefore, estimation of wastewater load for modelling purpose requires estimation of water demand. Kumar and Minocha (1999) have discussed State space *versus* multiple regression for forecasting urban water demand. Further, Kumar and Minocha (2000) have given an interesting discussion on applicability and limitations on use of regression models for analyzing stormwater loads. Traditional deterministic optimization models of river basin water quality management have been formulated to minimize total effluent treatment costs while satisfying water quality standards along the concerned river basin. ReVelle *et al.* (1968) have used linear programming technique for water quality management. Bayer (1974) used Non-linear programming in river basin modelling. Fujiwara *et al.* (1987) gave a chance constrained model for river quality management. Besides the mathematical programming techniques, fuzzy set theory has also been used for water quality management. Sasikumar and Mazumdar (1998) gave a fuzzy waste-load allocation model for water quality management of a river system using fuzzy multiple-objective optimization. Kumar and Minocha (1999) have discussed the applicability and limitations of the modelling as attempted by Sasikumar and Muzumdar (1998).

New Approaches in Water Quality Modelling

Many water quality management plans (WQMP) or river basin plans were developed more than 25 years ago when water quality data were insufficient to support sound modelling technologies. With advancement in computing science and technology and development of new modelling techniques, there have been emerged new approaches. As more data are available now, these new approaches are required to be applied in the area of water quality modelling.

In a case study Lung *et at.* (1999) found that because of limited water quality data and lack of skilled modelling professionals, the models developed previously to determine assimilative capacity of streams and waste load allocations were not efficient. It has been found that the quality of receiving waters had improved due to increased treatment of waste flows and generally met standards, although an old model had predicted non-attainment.

Of late water quality management problems have been addressed as multiple optimization problems by many researchers.

River Water Quality Management Models

With increased population as well as pollution and decreased water resources the management requirements arose. Therefore both river water quality and management were considered together and various water quality management models were developed which can be classified into one of the two broad categories:

(a) Models that minimize the waste treatment cost subject to the constraint that water quality standards are not violated at selected water quality checkpoints in the river system.
(b) Models that maximize the water quality subject to a constraint on water quality cost.

Modelling Methodology

The model is formulated to minimize the total cost incurred on treatment processes by the point load dischargers in a river basin and at the same time, maintain the water quality in the river by imposing certain effluent standards. The parameters involved in the selection of a treatment unit at the sources for a given target deficit are,

(i) amount of improvement in DO for varying treatment levels using QUAL2E of EPA (Brown and Barnwell, 1987) or WASP simulation model, or any other model.
(ii) The differential cost of treatment.

Data Requirements

Any simulation model requires some degree of modelling sophistication and expertise on the part of a user. The input data in QUAL2E can be grouped into following categories:

(i) The first step in preparing the QUAL2E inputs is to divide the stream into reaches, whch are stretches of stream that have uniform hydraulic characteristics. Each reach is then subdivided into computational elements of equal length.
(ii) The global variables include simulation variables, such as units and simulation type, water quality constituents, and some physical characteristics of the basin.
(iii) QUAL2E accommodates four types of hydraulic and mass-load forcing functions: headwater-inputs, point sources or withdrawals, incremental inflow/outflow along a reach, and the downstream boundary concentration (optional).
(iv) Local climatological data.

Utilizing the above information the water quality model simulates the constituent concentrations along a reach.

Table 29.1: Sample Effluent and River Data

Effluent Data					*River Flow Data*			
Discharger	*Effluent m^3/sec*	*Temp. °C*	*BOD Conc. mg/l*	*Do Conc. mg/l*	*Reach no.*	*Distance in km.*	*Velocity m/sec*	*Temp. (Av.)°C*
D1	0.49	30	125	1.62	RI	45	1.648	16.20
D2	1.24	30	225	2.15	R2	30	0.264	15.00
D3	0.96	30	140	2.03	R3	80	1.440	17.20
D4	0.74	30	323	1.80	R4	40	0.363	16.55
D5	0.63	30	250	1.23	R5	60	0.947	14,55

Table 29.2: Sample River Flow Data

River Flow Data					
River Reach	*Flow ($106m^3/day$)*	*Time of Flow (in days)*	*Deoxygenation const k_1 (1/day)*	*Reaeration const k_2 (1/day)*	*Saturation Do conc. mg/l*
RI	1.61	0.316	0.331	0.847	10.10
R2	1.25	1.312	0.328	0.743	9.85
R3	2.87	0.642	0.378	0.532	9.64
R4	1.98	1.281	0.410	0.831	9.78
R5	3.23	0.732	0.320	0.754	10.20
R6	4.22	1.218	0.357	0.670	9.90

Treatment Cost

The cost of each treatment level may be adopted on the basis of the treatment units employed, thereby eliminating the problem of cost functions and the complexity caused by the same in solving the problem. The cost adopted is the amortized cost (Table 29.3) of the treatment unit for another study. Unit costs of treatment are based on volume and BOD and are taken from the report on CETP by National Environmental Engineering Institute.

Table 29.3: Amortized Costs for Different Treatment Options (Dutt, 2001)

Source No.	*Treatment Cost, C (Rs.)*		
	Primary (I)	*Secondary (II)*	*Tertiary (III)*
D1	11,61,003	69,46,170	1,12,96,209
D2	58,42,009	2,99,59,020	4,87,20,856

The percentage of removal of carbonaceous BOD by a particular treatment unit varies with the characteristics of the waste. A set of percentage removal at each source corresponding to each of the treatment unit is assumed from the report by Dutt (2001).

To investigate the relationship between BOD load and the DO deficit, QUAL2E is run for stated values of percentage of treatment (Dutt, 2001) to find the DO sensitivity in the stream, for each source. There exist DO deficit maxima D_{ml}, D_{m2} and D_{m3}, at three locations in its stretch at a chanage of 520 km, 470 km and 380 km.

Table 29.4: Per cent Removals at Different Sources Corresponding to Each Treatment Level (Dutt, 2001) and Respective BOD Load

Source	*Treatment Levels*			*BOD Load at each level of treatment (kg/day)*			
	I	*II*	*III*	*Nil*	*I*	*II*	*III*
D1	30%	80%	95%	26,676	18,673	5,335	1,334
D2	20%	70%	80%	56,382	45,106	16,195	11,276
D3	25%	75%	'90%	78,617	58,963	19,654	7,862
D4	35%	80%	95%	46,413	30,169	9,283	2,321
D5	30%	75%	85%	91,541	64,079	22,885	13,731

Table 29.5 gives the values of D_{ml}, D_{m2} and D_{m3} after applying primary treatment at all the sources.

Table 29.5: DO Deficit after Applying Primary Treatment at All the Sources.

D_{ml}	D_{m2}	D_{m3}
8.60	9.07	9.79

After identifying the points of critical deficit, the next step is to find the sensitivity of DO, at these iocations for the BOD loads at various sources. The DO at different treatment levels for any specific source is found to be proportional to the percentage of treatment. But, since the percentage of treatment at different levels is varying with the sources, it is more emphatic to find the DO improvement resulting from upgrade to the higher levels of treatment.

To account for the varying treatment costs, a simple model may be developed, wherein the approach is to minimize the treatment cost for any given target DO deficit in the stream. For example, consider a target deficit of 6.0mg/1 is to be maintained in the stream. The values of DO deficit maxima after applying primary treatment (Table 29.5) are 8.60 mg/I, 9.07mg/1 and 9.79 mg/I. The most critical of the three is D_{m3} i.e. 9.79 mg/I.

REFERENCES

Bayer, M.B. 1974. Nonlinear Programming In River Basin Modelling, *Water Resource Bulletin*. vol.10(2).

Brown, L. C., and T. O. Barnwell, Jr. 1987. *The Enhanced Stream Water Quality Models QUAL2E and QUAL2E-UNCAS: Documentation and User Manual. EPA-600/3-87/007.* Environmental Research. Laboratory, Office of Research & Development, USEPA, Athens, Georgia, USA.

Dutt, Shubhra. 2001. *Optimal Resource Allocation Model Development for Water Quality Management in a River with Multiple Dischargers Using Mixed Integer Programming Technique*. M.E. Minor Project under the guidance of Vijay K. Minocha, Delhi College of Engg. Delhi University, India.

Fujiwara, 0., Gnanendran, S.K., and Ohgahi,S. 1987. Chance Constrained Model For Water Quality Management. Journal of Environmental Engineering, A.S.CE., vol. 113(5).

Kumar, A. and Minocha, V. K., 1999. State space *versus* multiple regression for forecasting. urban water demand: Discussion, *Journal of Water Resour. Plng. And Mgmt., A.S.C.E.,* vol. 125(4).

Kumar, A. and Minocha, V. K., 2000. Use of Regression Models for Analyzing Highway Storm-Water Loads: Discussion, *Journal of Environ., A.S.C.E.,* vol. 126(7).

Lung, W., and Sobeck, R.G. (1999). Renewed Use of BOD-DO Model In Water Quality Management. J. *Water Resour. Plng. and Mgmt., A.S.C.E.*, vol. 125 (4), 222-227.

ReVelle, C.S., Loucks, D.P., and Lynn, W.R. 1968. Linear Programming Applied To Water Quality Management, *Water Resources Research,* vol. 4(1).

Sasikumar, K. and Muzumdar, P.P. 1998 (March/April). Fuzzy Optimization Model for Water Quality Management of a River System, *Journal of Water Resource Planning and Management. A.S.C.E.*, vol. 124 (2).

30

UNSAFE CHROMIUM CONTAMINATION AND POLLUTION FROM COTTON ROLLER GINNING INDUSTRIES AND DESIGN AND DEVELOPMENT OF AN ECO-FRIENDLY ALTERNATIVE

G. Vijayan Iyer*

ABSTRACT

Most of the cotton ginning operations are performed by using double roller (DR) ginning machines which serve an important role in the ginning industries. The rollers used in these industries are made of Chrome Composite Leather-Clads (CCLC) covering fixed to a shaft. The CCLC coverings contain about 18,000 to 36,000 mg/kg (ppm) of chromium particles. When the seed-cotton is processed in DR ginning machine, the lint cotton is contaminated with chromium dust of about 140 to 1990 ppm as total chromium which is *carcinogenic substance* as against the safe limits of 0.1 ppm. Due to the constant dust-producing grinding action of CCLC rollers in DR gins, chromium leaks into environment beyond the permissible levels which threat to cotton processing workers and users of cotton garments. During the cotton ginning process due to persistent rubbing of rollers over stationary knife the chromium particles are carried into lint cotton such that the spun yarns contain about 100 to 200 ppm, which according to eco-standards for spun yarns should not be more than 0.1 ppm. The CCLC rollers used in cotton roller ginning machines gets powdered during the ginning process. As chromium is a specific dust, gin and mill workers users of cotton garments are directly exposed to this carcinogenic substance with the *carcinogenic risk* of 2020 chances per million and are vulnerable to health hazards. To offset this problem, pollution-free rubberized cotton fabric (RCF) washers and rollers both for laboratory and commercial studies have been fabricated and experimented. These eco-friendly chrome-less rollers are evaluated for their performances; with a particular reference to techno-commercial and environmental aspects. In this the roller covering is made of multiple layers of fabrics bonded together using a white rubber compounding which has a surface finish conducive to high ginning efficiency. On the basis of the design and development of various rollers with subsequent performance evaluation studies, pollution-free RCF washers and rollers have been demonstrated with a reference to techno-commercial and eco-friendliness in ginneries. These newly developed RCF rollers were successful in their operation. These were found more effective in functioning than CCLC roller ginneries. They have eliminated unsafe chromium related contamination and pollution problems. This cleaner

*Professor, Department of Mechanical Engineering , MNM Jain Engineering College, 23/30, Mariaman Koil Street, Phase-V, II Floor, K.K.Nagar(West), Chennai-600 078, Tamil Nadu, E-mail. gvijayan33@hotmail.com

technology is a *source-at-control* preventive measure such that the cotton ginning and textile industries meet the requirements of environmental standards being enforced by many countries and high quality yarns and fabrics meeting the international standards are produced. These ginneries have been tested commercially. This found better in all aspects with reference to cotton technological properties, dye-catch properties, physical and chemical properties. It could be used commercially as an alternative in the cotton roller ginning industries for the benefit of environment, society, gin and textile mill owners, traders, workers, employees and the government.

Introduction

India is one of the leading cotton growing countries in the world. It has 24 per cent of the world area under cotton cultivation and 11 per cent of the total cotton produced in the world. India ranks first in area with 11.1 million hectares of land under cotton cultivation and third in production. Cotton production has been estimated at 1700 lakh bales (each bale weighs 170 kg of lint cotton) during 2004-05 in the world. Out of total cotton production, 200 lakh bales was ginned from 5100 cotton ginning industries in India. The cotton roller ginning process is defined as the mechanical separation of cotton fibres from their seeds by means of one or more rollers to which fibres adhere while the seeds are impeded and struck off or pulled loose . Roller gins are of two kinds, viz. single roller (SR) gins and double roller (DR) gins. In these gins, spirally grooved leather rollers, pressed against fixed knives, are made to rotate at a definite speed. Crank or eccentric shaft close to leather rollers gives rise to oscillation action of moving knives. When the seed-cotton is fed to the machine, fibres adhere to the rough surface of the rollers and are carried in between the fixed knives and the rollers and the fibres are entangled in the process. The moving knives beat the seeds and separate the fibres, which are gripped, from the seed end. This process is repeated number of times and due to the 'push-and-pull' action; the fibres are separated from the seeds carried forward on the rollers to be dropped out of the machine. The ginned cotton seeds dropped down through the grid slots.

Problem Formulation

The roller constitutes an important element of roller gins. Until 1940, only Walrus animal hide was used as roller covering material in USA and UK. Later on due to the non-availability of Walrus, these countries did not allow this type of hides to be used and obsolete these roller gins. Sheep and Buffalo chrome tanned hides, were used as substitutes in the roller ginning machines, though the interfibrillary action was not satisfactory compared to walrus hides. The roller materials, viz. ordinary leather, newspaper, corkboard, and coconut coir were also tried, but have not been found suitable. Since 1940, *chrome composite leather-cladding* (CCLC) material has been under use for making rollers of roller gins till now in India and Africa. The CCLC rollers have not been used in USA and UK, since many years. The CCLC roller is made of CCLC washers, which are filled and compressed in a shaft to a high pressure. One of the important property pertains to roller covering material is interfibrillary action to be considered in the development of roller ginning process.

Since the semi-finished chrome leather washers contains 3 to 4 per cent as total chromium that is around 36,000 ppm (mg/kg) of chromium and are being used by roller ginning industries, attention has been drawn to view the contaminating and polluting aspects during the complete process. The chrome specific dust pollutes the gin-house air and the cotton processing workers suffer from chromium bound diseases and physiological disorders due to the occupational exposures. The chromium adsorbed into lint causes allergic symptoms, cancer incidence, brain damage, chronic ulceration and perforation of nasal septums to cotton processing workers. Toxic effects are produced by prolonged contact with airborne or solid or liquid chromium contamination and pollution even in small quantities. Several CCLC rollers are required in an year due to tremendous wear and tear action arising out of grinding action. This also results in *chrome specific dust* (CSD) production during the process of ginning operation, which is the major unsafe environmental chromium contamination and pollution problems from all 5100 roller ginning industries in the nine cotton growing States of India. The CSD contaminates the lint cotton, spun yarns, woven fabrics and cotton seeds during the cotton roller ginning process.

As far as the roller ginning machine construction is concerned, roller is the major component of DR gin. The gin roller length varies from 1025 to 1148 mm with a diameter varying from 178 to 180 mm suitable for operation. The roller consists of 78 to 80 numbers washer disks. Each washer disk is made of dimensions having the diameter 180 mm and thick 1 mm comprising of 18 numbers CCLC flaps stitched and bonded together. Basic Chromium Sulphate (BCS) $Cr(OH)SO_4\ nH_2O$ and impure chromate having 45-50 per cent basicity are used during chrome leather tanning process for making such CCLC flaps. The various unit operations involved in making washers to final shape of the roller are (i) The washers are filled in a steel shaft having square crosssection of 50 mm^2 or hexagonal section of 50 mm E/E to form a roller, (ii) The filled washers are compressed to a pressure of 14 kg/cm^2 by using a conventional pressing machine. The roller is to be pressed on both sides by adding required number of washers on each side, (iii) The pressed roller is turned and finished to diameter 180 mm in a centre lathe, (iv) Spiral grooves are made on the surface of the finished rollers. The finished roller is ready for grooving operation by using band saw; initially by marking 'U'-shaped spiral grooves, fixing in the grooving machine and lastly spiral grooves are made on the roller surface by band saw or circular saw cutting machine. The ginning efficiency primarily depends upon the surface speed of the roller and number of working strokes on the moving knife. While operation of these rollers in the ginning machines, the rate of ginning goes on declining when the roller diameter is reduced. At the end of cotton season having three months duration, that is after the diameter is reduced to 114 mm, the washers are removed from the shaft. Again the new washers are recovered and cladded in shaft. The worn out and consumed washer disks after considerable period of usage for about three months.

Cotton seeds get contaminated with chromium from the source. Present work was aimed initially to identify and study the environmental and health related problems faced with currently used Chrome Composite Leather-Cladding (CCLC) rollers. Some experimental results of chromium analysis and relevant eco-standards, quantity of respirable and suspended particulate matter samples, health survey observations and cotton technological reports of eco-friendly lint and chrome contaminated lint are presented for assessing the hazards of chromium contamination of lint, yarn and fabrics and air pollution problems due to chrome specific dust (CSD). After realizing the hazards of chromium contamination in lint cotton, spun yarns, woven fabrics, cotton seeds, textile effluent and chromium air pollution due to dust-producing grinding CCLC rollers used in ginning environment, there is a need to eliminate this chromium contamination and pollution problem from cotton roller ginning and textile industries.

Hence, the work has been carried out with the following principal objectives:

- To identify and study the environmental related problems existing with the present chrome rollers employed in cotton roller ginning industries,
- To design and develop eco-friendly chrome-free roller and evaluate its performance with a particular reference to environmental and techno-commercial aspects in ginning industries.

With the author's research background and practical experience in ginning and textile industries, present study is attempted to eliminate this problem, through a suitable design and development of an eco-friendly, chrome-free roller for cotton roller gins. An eco-friendly roller ginning process is developed for replacing conventional CCLC roller ginning process to eliminate the unsafe chromium contamination and pollution from roller ginning industries so as to meet the requirements of the environmental standards while maintaining high quality spun yarns and woven fabrics meeting the international standards.

Environmental Impacts of CCLC Rollers

Environmental impacts of CCLC rollers are assessed from the pollutants, viz. cotton dust and chrome specific dust (CSD) in the mill atmosphere. The cotton dust released in the ginning process is a complex and variable mixture of cotton fibres, undeveloped ovules, cotton plant debris including twigs, bract and pericarp particles left after the ginning process together with soil particles, bacteria, fungi and residues from pesticides. The visible and invisible dust in the mill atmosphere is known as 'Fly'. The ambient air particles of about 2.5mm are classified as cotton dust in ginning environment. Byssinosis is a disease due to the inhalation of cotton dust over long period of time (Shirley, Vol. II, 1982). It is a permanent disabling lung disease. The symptom is chronic cough ending in chronic bronchitis (respiratory disorder). India has a large number of ginning and textile mills employing 48 per cent of all the factory workers. About 55 per cent of mill workers suffer from byssinosis disease. As per the rough estimate during field survey/discussions with ginning industry management, presently, there are about 2,13,000 CCLC rollers, which comprise of 1,70,40,000 CCLC washers are used for a cotton season of three months in our country. At present, there are about 7,60,000 people working in 5100 roller ginning industries and 15,90,000 people working in 6,000 textile industries who are occupationally affected due to the unsafe chromium and pollution problems.

Materials and Methods

For the present study, roller wearing and compaction rate study were conducted in Roller ginning industries at Belgaum for the cotton seasons 1996-97 to 1998-2000 and 2001-02 including 2004-05. The roller gins are adjusted using spacers as per the standards. Grooving is done regularly at the start of each shift as per cotton varieties. To study the heavy metal as total chromium mg/kg (ppm) in cotton lint samples, seed samples, seed-cotton samples, CCLC roller samples, CCLC roller samples collected during grooving operation, soil samples from the region of investigation is made, root of the plant for bio-availability, fibre, yarn, fabric samples, textile effluent samples, the standard American Public Health Association (APHA) method is followed for chromium (as total and hexavalent) analysis using Atomic Absorption Spectrometer (AAS). Respirable and suspended particulate matter quantity in gin house air is monitored using High Volume Air Sampler (HVAS) cascade impactor with appropriate glass fibre filters. The quantity of pollutants are collected in HVAS as 8 hours basis and analyzed for chromium. The worker dose and exposure time are found using a personal sampler. Cotton technological parameters were tested using High Volume Instrument (HVI) and Scanning Electron Microscope (SEM) for chrome roller ginned lint and eco-friendly roller ginned lint and were confirmed to the international standards.

A health study is conducted by the author at Guntur, Bailhongal, Sendwa, Surendranagar, and Tanzania, where maximum number of ginning factories are situated to survey the conditions of health effects and occupational health hazard exposures. In view of the environmental problems existing with the present CCLC rollers employed in ginning industries and subsequent to literature survey made, it was decided to develop environmentally free alternatives.

The ginning investigations were carried out at Central Institute for Research on Cotton Technology (CIRCOT), Mumbai. The laboratory rollers for GRED were designed and fabricated at Calcutta at a local manufacturing firm. Experiments with the designed rollers were conducted at CIRCOT, Mumbai along with the cotton technological parameters. After the initial tests, pilot model rollers were designed and fabricated which were tested in ginning factories at Bailhongal and Sendhwa. Environmental analysis was done in Centre of Mining Environment, Indian School of Mines, Dhanbad, and Eco-Textile laboratory, Mumbai. Mechanical properties were analyzed in various mechanical engineering laboratories.

Problem Solution

The second part of the present study pertains to nullify the unsafe chromium pollution and contamination problems in cotton ginning and textile industries. An exhaustive study is needed for the design and development of eco-friendly chrome less roller, which can be alternatives to the existing CCLC rollers.

An exhaustive material studies are done for the suitable material's selection of the gin rollers. These were made of Walrus animal skin, Spider tuck packing, coir-board, rubber packing, metal cylinder, rubber roll, fabric and rubber packing, leather, cotton, rubber and cork, plastics and fluorinated ethylene propylene. The peculiar gripping action or adherence of the cotton fibres to the roller surface was considered while designing the rollers. The leather surfaces possess interfibrillary action, which enables to adhere the fibre on the surface. This particular property was studied extensively for the different materials and combination of different materials so as to design and fabricate laboratory gin chrome less rollers for gin roller experimentation device (GRED) and prototype eco-friendly chrome less rollers for existing DR gins.

Results and Discussions

An experiment is conducted to find out the wearing and compactness rate of CCLC rollers used by roller ginning industries for a season lasting three months. At the start of season the diameter of rollers are 180mm. At the end of season the roller dimensions are noted at left, middle and right positions for all the roller gins in the factory, that is 18 ginning machines. Apart from the wear and tear rate, the quantity of pollutants generating during the operation, viz. chromium, leather powder, cotton dust and chrome specific dust. It is found that the wearing rate is 0.033mm/hour and the percentage material removed per roller 43.8 per cent. The final diameter at the end of study is nearing to 140mm. The compaction rate is 0.050 mm/hour. Chromium analysis report of cotton lint samples, seed and seed linter, seed-cotton samples, fibre, yarn and fabric samples are tabulated.

The CCLC roller contains 18,077 mg/kg (ppm) to 30,780 mg/kg (ppm) as total chromium (3 to 4% as total chromium). This included trivalent and hexavalent chromium. During the ginning operation, lint adsorbs chromium particles which contains 143 mg/kg (ppm) to 1994 mg/kg (ppm). The CCLC roller is grooved at the start of each shift and filing or turning of the roller for levelling is done to get uniform diameter at start of each season. At that time, the wearing of roller is more and presence of chromium to the extent of 1994 mg/kg (ppm) with lint. The total weight of chromium removed during a cotton season of 16 hours per day is 450 to 600 grams per gin roller gin. The chrome specific dust from one ginning machine enters in to environment and being adsorbed in lint stage having the level of 143 ppm. The environmental standards for chromium in spun yarn is 2 ppm and Cr (III) for baby clothing and fabric is 0.1 ppm and nil for Cr (VI). The traces found contain hexavalent chromium being adsorbed from contaminated lint, yarn to fabrics and subsequently cannot be removed in fabrics. There is evidence that the toxic effects on humans due to Cr (III) and Cr (VI) of its carcinogenecity and corrosiveness. The analysis show that traces of Cr (VI) are found in even analar grade trivalent chromium compounds and complications do arise due to the reducibility nature of these traces that affect the organic tissues of the body. This regenerating effects occur rapidly and dependent of the worker dose and exposure time.

Results of Design and Development of Eco-friendly Technology

The manufacturing technology, design engineering features and assembly drawings show that the conventional fabric and rubber roller gin covering material is selected with the following characteristics, namely,

(1) Hardness of 90 (type DO durometer),
(2) 9 to 10 layers of fabrics 20 mm length,
(3) Thickness of fabrics 1.2 mm,
(4) The rubber compounding is resilient, and
(5) 0.76 mm of fibre bristles protrude beyond the rubber surface is maintained in spite of wear.

On the basis of the design and development of various rollers with subsequent performance evaluation studies, chrome-free RCF roller has been demonstrated with reference to techno-commercial and eco-friendliness in ginning industries. The newly developed RCF rollers are successful and effective in functioning and in ginning out the seed-cotton. Cost economics study reveals that eco-friendly RCF roller ginnery sounds better in all aspects with reference to environmental, cotton technological and techno-commercial aspects.

Estimation of Cancer Risk due to Occupational Exposure

Estimation of the cancer risk for a 60-kg worker who has worked in a ginnery and textile mill exposed to a particular carcinogen under the following circumstances. Exposure time was 6 days per week, 30 weeks per year, over a period of 20 years. The worker was to breath heavily for 2-hour per workday at the rate of 1.5 m^3/hour and 10 hour per workday at a moderate breathing rate of 1 m^3/hour. The carcinogen has a potency factor of 0.02 $(mg/kg/day)^{-1}$ and an absorption factor was estimated at 80 per cent and its average concentration in the air 2.05 mg/m^3. The data were obtained from chromium concentration in ambient air.

The total amount of air breathed per workday =

$\therefore$ Daily intake rate = $1.5 \times 2\ (m^3/day) + 1\ (m^3/day) \times 10 = 13\ m^3/day$.

The total dose = Contaminant concentration × Intake rate × Exposure time × Absorption Factor

$13\ m^3/day \times 6\ days/week \times 30\ weeks\ per\ year \times 20\ years \times 0.8 \times 2.05\ (mg/kg/day)^{-1}$

$= 76752$ mg

Using a standard estimate of 70 years for a lifetime, the chronic intake daily (CID) was

$$\text{CDI in mg/kg/day} = \frac{\text{Total dose (mg)}}{\text{Body weight (kg)} \times \text{Lifetime (days)}}$$

$$= \frac{76752}{13 \times 70 \times 180\ \text{days/year}}$$

$$= 0.1\ \text{mg/kg/day}$$

$$\text{The cancer risk} = 0.1\ (\text{mg/kg/day}) \times 0.02\ (\text{mg/kg/day})^{-1}$$

$$= 2.02 \times 10^{-3}\ (\text{2020 chances in one million})$$

Conclusions and Suggestions

The CCLC rollers used in ginning industries get powdered during ginning operation and enter the environment as Chrome Specific Dust. It was observed that the chrome specific dust contaminates cotton and its products. Since the semi-finished chrome leather washers contains 3 to 4 per cent as total chromium that is around 36,000 ppm (mg/kg) of chromium and are being used by roller ginning industries, attention has been drawn to view the contaminating and polluting aspects during the complete process. The chromium adsorbed into lint causes allergic symptoms, cancer incidence, brain damage, chronic ulceration and perforation of nasal septums to cotton processing workers. Toxic effects are produced by prolonged contact with airborne or solid or liquid chromium contamination and pollution even in small quantities. India has a large number of ginning and textile mills employing 48 per cent of all the factory workers. About 55 per cent of mill workers suffer from byssinosis disease. As per the rough estimate during field survey/discussions with ginning industry management, presently, there are about 2,13,000 CCLC rollers, which comprise of 1,70,40,000 CCLC washers are used for a cotton season of three months in our country. At present, there are about 7,60,000 people working in 5100 roller ginning industries and 15,90,000 people working in 6,000 textile industries who are occupationally affected due to the unsafe chromium and pollution problems.

The chromium contamination levels for cotton and its products were abnormal for all the samples except that the cotton samples obtained from RCF roller gin rollers, i.e. eco-friendly ginning industries. As per the environmental standards, chromium content in cotton and its products not to be more than 0.1 ppm. The samples, namely lint cotton, yarn, fabrics, seed, linter, edible oil and oil cake were found contaminated and their levels were in the range of 110 to 1990 ppm obtained from the source of dust-producing grinding CCLC rollers sample which contained 18,077 to 30,783 ppm. The ginned lint cotton gets contaminated to an extent of 143 to 1990 mg/kg (ppm) of chromium and the woven fabrics to the tune of 17 to 45 ppm of chromium against the safe limit of 0.1 ppm. Hence, chromium leaks into environment beyond the permissible levels which threat to cotton processing workers and users of cotton garments. The cancer risk of occupational exposures was found to be 2020 chances in all mills. The chromium was not detected from RCF ginned lint cotton as there is no chromium in the source, which confines to the eco-standards. The air pollution due to CSD and cotton dust, which is responsible for synergistic (augmentative) health complications of chromium based diseases and byssinosis diseases on ginning industry workers. Almost most of the mills in India are not provided with dust control systems. Nor they provide personal protection devices to the workers. It is mentioned that the ginning industries are located in and around cotton growing areas and employ women in the age group of 21 to 40 years for menial jobs and male workers in the age group of 18 to 50 years. The women often come along with their children for performing their jobs, like (i) feeding seed-cotton (or *kapas*), (ii) collecting the lint cotton, seed and floor sweeping, (iii) cleaning and grading the seed-cotton and (iv) light activities. The children are exposed directly to CSD. The health effects and reports of the workers have not come out into public, because (i) almost all the workers are not in regular employment, (ii) the cotton ginning industry functions seasonally for 6-8 months in semi-arid zones and 8-10 months in rain fed areas in an year, (iii) the workers are reluctant to go for their medical checkup because of their negligence and fear, and (iv) they are economically not sound enough to go for their medical treatments.

On the basis of the design and development of various rollers with subsequent performance evaluation studies, chrome-free RCF roller has been demonstrated with reference to techno-commercial and eco-friendliness in ginning industries. The newly developed RCF rollers are successful and effective in functioning and in ginning out the seed-cotton. Cost economics study reveals that eco-friendly RCF roller ginnery sounds better in all aspects with reference to environmental, cotton technological and techno-commercial aspects. This improved technology is amenable for commercialization to the industries.

Though the initial cost of the RCF roller is 11 times more than the life of CCLC roller, the high price is compensated, as it is durable up to an estimated life of seven years than more of a few months of CCLC rollers. Besides, it ensures the following advantages.

(1) There is negligible wear and tear and also zero maintenance.
(2) High ginning efficiency and output of about 1.25 times more than the CCLC rollers because the developed roller made up of rubberized cotton fabrics has a surface finish conducive to high ginning efficiency.
(3) 50 per cent reduction in the weight of the rollers consume 25 per cent less in energy consumption that is power saving of three times less compared to CCLC roller ginneries.

31

SAFETY IN STEEL PLANT MAINTENANCE

Shashi Kant*

ABSTRACT

Steel industry poses one of the most difficult challenges in the area of safety. And this paper attempts to highlight these challenges and suggest to mitigate them. Taking stock of hazards in steel industry, it emphasizes on proactive safety management and better maintenance procedures for rail and road movement, heavy earth moving equipments, crane movement and convèyor operation.

Steel Industry no doubt, poses one of the most difficult challenges in the area of safety, health and environment when compared to many other industries. This challenge is gigantic task for SAIL in the wake of liberalisation, globalisation and disappearance of geographical boundaries. Steel market is getting more and more competitive within the country and outside today. Cost of safety may be high. But cost of not ensuring safety is still higher. It has been proved statistically that an industry, which has a good record of safety also, has a good record of production.

Preamble

Steel industry is generally based on the cascading effect of various unit operations. In order to maintain these operations proactive maintenance practices should be adopted. Planned Maintenance and Autonomous maintenance should be adopted and integrated with safety aspects. The overall objective of the maintenance process is to reduce downtime, increase efficiency, prevent deterioration of equipment and ensure that there are no accidents/incident in the workplace due to mechanical failures. The process industry has witnessed some of the major disasters in the world. The prominent among them were Bhopal (India, 1984), Flixoborough (England, 1974) and Seveso (Italy, 1976). The major reasons for the above disasters were due to negligence, not working in accordance with engineering practices and

*Executive Director (Safety), Steel Authority of India Limited, Doranda, Ranchi

inadequate maintenance plan. Bhopal wouldn't have happened if the protective equipment were in place. Also the Refrigeration system was shut down without any plan and the flare system which could have arrested the vapors by burning them was out of order. All these events led to the domino effect which caused the release of MIC (Methyl Isocyanide) thereby creating havoc.

The above incident confirm that the accidents happened due to mechanical failures coupled with operational errors so there is a need for proactive approach on maintenance of unit operations.

Introduction

Steel making was largely a development of the 19th century, with the invention of melting processes; the Bessemer (1855), the open hearth, usually fired by producer gas (1864); and the electric furnace (1900). Since the middle of the 20th century, oxygen conversion, pre-eminently the Linz-Donowitz (LD) process by oxygen lance, has made it possible to manufacture high quality steel with relatively low production costs.

Today, steel production is an index of national prosperity and the basis of mass production in many other industries such as ship-building, automobiles, construction, machinery, tools, and industrial and domestic equipments. The development of transport, in particular by sea, has made the international exchange of the raw materials required (iron ores, coal, fuel oil, scrap and additives) economically profitable.

Hazards in Steel Industry

In the iron and steel industry, large amounts of material are processed, transported and conveyed by massive equipment. Steel works typically have sophisticated safety and health programmes to address hazards in an environment that can be unforgiving. An integrated approach combining good engineering and maintenance practices, safe job procedures, worker training and use of personal protective equipment (PPE) is usually required to control hazards. Maintenance is particularly important for accident prevention. Its purpose is to ensure the efficiency of the equipment and maintain fully operative guards, because failure may cause accidents. Adhering to safe operating practices and safety rules is also very important because of the complexity, size and speed of process equipment and machinery.

Proactive Safety Management

The recent spurt in fatal accidents in steel plants and mines has forced us to put on our thinking caps and find out measures to arrest this trend. A fatal accident occurs only after repeated violations and following unsafe practices repeatedly over a period of time. The safety triangle indicates that for one fatal accident to occur a few thousand violations take place. Fatal accidents can only be controlled if we shrink the base of the triangle. One of the ways to trim the base of the triangle is through safety audits, training and mental attitude of the concerned employees and follow standard operating and maintenance practices. We have got to have faith that "All accidents are caused and can be prevented" and prevention of accidents is managerial responsibility. If we go by the above principle through lot of managerial actions, accidents will be prevented and precious lives saved and injuries reduced considerably. Being very religious we sometimes have so much faith in 'God' that we tend to avoid our responsibilities and term accidents as 'God Sent' or an act of God. This brings us a sense of complacence and no worthwhile improvement takes place once we get complacent. A large number of near miss cases do occur and there is ideally no difference between an accident and a near miss case. These near miss cases if investigated and remedial action taken, lot of accidents can be averted.

Many experts (like Henrich/Bird) has opined that there is a relationship between the number of minor incidences and the fatality which occurs in an industrial environment. This has been observed in our plant working also. If we are able to prevent small incidences certainly the serious accidents and fatality will come down. This is an old saying that "prevention is better than cure", but in an industrial environment this is more relevant and its effect is clearly visible. The wide variation between the accident and near miss case is more appreciable. Our main approach is to bring more awareness to the employees to understand this dictum and take all preventive actions rather than face accident and suffering in different forms. Many times our employees have shown their belief in this theory and got benefited. We are going to describe some of the incidences where the employees' preparedness could save accident and bring benefit to company and employees.

Analysis of Accidents

A huge quantity of raw materials, semis and finished products are being transported between departments, from mines to plants and to stockyards from plants by means of heavy equipments like rail, dumpers, trucks, etc. the large workforce, regular as well contractors employees uses powered two wheelers, cars, jeeps, etc. and bi-cycles. The movements of materials and men keep the workplace busy. These movement calls for monitoring. A quite number of accidents had taken place and some of them are narrated here with the precautions taken to avoid such incidents in future. Thrust has been laid in the following areas to avoid accidents.

(a) Rail and Road movement including heavy earth moving equipments
(b) Crane movement
(c) Conveyor operation

Rail and Road Movement

The materials are being moved from one place to another through Rail network, which is around 60-70 per cent of the total movement. The vast network of rail are being used to transport raw materials like iron ore, coal, etc., liquid hot metal, semi and finished goods.

— One sticker BOF slag ladle kept under water-cooling for about 15 hrs., was being taken backward for shunting to the parallel line for placing inside the BOF shed. Before entering the shed, the slag in the ladle exploded during movement. The victim, who was standing outside the diesel loco and engaged for withdrawing the ladles, received severe burn injury on his body and hand.

- While the victim was crossing the rail track near No.1 S.W. Tower of No.3 Blast Furnace, he was run over by Light Loco, which was coming after placing slag ladles at No.1 BF, leading to his death on the spot.
- While trying to bring back to position/straighten a half tilted slag ladle containing liquid slag which had been sent back from slag yard with tilting problem in half tilted condition. During the process of straightening when the disengagement of the drive mechanism from the ladle was being carried out (disengagement of worm wheel the driving worm shaft). The ladle jerked and tilted further rather than straightening. In this process the remaining hot slag spilled on to the ground. The resultant flames and liquid slag caused severe burn injuries who subsequently succumbed to his injuries. The jamming of the rack against accidental movement of the slag cup was inadequate and improper with a slag boulder, which crushed during the process of ladle rolling and caused this serious accident.

Recommendations emerged after enquiries are being implemented and monitored continuously to prevent such occurrences. Some of them are:

1. Joint inspection of locos, roads and rail crossings.
2. Inspection of trucks and trailers and illumination level to be done.
3. SOPs to be prepared by the concerned department for respective workmen and the safety department to ensure the implementation through extensive awareness programme.
4. In no case loco should be driven by only one person. Another person must be on loco to see the opposite side of driver's side and to direct the loco driver.
5. Surprise check to be carried out by SED to ensure proper running and manning of loco.
6. Speed of loco should be maintained at specified speed.
7. Red blinker light to be provided on loco for use on the crossings.
8. Insertion of Buffer car in between Loco and slag pot carriage.
9. Clearance for movement of water-cooled slag pot only after complete stoppage of steaming and physical checking of coldness.
10. Crews to be inside cabin while on movement with water-cooled slag pot.
11. Heavy vehicle movement to be restricted during shift changing hours.
12. Medical examination of drivers to be done regularly.

For Decoupling Operation

1. Two porters must be present at the time of decoupling and coupling—one for signalling and one for operating the coupling if required.
2. Loco Operator must be made responsible to ensure the presence of both porters at the time of coupling/decoupling jobs.
3. Rail traffic should examine the health of automatic couplings where it should work automatically.
4. If possible decoupling/coupling should be avoided at curve area of the rail track.
5. Proper type/size stopper to be provided with the wagons need to be shunted.

6. Stopper must be put on both sides of the decoupled wagons.
7. Persons handling such type of jobs should be trained and made alert.

Another concern area is accidents in roads. In steel plants, large workforce work round the clock and a huge quantity of materials is transported by trucks, dumpers and other heavy vehicles from one place to another throughout the day. These movements of men and material keep the roads inside the plant busy. Narrated here are some of the accidents happened inside the plant premises.

— After getting down from duty bus, the victim was going to his workplace. The bus while reversing hit and run over him.
— While coming to attend "A" shift duty by Moped, the victim collided with a cyclist and fell down on the road and succumbed to his injuries.
— While the victim, coming from SMS-2 by motor cycle on road at the down gradient chowk of modernization over bridge was hit by a dumper coming in same direction causing severe head injury.
— While travelling on truck with materials at SMS stage, the victim fell down from the moving truck due to heavy jerk and was run over by the rear wheel.

The analysis shows that during last 12 year, 58 fatalities happened due to road movements. The causing agency-wise analysis shows that majority of accidents is due to dumper movement and truck movement.

- Dumper : 27%
- Tipper : 5%
- Tyre mounted crane : 9%
- 2 wheelers : 9%
- Bus : 7%
- Dozer : 4%
- Truck : 25%
- Trailer : 7%

A large step has been taken by banning of entry into the plant with out the use of crash helmets but movement within the plant needs to be improved much further as regards driving speed and use of crash helmets within the plant. Many times the user takes off the helmet after leaving the plant. This calls for attitudinal change and continuing our efforts endlessly and repeatedly. And the other measures taken are

- Relevant Sign Boards, Caution Boards, Red flags and Slogans to be prominently displayed to caution the persons concerned.
- A special drive to be taken to educate the people of different agencies deployed at the construction site for follow of safety precautions.
- Periodic drive to be undertaken to ensure 100 per cent usage of Crash helmet during movement in work areas.

Heavy Earth Moving Equipments

In surface mining large capacity Draglines, Shovels, Drills and Dumpers are being increasingly used in coal mines. In the Metalliferrous sector also intensive mechanization have taken place in open cast mines with thrust on higher capacity mines. Out of these machines, Wheeled Trackless Equipment like dumpers, tippers/trucks are primary means of transport in mechanized open cast mines. Although these machines are costly and robustly built and safe with many state-of-the-art safety devices, it remains chief cause of concern with a contribution of over 50 per cent of all the fatal accidents in open cast coal mines, 26.4 per cent in non-coal mines and 34.6 per cent of all fatal accidents in overall open cast mining operations. Narrated here are the common types of accidents and caused sometimes fatalities.

— While negotiating the gradient the victim would have lost control over the Dumper and it toppled down from Bench Nos. 6 to 7 from a height of about 12 mtrs.
— One Dumper Sl. No. 14 fell from dumping area at HK quarry from a height of 45 m (approx.) on Coal Bed. The body of the Operator of the vehicle Sri Jawahar Prasad was found lying outside the cabin of the Dumper.

In one of the road accidents which is of recent origin a scooterist has been run over by a dumper while trying to overtake. We must remember that Dumpers are off highway equipment and cannot be treated at par with other road vehicles like trucks or buses. The driving cabinet is on the left hand side whereas the visibility of the driver on the right hand side, i.e. the side people are trying to overtake is not there at all. If it is so, it is from the rear view mirror (located on the right side) which at times may be maintained or not or may be dirty. Have we a practice of ensuring the reliability or workability of the rear-viewing mirror on the right side because this mirror can save a life it need to be treated with the same significance as headlight or brake.

Most of the accidents during transport by trucks and dumpers have resulted either due to carelessness or indiscipline. Some of the major causes are:

(i) Inadequately maintained haul roads like absence of provision of proper haul road width, bundh/parapet wall, escape lanes, road signs along curves and crossing and other statutory provisions.
(ii) Poor maintenance of vehicles and their mechanical failure, for example, Non-functioning of safety warning devices like light, horn and other gauges in the opposit Cabin, ineffective brake, emergency steering, AVA and missing or broken or dirty wind screen.
(iii) Reversal of vehicles mainly due to absence of spotters and stop blocks while tipping at stockpiles, waste dump and inadequate supervision.
(iv) Unauthorized riding of vehicles particularly in hilly area where workers may try to take a lift for their personal work avoiding a long walk.
(v) Loss of control of vehicle in hilly terrain due to carelessness and driving at excessive speed.
(vi) Improper artificial lighting at dark hours and during thick fog.

Other most important accident-prone area is dumper/trucks maintenance/repair shop. The main contributing causes are:

(i) Non-adherence to SMP's in the work areas.
(ii) Lack of safety awareness and inadequate use of PPE`s.
(iii) Poor access to working spaces which forces the employee to take unsafe positions.
(iv) Unauthorized driving of dumpers by inexperienced khalasis/mechanics.
(v) Choice of wrong tools/ancillary equipment for the job.
(vi) Inadequate method of handling heavy parts.
(vii) Poor housekeeping of repair shed/workshop.

To overcome accidents to HEM vehicles, the following aspects are being taken up with the concerned for avoiding such recurrences.

1. Dumper Operators to be imparted Structured Training in accordance with the revised training module of DGMS. Specialized Training of Dumper Operators to be arranged in consultation with the original equipment manufacturers.
2. Heaps of loose dumped materials along the edge of the haul roads for serving the purpose of a temporary barrier.
3. Suitable berm of either height not less than 1m or as high as size of Dumper tyre to be ensured as per statute along the haul roads which are of permanent nature.
4. Provision of seat belts to be made in the Dumpers wherever not yet provided.
5. Dump yard should be properly marked to restrict the Dumpers reversing beyond the safe limits and also proper beam to be maintained.
6. Sufficient speed limit board to be placed on the haul roads to control the speed of the Dumpers by the operators.
7. The concerned operator should fill up the logbook of the machine. The defects indicated by the operator should be properly attended to keep the machine in healthy condition.

Crane Movement

E.O.T crane is one of the main equipment in a steel plant and is one of the most important modes of material transport in a Steel Plant. It is capable of handling much more than any other mode of transport can handle. We can't think of smooth/uninterrupted production without proper reliability and availability of E.O.T. Crane. Due to its regular use sometimes it poses a safety hazards.

We lost few lives in crane operation due to limit switch failure, improper communication, pressed between crane and gantry, rope failure, etc. presented here are some of the accidents happened in crane operation and lost few lives and some near misses cases also.

— While changing main hoist rope of a hot strip mill crane, the free end of the rope fell down on the head of Mr. X who was working with rope changing group. The approximate weight of the rope was 15 kg and it fell from a height of about 20 metre. With normal direct impact it is sufficient to break the head of any human being. The concerned employee was wearing safety helmet and his head could be saved fully without any injury.

— There was metal spillage in ingot mould Foundry/Slag Pot area due to failure of crane hoist mechanism causing fatal injuries due to burn injuries. Out of two hoisting ropes on the SMS side one rope (which was hemp core) snapped and subsequently other rope also snapped not being able to take full load, thereby causing the traverse as well as ladle to tilt on the SMS side, which has resulted in the spillage of molten steel on the ground and causing the accident.

Suggestion for preventing recurrences

1. All cranes which are handling hot metal/liquid steel, must have a very rigid and explicit maintenance schedule and implementation of this maintenance schedule must be recorded in register/file.
2. The preventive maintenance schedule must include periodic checking of the diameter of rope, being used, at number of places and at various cross-sections in addition to examining the overall condition of the rope and rope strands as per relevant standards. This should be recorded and monitored regularly.
3. Whenever any major repair/replacement/breakdown rectification is done on such important cranes, the same must be recorded in the history register with the signature of the person getting that job executed.
4. Before changing ropes, it must be ensured that the rope being used is of proper specification.
5. Provision for proper lubrication of the wire ropes and periodic application of cadmium compound, as part of maintenance must be ensured.
6. Load testing of all hot metal cranes must be done as per factory act/rules.
7. All employees working in hot metal handling area must wear required PPEs including fire retardant dresses.
8. No office/sitting accommodation should be allowed to be built or used in the near vicinity of such places, where hot metal is being handled/poured. These must be located at a safe distance as per site condition and taking in to consideration all eventualities with proper risk analysis.
9. Suitable escape routes (minimum two numbers) on opposite side, where hot metal is being handled, must be provided. This should be valid for all platforms and working area at which employees are supposed to stand during operation.
10. In such areas where hot metal is being handled and likely to spill over by any chance, no inflammable material should be allowed to be stored. Parking of vehicles should be away from work places.

Conveyor Safety

In industry, conveyor safety is generally considered as a part of general safety. This practice may not give sufficient attention to the nuances required to provide a safe conveyor system. The responsibility for the planning, design and implementation of a conveyor system is often fragmented, lacking central responsibility. The majority of conveyor accidents are a direct result of an original design or implementation error. Conveyor safety is a critical factor in any material handling operat[illegible] Typical conveyor hazards occur at:

- Power transmissions
- Shear points
- Spill points
- Transfer counterweights
- Passage areas under conveyor
- Nip points
- Pinch points
- Areas under counterweights
- Transfer mechanisms
- Under special circumstances

Number of precautions had been taken in steel plants and mines with respect to conveyor operations, still accidents occur due to negligence and over confidence. Mentioned below are some of the cases happened.

— Two DPR Attendant were deputed for operation and cleaning job at Conveyor Nos.1 and 1A and at about 1.45 PM one of the DPR was lying under the tail end drum of Conveyor-1 and was lying below the conveyor in badly injured condition.

— Victim was deputed for operation of Shuttle Conveyor Belt # 20 at Coke Sorting Plant. It was reported that shuttle belt conveyor was stalled. Body of the victim was found entangled with the conveyor belt/roller.

— The chain conveyor was in operation and the workman was by the side of conveyor head. All of a sudden the head of the conveyor tilted and fell along the side where victim was standing. He was pressed between the conveyor head and sand barricade.

The various committees constituted for enquiry recommended various recommendations and are being followed up for implementation.

1. No person should work or move along running conveyor wearing loose garments/ *gamachha*.
2. Spillage cleaning should be done preferably after taking shut down.
3. No person should work alone in an area where if any accident occurs.
4. Providing suitable (detachable) guards of adequate strength at three sides of the tail end and maintaining the same in position to prevent inadvertent contact with the moving parts of the conveyor.
5. Adequate means of communication by telephone or ringing bell may be provided near the face conveyor wherever interlocking is not possible.
6. Strict supervision by departmental employees should be ensured for the contractual jobs.
7. Prominent display of SOP, Safety instructions/Caution boards and Do's & Don'ts at vulnerable locations.
8. Undertaking special drives for ensuring functional operability of safety switch/Pull cord switches.

Now-a-days, stress has been laid to maintain the following devices with respect to conveyor safety.

1. Emergency stops/Pull chords
2. Clearance between equipments and Crossover steps
3. Illumination and Earthing of motors
4. Start/Stop buttons in loading/unloading areas and knowledge about the same to operator
5. Conveyor over run arrest arrangements and overloading protection device
6. Fall of material protection from chutes/hopper
7. Cleaning system of belt (scrappers/rotating brushes) and walkways/dashboards

How Maintainers Can Work More Safely

Maintenance departments are often recognized as caretakers of company safety initiatives, they are inundated with all things safety: messages, procedures, meetings, checks, equipment, training and permits, etc. Despite their good intentions, however, some maintenance professionals still get hurt on the job and mentioned below are some of them.

— An explosion took place in the Basic Oxygen Furnace in Steel Melting Shop due to water accumulation where seven persons were injured during prolonged maintenance of tilting mechanism of the furnace.

Following recommendations emerged after investigation has been put into action and they are being monitored on daily basis.

1. Incorporation of all safety procedure in SOP's and work instruction like while tilting the converter. Supervisor should ask the employees to move to safer place and held up converter with liquid steel should not be rocked beyond 45 degree. On either side whenever required to ignite the bath.
2. Prepare DO's and DON' Ts and display at the hazardous places.
3. Make focused safety awareness Training manual for each operation and impart training in shifts.
4. Safety talk should be given to concerned employees by shift in charges at the beginning of the shift.
5. Safety personnel to be associated with during prolonged breakdown, maintenance critical equipment.

In another instance, during partial relining of Blast Furnace (BF-3), cleaning of stoves were taken up. A Brick Layer Supervisor while working inside the Stove was affected by BF Gas. The recommendations emerged are as follows:

1. A main cooler fan of high capacity to be placed on the stove near the dome to prevent any gas entering into the stove from outside. Additional onc, i.e. in total two numbers of c[illegible]ed air line to be introduced through dome opening in order to enrich ai[illegible]he stove.

2. To make arrangement of breathing apparatus of different type (smaller size) so that the person carrying the instrument should not find it difficult to enter in a confined place or elsewhere.
3. To ensure rotation of workers, supervisors, every half an hour, while working in a confined place, no one should work more than half an hour at a stretch.

In another shocking incident, one Sr. Technician was coming out of the Mill to return back to his home after performing his Night Shift. The Crop-Shear pit was full of hot boiling water having a temperature in the range of 95° - 100° C. Lots of vapour were coming out of the pit making the visibility poor in the area. The victim, while moving towards the Crossover on the pit to reach the other side, misjudged the crossover due to poor visibility and fell into the Crop-Shear pit and succumbed soon after he fell into the pit.

The investigation committee had recommended the following aspects to be looked into while maintenance jobs are being planned in mills area.

1. Mill should not be started without ascertaining and obtaining the written clearance that the pit covers and railings have been restored back.
2. After the cleaning is over the covers and the railings of the pit should be restored immediately.
3. When the covers are removed for cleaning the pit, the area should be barricaded with red flags, temporary railings, florescent red ribbon etc. with caution board.

Considering above stated incidents, the following aspects should be considered with respect to Maintenance Procedures:

- Human factors;
- Poorly skilled workforce;
- Unconscious and conscious incompetence;
- Good maintainability principles;
- Knowledge of failure rate and maintainability; and
- Clear criteria for recognition of faults and marginal performance.

Conclusion

Many recommendations would come after the accident but how are we going to follow the same is the big question. Generally for a few days they would be adhered to but after some time people would come to their old behaviour and all recommendations vanish away. Safety is an emotive word having different connotations for different people in different circumstances, at different times and at different places. In common parlance, safety means "Being safe, freedom from risks and dangers", when put in industrial practice, it means removal/ reduction of different types of accidents. To a conceptualist, it means loss prevention in totality.

32

BEHAVIOUR BASED SAFETY
Save Lives by Involving Employees

Rajesh Goel*

ABSTRACT

B.F. Skinner says that their behaviour could be modified by giving them certain stimuli and either rewarding them or punishing them. Behaviour based safety programmes seek to do the same thing. Identify unsafe behaviour and punish those who practice it and identify safe behaviour and reward those who practice it. We know people are not rats so how does human society differ and make this idea of behaviour based safety programmes more complex than it may seem at first glance?

An interesting variation of this approach does not require the observer to give one-to-one feedback: a single observer counts the number of instances of behaviour like not wearing goggles, dust mask, hearing protection etc. If observations are done at regular intervals trend data can be prepared and mere fact observing and reporting on the behaviour is often enough to generate improvement.[1]

More recently, the emphasis in safety has shifted towards the behaviours and attitudes of employees at all levels of the organization—so-called *Behaviour-Based Safety*. It works on the premise that most accidents are a result of the risk-taking behaviours of employees. This approach wants to minimize the so-called at-risk behaviours and increase the safe behaviours of workers.

Safe behaviour strategies are aimed at culture change. Behaviour safety is the application of behavioural psychology to promote safe behaviour in the work place using involvement of employees. Encouraging desirable behaviour, however, by positively acknowledging safe behaviour is more likely to be successful, as it does not have to be given immediately and every time. Evidence indicates that one of the most powerful methods of encouraging desirable behaviour is to provide social rewards in the form of praise or recognition.

We can conclude that the unsafe behaviour can be predicted and analyze and can be improved also by culture change. Behaviour safety creates employees ownership by working with a team of employees to plan and implement the process.

*Assistant Manager, OSRAM India Pvt. Ltd., Delhi Road, Sonepat-131 001, (Haryana).

Introduction

The first way in which we differ from the assumptions made by Skinner is, of course, in the way society is organized. It is, in our social and economic system, hierarchical. Within the workplace this is translated into those who own and control the workplace and those who work there. And it is this concept of control, which we must understand if we are going to make safer workplaces. For those who control the workplace must, by definition, have more influence in whether it is safe or unsafe. And those people who control the workplace are of course employers. Those who work there are workers and they, in our social and economic system, have extremely little control over the workplace. So the question really is, whose behaviour needs to be modified? How do we get the biggest "bang for our buck"?

Is the idea of behaviour-based safety not really just a less fatalistic version of "accident proneness" that wonderful theory that says that some people just can't stop hurting themselves? By this logic, of course, we see who the most accident-prone people are in our society. They are underground miners, construction workers, and loggers. Office workers must, of course, be extremely careful and safe people because they get hurt far less often.

And it is in these rather extreme examples that we see the fundamental flaw in the logic of the concept of behaviour-based safety. It assumes that the nature of the workplace and the work itself is less important than people's behaviour. I say that is nonsense. Some workplaces are extremely hazardous and those are (usually) where people die at work. Others are less so and those are where people tend to avoid death and serious injury. It isn't their behaviour that determined their safety; it's the nature of the work.

Employers and Workers

But that as well of course is also too simplistic. Because someone controls the workplace. Somebody (or a group of people) has made decisions about how work is organized, regardless of the industry. Somebody has decided whether the coal mine will be open pit or underground. Someone has decided whether there will be working methane monitors, sufficient ventilation, sources of ignition allowed underground, proper supports for the tunnels, mine rescue stations.

What about Clifford Frame? He owned the mine. He made the decision to mine the Frood seam, the notoriously gassy coal seam that had been known to be extremely dangerous for decades. He made the decision to hire management with a clear preference for production over safety. And the mine manager and the rest of the management team carried out that preference to the hilt. They made the decisions not to worry about whether the methane monitors worked or not. They made the decisions about whether to use rock dust or not to quell the coal dust. They made the decisions about allowing sources of ignition underground. Whose behaviour needed to be modified? It was Clifford Frame's and other members of management who needed their behaviour modified.

Know About Unsafe Behaviour

Major accidents can frequently be traced to failures in safety management systems. Even when enormous effort has gone into perfecting these systems, it seems they remain fallible. It is largely for this reason that the concept of safety culture is now receiving widespread attention. This is not to say that systems are irrelevant, but rather that they will function better in organisations which have developed a culture of safety.

The most common "programmes require front line staff to carry out behavioural safety observations on their colleagues" and feed the results back on a one-to-one basis. The feedback process requires sensitivity and observers need to be trained to do this effectively. An interesting variation of this approach does not require the observer to give one-to-one feedback: a single observer counts the number of instances of the behaviour, for example, cases of workers not wearing hearing protection, and reports the statistics. If observations are done at regular intervals trend data can be prepared and behaviour can be predicted to generate improvement.

Prediction of Unsafe Behaviour

Traditionally, occupational health and safety (OH&S) interventions have centred on controlling the physical work environment and work procedures of employees in an effort to prevent errors and accidents. Examples include the documentation of detailed procedures designed to provide the safest way of completing tasks, procedures for handing over uncompleted tasks to colleagues, strict safety guidelines for the operation of machinery, and the wearing of personal protective equipment. A complementary approach to human error focuses on the human factors in work accidents. This approach takes into account the inevitability of human error and seeks to contextual behaviour so that a greater understanding can be realised. Where strict procedural guidelines attempt to mechanise and standardize behaviour, a human factors perspective acknowledges individual differences and focuses on psychological pressures and factors that influence behaviour. The present study is concerned with the human factors that contribute to violations.

Models of Unsafe Behaviour

A number of researchers have attempted to categorise safety climate variables with the aim of constructing models to explain the interactions among the variables. Variables are classified at either the organisational, group, or individual level. The study investigated the influence of organisational, job, and individual variables on errors using Structural Equation Modelling (SEM). The results suggested that the effect of organisational level factors on errors is mediated by individual level factors, such as health and stress those leads to unsafe behaviour. Fogarty and Neal (in 2002) combined these two variables in their work on the causes of both violations and errors in the construction industry. The authors hypothesized that safety climate variables would predict violations, whereas individual level variables would predict errors. The relations among the variables used by Fogarty and Neal are modelled in Fig. 32.1.

As illustrated by the model, safety climate variables are seen as impacting directly on violations and the personal resources of employees. The employees' resources influence the psychological strain they feel, which is directly responsible for following procedures. Perceived behavioural control suggests there are times where, despite best intentions to act in a certain manner, individuals feel incapable of fulfilling a planned activity. In the same way, employees may feel that they are not able to complete work tasks according to procedures

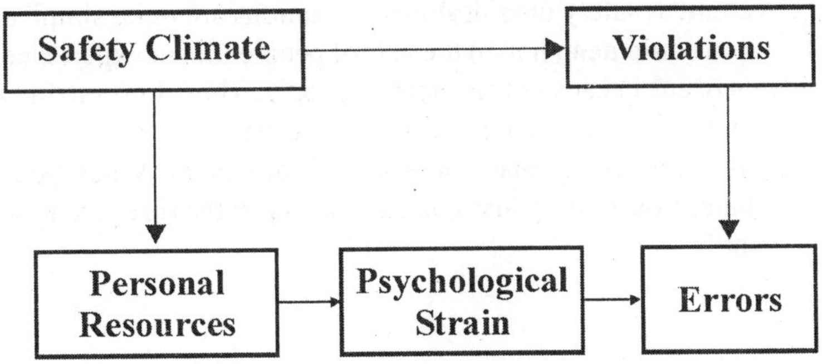

Fig. 32.1: Fogarty and Neal's (2002) model of the organisational and individual level variables that act as predictors of violations and error in the construction industry

and rules because of external factors that are beyond their direct control. Examples of these external influences include lack of equipment, lack of personnel, lack of time, and production pressures.

In safety literature these factors are often combined under the construct of workplace pressures, elements of work that are beyond the control of individual workers. Consequently, it is suggested that workplace pressures will be associated with employee intentions to violate and actual violations of procedures. Thus, the TPB maps quite nicely onto models generated by some studies of safety climate (Fogarty & Neal, 2002). This study will extend the TPB by suggesting management attitude to safety is responsible for the spurious link between these unanalyzed variables

The importance of management attitudes to safety is well documented, indeed, it extends back to Zohar's (1980) initial study of safety climate. Zohar found that an employee's perception of his or her manager's attitudes toward safety was the most important predictor of safety climate. Since then, studies applying safety climate and culture to mining accidents, the aviation industry, and construction workers have all highlighted the important role played by management in ensuring the safety of organisations. We suggest here that management attitudes will exert an indirect influence on violation behaviour via own attitudes, subjective norms, and perceived control. We also suggest that subjective norms will have a direct effect on own attitudes. The resulting model is shown in Fig. 32.2.

Modification Strategy of Unsafe Behaviour

After prediction of unsafe behaviour in organization the approach to be discussed here is the use of behaviour modification strategies to promote safe behaviour. The popularity of the behaviour modification approach stems from the widely held view that "human factors" contribute to the great majority of accidents. A conclusion which is frequently drawn from this observation is that the focus of accident prevention efforts needs to be shifted from engineering solutions to ensuring compliance with safe work practices. As the General Manager of DuPont, Australia once said,

"Both government safety organisations and unions are quite simplistic on safety. They focus on equipment, not on the acts of people. In our experience, 95 per cent of accidents occur because of the acts of people. They do something they're not supposed to do and are trained not to do, but they do it anyway. Changing this behaviour is much harder than focusing on equipment. When you've done the technical things you've only just started. That's just the tip of the iceberg of safety management."

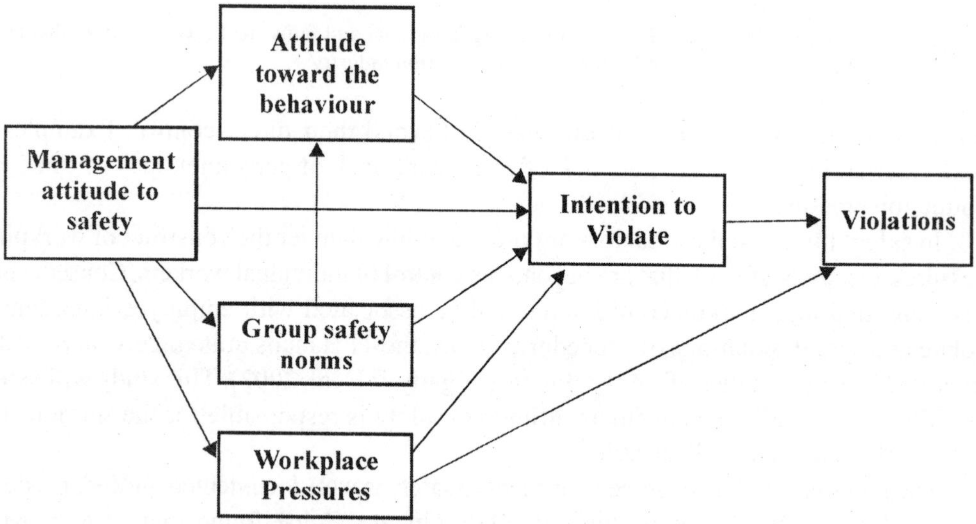

Fig. 32.2: A model of safety behaviour based on the TPB.

More recently, the emphasis in safety has shifted towards the behaviours and attitudes of employees at all levels of the organisation—so-called *Behaviour-based safety.* Behaviour-based safety refers to an approach to safety management that focuses on the "people factors" in safety. It works on the premise that most accidents are a result of the risk-taking behaviours of employees.

Risk-based safety refers to the abilities and capabilities of the human being to identify, assess and understand risk in their work environments. In any given work situation, there will always be an element of risk that has to be taken. How well these are taken is what is defined as risk competency.

Many companies are seeking to inculcate risk-awareness into their employees. Esso, for example, has a "step back five by five" program. The idea is that before starting a new job, the employee should take five steps back, metaphorically, and take five minutes to think about what might go wrong and how this might be avoided.

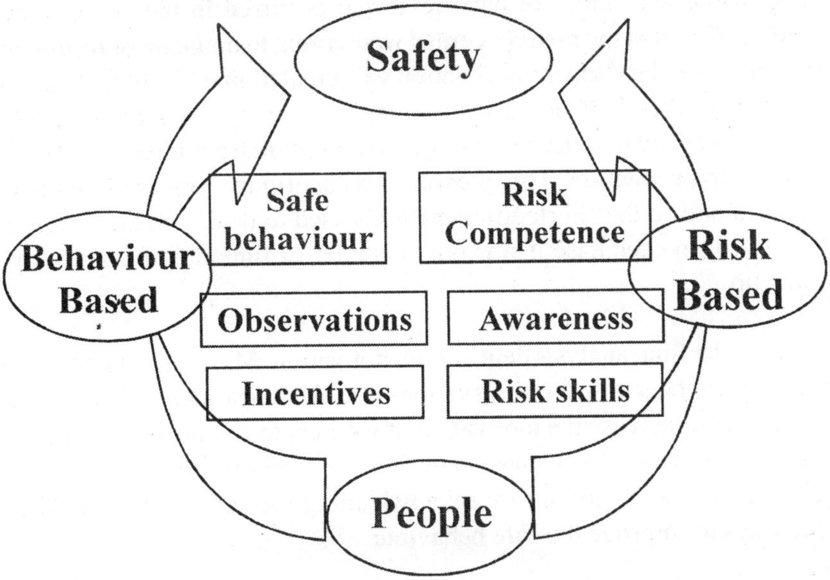

To achieve risk competency human error as shown in the model below has to be overcome.

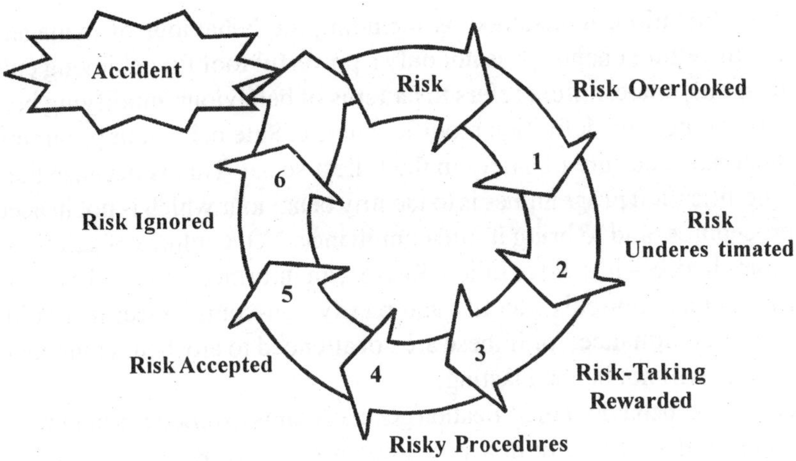

Modifying the Employee Behaviour by Involving Them

Safety today...is not as much an environment problem as it was. Our primary job is not to control "things". Safety today is a people problem. We are in a situation where we must learn to understand people and where we must learn to control people's behaviour...

When we talk about the management control of accidents, we really mean to talk about how management can control the behaviour of people. When we talk of how management can control the behaviour of people, we really mean, for all practical purposes, how management can get those people to want to operate in the manner management wants. In the study of accidents in British factories, the results of which were published under the title Safety or Profit, authors Theo Nichols and Pete Armstrong have this to say:

> Each of the accidents we have reviewed occurred in the context of a process failure and whilst the men concerned were trying to maintain or restore production. In every case the dangerous situation was created in order to make it quicker and easier to do this. In every case the company's safety rules were broken. The men acted as they did in order to cope with the pressure from foremen and management to keep up production. This pressure was continual, process failures were fairly frequent and so the shortcutting methods used to deal with them were repeatedly employed. In each case it was only a matter of time before somebody's number came up.

All safety in the final analysis depends on motivation. Most important tools being used in this approach are peer-to-peer observations and feedback and incentive systems. The focus with these is to increase the motivation of workers to engage in safe work. To engage the employee in safe work the management should control the unsafe work practices and encourage the desirable behaviour by acknowledging the safe behaviour. The following could be the ways to improve the safe behaviour.

Organisation Culture Change Through Participation

Behaviour modification, understood as including the behaviour of managers, is closely connected with culture change. It is not only a powerful tool for achieving culture change, but more than this, since culture refers to patterns of behaviour, modifying behaviour in an organisational context by definition modifies culture. Safe behaviour programmes will not, however, generate a culture of safety in the fullest sense. This is because the aim of most behaviour modification programmes is to identify behaviour which is not in accord with safe working procedures, and to bring it into compliance. Yet a culture of safety is more than a culture of compliance—it is, if we follow Reason, an informed culture. To be fully informed, an organisation must not only identify and rectify non-compliance; it should identify the reasons for non-compliance, for if these are not attended to any behaviour change achieved in the short term will not be long lasting.

To summarise, behaviour modification generally aims to produce a culture of compliance. This is a culture of safety only in a very limited sense. In Reason's work, the culture of safety is a far richer concept, lying beyond the reach of conventional behaviour modification techniques.

Safety Awards

Safety awards in some form or another are almost always part of the behaviour based safety systems. They are the "carrot" to reward good behaviour. But that behaviour is almost always based exclusively on not reporting injuries, not going off on lost time. The BC Board recognized this deficiency in 1974 when it cancelled its safety award system based on reported injuries and accepted claims rates and replaced it with a useful safety awards programme based on recognition for innovation, for coming up with a new idea to make workplaces safer such inventing a movable guard on a pit in a service garage.

What can be Done to Control BBS?

Behaviour-based safety programmes weaken hard-won protections and discourage workers from taking a more active role in the union. A number of unions in Canada and the United States have issued policy positions opposing "blame the worker" approaches to health and safety. In order to combat BBS programmes unions are advising their members to do the following:

- Use their health and safety bargaining rights to negotiate against use of incentive programmes;
- Draft policies and position papers against BBS programmes;
- Communicate to their members (workshops, leaflets, brochures, buttons etc.) the hazards of BBS and the real sources of injury and illness thus helping to dispel myth of "careless worker"; and
- Press government for improved health and safety laws and enforcement of existing legislation.

Joint Health and Safety Committees are Being Encouraged to:

- Exercise their right to regularly inspect the workplace;
- Recommend establishment of a health and safety program;
- Press for hazard awareness training for all workers; and
- Press for certification training for all committee members.

Workers and their Representatives are also advised to:

- Report all workplace hazards;
- Report injuries and illnesses;
- Refuse unsafe or unhealthy work.

Conclusion

Safety culture is an attractive idea because if promises a way to overcome the limitations of safety systems by involving the employees. Many of those who refer to safety culture have in mind an organisation whose members are all individually safety conscious. Reason, on the other hand, while acknowledging that safety culture has implications for the behaviour of individuals, insists that the concept be used to describe truly organisational phenomena and not simply the aggregated behaviour of individuals. The TPB draws our attention to the role of subjective norms and mediating construct of intentions, both of which are useful additions to the network of variables that interact to influence safety behaviours. The theory of planned behaviour could be extended to include the management attitudes.

Safe behaviour strategies are aimed at culture change by involving all level of employees. Encouraging desirable behaviour, however, by positively acknowledging safe behaviour is more likely to be successful. Another hallmark of most behaviour based safety programmes is safety incentives or safety awards programmes. Safety incentive programmes offer "prizes" or "rewards" to workers or groups of workers to encourage them to work safely.

NOTE

1. Whiting, J.(1993), "On Safe Behaviour", 64 *Australian Safety News,* 7, pp. 43-44.

REFERENCES

Dan Petersen, *Safety Management: A Human Approach,* (Englewood Cliffs, N.J.: Aloray, 1975), pp. 24-5.

Fogarty, G., & Neal, T. (2002). *Explaining safety violations and errors in the construction industry.* XXV International Congress of Applied Psychology, Singapore, July 7-12.

Hofmann, D.A., & Stetzer, A. (1996). A cross-level investigation of factors influencing unsafe behaviors and accidents. *Personnel psychology,* 49, 307-339.

McSween, T.E., *Values-Based Safety®: Improving Your Safety Culture with a Behavioral Approach,* 2003, John Wiley and Sons, NY.

Theo Nichols and Pete Armstrong, *Safety or Profit: Industrial Accidents and the Conventional Wisdom,* (England: Falling Wall Press, 1973), p. 20.

Zohar, D. (1980). Safety Climate in Industrial Organizations: Theoretical and Applied Implications. *Journal of Applied Psychology,* 65(1), 96-102.

33

INTEGRATION OF OH&S WITH OTHER MANAGEMENT SYSTEMS IN CORPORATE SECTOR

S.K. Sharma*

ABSTRACT

Implementation of Integrated Management System (IMS) by organisations is at a nascent stage in India, and only a few units in the organised sector have adopted the system. It is now evident that IMS is a time saving device and enhances the effectiveness and efficiency of the organisation. But there are certain constraints in the development and implementation of this system, and unless the certification bodies follow common guidelines for examination of documents such as ISO-Guide 72, and the industry consider the quality, environment aspects and impacts, risk associated with the operational activity as its integral part, it would be difficult to implement the system in its true perspective.

"The first duty of business is to survive with good health, and the guiding principle is not maximization of profits but to avoid the loss," said Peter Drucker.

Health and Safety are one of the most important aspects of a successful organisation's effective functioning which ensures an accident free and sustainable environment.

It is now accepted that legislation alone cannot guarantee an accident free environment. A number of legislations have been prescribed on OHS regarding health and safety of the work force, but it is an accepted fact that accidents have occurred in the past, occur presently and will occur in the future also. Accidents are universal but can be reduced. It is, therefore, required that we should make efforts and should have the will power to reduce the accidents and health problems caused by occupational hazards. The mission of OH&S is to protect the health and safety of persons at work place. Worldwide data about health and safety at work place are alarming as per ILO report. This demands adoption of a structural approach for identification of Hazards, their risk involved and control of risks, but it is necessary to align or integrate the OH&S Management system as well as the other systems into the overall business management process so as to avoid the conflicts between the management systems and other organisational priorities. All internal decisions should be taken keeping in view the requirements of quality, environment and occupational health and safety.

*Ex. Director (S.G), BIS, Presently, CEO - Quality & System Consultants, Faridabad

Requirement of Integrated Management System

IS 18001:2000 clause 4.1 describes a provision for development and integration of the requirements of the OH&S management system with overall business management process. It also aims at developing the capability to balance and resolve conflicts between OH&S and other organisational objectives.

The present scenario in the corporate sector organisations is that most of the organisations have obtained the certifications for ISO: 9001:2000, ISO: 14001:96, OHSAS: 18001:99 and even 5 S standard, but invariably it could be seen that the systems fail to merge with general management practices. Reason being, that our priorities are different for different subjects. Our most priority subjects are production, finance and profits, other management systems are considered either tools of marketing or a status symbol, or have been imposed upon by our customers and legislative requirements. In fact ISO: 9000 and other management systems have become rituals and prestige symbols for the industry and service providers, who hardly take the advantage of continual improvement through these systems, or try to achieve the excellence model. Each management system and the practices followed by the organisations have their own priorities. In one of the largest power plant in India, while conducting an internal audit training programme, I asked one of the senior officers of the plant to narrate the OH&S Policy of the organisation, the answer was that, "there are so many systems in our plant such as ISO: 9001, ISO: 14001, OHSAS: 18001, 5 S, SA-8000 and lastly TQM, and therefore, how many policies one could remember?" He further added, that "in order to implement the systems we are required to carry out management reviews for each system periodically, and have to face the auditors for one system or the other at frequent intervals, in such a situation should we devote our time in our own productive activities or in learning all these systems at this age?" I was stunned to hear this answer from a responsible officer, but it is a reality in most of the organisations.

Similarly, during an audit at one of the Banks, I found that records maintained by the Bank were missing from the list of records to be maintained as per ISO requirement. On asking as to why the information about the records maintained and their retention period are missing from the list of records, the answer was that "we are required to maintain the list of records required by ISO only." It clearly showed that the Bank's working didn't adopt Quality Management System and it was considered as a separate identity and not a part of working practices. Now who is to be blamed for such a situation, which is prevailing in most of the ISO certified organisations in our country, obviously, the certification bodies and their auditors, who do not verify the real implementation of the Quality Management System in actual functioning of the organisation.

Such a situation prevails in case of different management systems also. Each system maintains a barrier of separate policies, objectives, records and monitoring of the systems. It is necessary that a change in our thinking has to be brought into integrate all the systems with main functions of the organisation, in order to achieve required effectiveness and efficiency of the organisation, and therefore, an IMS (Integrated Management System) should be developed in a structured way focusing on quality performance, environment, health and safety, finance and HRD.

There should be one system in an organisation where activities are performed taking into consideration their overall performance, economy, quality, environmental, health and safety issues together. The compatibility of all the three standards serves an advantage to the organisation in this regard. Integrated system wherever adopted have given good returns in the form of reduced operational costs. Controlled management processes take less time. Main advantages of IMS include reduction of duplication of management review meetings, internal and external audits, mitigation of risks, and thus increased profitability. IMS, if implemented in its true perspective, also leads to reduction in conflict of priorities, elimination of responsibilities and authority conflicts, which creates a conducive work environment and ultimately leads to employee satisfaction.

How to Develop an Integrated Management System

At present there is no national or international standards on integration and consultants have developed their own methods depending upon the needs of the organisation by considering ISO: 9001:2000 or ISO: 14001:2004 as a base standard and alignment of the similar clauses of other standards. However, there are difficulties in this regard, that some of the EMS or OH&S requirements don't match with QMS requirements, such as study of aspect and impact in EMS and identification of hazards, risks analysis, legal requirements and disaster management plan etc. in EMS/OH&S system.

In order to assist the organisations to adopt uniform documentation for IMS, ISO have brought out a guide ISO-72 about integration of different systems, based on PDCA cycle. EMS & OHS activities such as aspect and impact analysis, identification of hazards, risk analysis, and resource management have been put in the in the planning clause, thereafter come the operational functions and implementation. Internal/external audits for monitoring and performance appraisal, corrective and preventive actions followed by management review for continual improvement. This is a guide and must be accepted by certification bodies for smooth transitions to IMS. The details of the guide in brief are given in the table below.

Integrated Management System Framework (ISO Guide 72)

Organisations wishing to implement a highly effective management system integrating ISO9001, ISO14001, and OHSAS18001 may consider the adoption of the system framework as presented in ISO Guide 72, Guidelines for the justification and development of management system standards.

Constraints for IMS

1. There is no balanced approach in attending the priorities of the jobs, which needs to be looked into.
2. Another difficulty in integration of different management systems lies in the fact that there is no provision for covering all the standards in one certificate. The reason being that there is no ISO standard for OH&S, and as long as the separate licences are to be obtained by the organisations, there cannot be full alignment of the management systems with the practices of organisations. However, if the documents are prepared as per ISO guide 72, there could be better degree of

The following is a tabulation of ISO Guide 72 management system common elements *versus* the requirements of ISO9001, ISO14001 and OHSAS18001

Common Elements	*ISO9001 (2000)*	*ISO14001 (2004)*	*OHSAS18001 (1999)*
B1 Policy			
B1 Policy and principles	4.1, 5.1, 5.3	4.1, 4.2	4.1, 4.2
B2 Planning			
B2.1 Identification of needs, requirements and analysis of critical issues.	5.2, 5.4.1, 5.4.2, 7.2.1, 7.2.2	4.3.1, 4.3.2	4.3.1, 4.3.2
B2.2 Selection of significant issues tobe addressed.	5.4.1	4.3.3	4.3.3
B2.3 Setting of objectives and targets	5.4.1	4.3.3	4.3.3
B2.4 Identification of resources	6.1	4.4.1	4.4.1
B2.5 Identification of organisational structure, roles, responsibilities and authorities	5.5.1, 5.5.2	4.4.1	4.4.1
B2.6 Planning of operational processes	7.1	4.3.3	4.3.4
B2.7 Contingency preparedness for foreseeable events	7.1, 8.3	4.4.7	4.4.7
B3 Implementation and Operation			
B 3.1 Operation control	7.2, 7.3, 7.4,7.5	4.4.6	4.4.6
B3.2 Management of human resources	6.2	4.4.2	4.4.2, 4.4.6
B3.3 Management of other resources, purchase, stores, maintenance, and housekeeping.	6.3, 6.4	4.4.1	4.4.4, 4.4.6
B3.4 Documentation and its control	4.2	4.4.4, 4.4.5, 4.5.4	4.4.5, 4.5.3
B3.5 Communication and Reporting	5.5.3, 7.2.3	4.4.3	4.4.3
B3.6 Relationship with customers	7.2.3		
B4 Performance			
B4.1 Monitoring and measuring	7.6, 8.1, 8.2.1, 8.2.3, 8.2.4	4.5.1, 4.5.2	4.5.1
B4.2 Analyzing and handling Non-conformities	8.3, 8.4	4.5.3	4.5.2
B4.3 System audits	8.2.2	4.5.5	4.5.4
B5 Improvement			
B5.1 Corrective action	8.5.2	4.5.3	4.5.2
B5.2 Preventive action	8.5.3	4.5.3	4.5.2
B5.3 Continual improvement	5.4.2, 8.5.1	4.2, 4.6	4.2, 4.6
B6 Management Review			
B6.1 Management Review	5.6	4.6	4.6

integration of the different systems, which could be synchronized with normal practices adopted by the organisation. But the ideal system of IMS will be implemented when there will be one licence for at least three management systems. 5 S requirements can be covered under clause Identification and traceability, if required.

3. Also, as per Indian Factories Act, 1948 the Chief Executive is designated as the occupier, who is responsible for health and safety of persons at work place in the organisation, but this authority is delegated to a lower rank person in most of the organisation and hardly any CEO is directly responsible for health and safety of the workforce. As a result of this the subject is considered as of less priority and not as a fully aligned part of the overall management of the organisation.

Social Responsibility of the Corporate Sector

At present there is no ISO standard on the subject, however, guidelines about the social responsibilities of the corporates have been prepared by an American corporate research institution. SA: 8000:2000 includes ILO conventions regarding forced labour, occupational health and safety, appointment of a management representative to look into health and safety of employees, making availability of the good working environment including utilities etc. ISO has taken up the task for evolving a standard on social responsibility which is likely to be finalised by the end of year 2008. Till such a time the standard is published, corporate sector must adopt the requirements of SA-8000 at the same priority as given to other areas of production, finance, marketing etc. This will result in increased effectiveness and efficiency of the organisation.

Conclusions

1. Management systems such as QMS, EMS, and OH&S should be considered as an integrated part of overall management;
2. Integrated Management System will reduce duplication of work, which will not only result in time saving, but also cost reductions and increased profitability;
3. ISO Guide 72 should be followed for preparation of documents for IMS, which is based upon PDCA cycle; and
4. IAF should allow a provision for certification bodies for grant of a single certificate for all the three systems, i.e. Integrated Management System.

REFERENCES

ISO-Guide 72

ISO: 9001:2000

ISO: 14001:2004

IS: 18001:2000

34

INDUSTRIAL SAFETY

Shivaji Singh*

ABSTRACT

In present global competitive edge and major industrialisation, time has come to re-examine our present framework of thinking in occupational safety and health and modify it to achieve our goal, namely zeroing on accidents. It is high time to look into at some newer principles that have appeared in safety field in the industrially developed countries in the West.

The total impacts of the losses are not being recognised by the Corporate Management and even the executive management at the Plant/work site level. However, when there is major property damage or serious injury to the personnel, it is common to collect and record certain concerned facts and figures, to enable subsequently to recover a portion of loss through insurance or perhaps through legal action if desirable.

There are losses of varying degree and frequency, in all business activities, that are not realised as controllable. They generally pass of as a normal part of operational activities.

Every loss, regardless of extent indicates a breakdown in management control at the shop floor level who are most closely associated with the activities where loss occurred.

An important development in Industrial safety in recent years has been that of total loss control (TLC) and of damage control.

The term 'total loss control' is identified here as all-inclusive functions, considering the control conditions that can contribute to or be totally responsible for the interference with orderly progress to attain the objectives of an organisation. Losses are generated as the result of an accident or an unintended interruption of an activity involving damage to equipment/machinery, facilities, materials, and environment and occasionally injury to people, besides many other causes.

Cost conscious organisation finds that they must control accident cost if they are to do business in a highly competitive global market. They do not consider the safety a nuisance, but a necessity not a fringe benefit, but a prerequisite.

*Engineer (Indl. Safety), Tehri Hydro Development Corporation Ltd., Tehri, Uttaranchal

Paper Details

Present Safety Concepts

Heinrich's Theory-1931—Foundation of a major injury

If a person commits an unsafe action of the same kind repeatedly say 330 times, the results of these 330 unsafe action are :

1	—	Major injury accident
29	—	Minor injury accidents
300	—	No injury accidents (property damage accidents)

If one repeats an unsafe action of the same kind, then it is not possible to avoid minor and major injuries.

Frank Bird Theory-1969 (New)

If a person commits an unsafe action of the same kind repeatedly say 641 times, the results of these 641 unsafe action are:

1	—	Major injury accident
10	—	Minor injury accidents
30	—	No injury accidents (property damage accidents)
600	—	Near miss incident

Before any unsafe act results in a minor injury, the person committing it, must have performed many-many unsafe actions while did not result in any injury—perhaps resulted in property damages. Such repeated performance of unsafe actions prior to an injury accident although seen or noticed by the supervisor or his superior, are not objected to and therefore, the person gets encouraged to repeat it as often he wants to and in the cyclic order, the repeated unsafe acts does at a certain stage result into minor injury and subsequently into major injury.

Hence to prevent or reduce injury accidents we should take a serious notice of causes of minor injuries accidents, first aid injuries accidents, property damage accidents and near miss incidents and find out their root causes—just not immediate cause or proximate causes.

Industrial Safety Gives More Emphasis on Freedom from Losses Due to Accidents

Following things to be made free from accidents in industries:

- Human beings employed.
- Machinery and equipments.
- Work place environment.
- Buildings and structures.
- Raw materials and finished products.

Losses to the above can be controlled by total loss control (TLC) approach. Industrial works are mainly full of hazards. We have to take precautions with a view to ensure safety of men, machineries/materials and environment.

Accidents in the industries are very costly, and also effect the ultimate business objectives and corporate image. Losses due to the accidents to the human beings cannot be compensated.

Accident may lead to-

- Injury or even death of the employees/workers.
- Damage to the plant/equipments.
- Disruption of production schedules.
- Lowers morale of personnel.
- Prolonged absence of trained workers etc.

We know that the accidents mainly occur due to

- Human error or unsafe action - 88%
- Mechanical failure or unsafe physical conditions - 10%
- Natural clematis due to earthquake, flood, cyclone etc. - 2%

It is prime requirement to eliminate unsafe action and unsafe physical conditions at shop floor to avoid any accident.

Safety is a factor, which influences the behaviour and attitude of the employees towards doing safe work and eliminating an unsafe action and unsafe conditions at the work site/ plant.

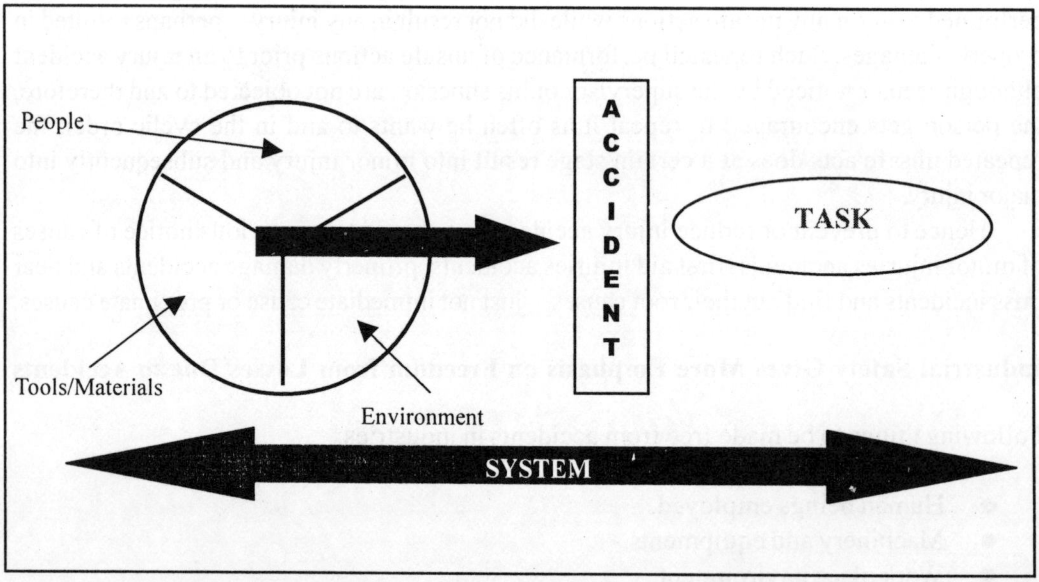

A system is an orderly arrangement of components that are interrelated, act and interact with each other to perform task or function in a particular environment. An accident results from deficiencies in people, tools or materials and environment and sometimes deficiencies in the task.

Hazard Identification Techniques

***J.S.A.* [Job safety analysis]** - to identify potential hazard of job and establish corrective and preventive action to control the hazards

Purpose

- To analyse the risk and hazards associated with work activities (job)
- To suggest safe work procedure [SOP] to control the hazards.

HAZOP [Hazard and operability] Study - Hazards and operability related problems.

Purpose

- Hazard identification
- Aid in design of safe plant
- Training aid to plant personnel
- Preparation of operating manual/sop
- Conformity with codes of practices
- Demonstration of commitment to safety

Safety Audit - Co-operative effort to improve overall safety performance.

Safety Inspection

F.M.E.A. - To know the hazards and their consequences.

Technique of Accident Prevention

Accident prevention is relatively simple. Because they are caused, accidents are preventable. For effective Industrial Safety programme we need three E's

- Engineering
- Education
- Enforcement

The plant/processes/works must be engineered for occupational safety and health at initial stage, starting from site selection, layout, design etc. During operation of the plant, attention should be given to their proper maintenance and upkeep by education of the employees and emphasing on the line managers about their roles and responsibilities in enforcing the safety norms/procedures/rules etc. at shop floor/work site.

Sequence of Accident Occurrences

Basic theory of accident occurrence may be stated as under

An injury occurs only at the result of an accident

↓

An accident occurs only due to unsafe action and unsafe condition or both

↓

Unsafe action and unsafe conditions exist because of human faults

↓

Faults of a person are inherited or acquired from the social environment and the cause for these defects are

↓

- Physiological or psychological unsuitability
- Lack of knowledge or skill
- Improper mechanical or physical environment

Thus when an accident occurs, there is always a chain of events, which occurs, in a logical and fixed order. Each link in the chain is dependent upon the preceding link.

Occupational Health and Hygiene

(Adoption of work to man and of each man to his job)

It is essentially a preventive medical science, which aims at prevention of diseases of occupational origin and maintenance of the highest degree of physical, mental and social well-being of workers in all occupations.

A human being at work place has to adjust himself with materials, machineries and physical environment with an objective to achieve the production target and ultimately business goal. But maladjustment with either of these would lead to bodily injuries, equipment/ machineries damage, and discomfort and health impairment. The effected body parts of a person due to occupational exposure/injury/accident may be cured, but the occupational diseases caused due to poor working environment are not curable. Moreover, the productivity in an industry may also effected because of impairment caused to the employees due to accident. Therefore, it is moral and legal responsibility of industrial management to protect and preserve the health of the workers due to polluted work environment.

Careful planning of location, design and layout of plant is essential to ensure efficiency and safety of industrial operations. If appropriate measures are taken at the initial planning stage, numerous accidents, occupational diseases, fire and explosion etc. can be prevented. Safety and efficiency go hand in hand. The best way to avoid accidents is to promote safety consciousness to inculcate safe working habit and to identify and eliminate operational and maintenance work procedure.

As per the Section 7A(1) of the Factory Act, 1948, every employer shall ensure the health, safety and welfare of all workers while they are at work in the industry/factory, and Section 7A (2)© emphasises the provisions of maintenance or monitoring of such work environment in the factory for the workers, that is safe, without risk to health.

In this regard industrial management should establish work environment monitoring facilities in their factory/plant and periodical work environment monitoring should be done to ensure any deviations in the system and process and their proper records should be maintained. The periodical monitoring in the industries would also reveal the compliance of the provisions laid down in Section 41F (2nd schedule) of Factory Act, 1948. The national data on work place monitoring will help to update and develop permissible limit of exposure to various contaminants.

A Better Occupational Health and Hygiene Practice Must Include

- Identification of health hazards
- Remedial and contro measures of health hazards through

    (a) Engineering methods like
        1. Substitution
        2. Segregation
        3. Wet method
        4. Exhaust ventilation etc.

    (b) Administrative like
        1. Enforcement of PPEs
        2. Relocation of effected employees.

It has been observed that occupational health and hygiene is not the priority of the most of the industries. Conscious level has to be increased among the workforce to realize that employees are the most valuable assets of an organization. If work environment is not made safe, later chronic effects would be seen in the impaired productivity.

Evaluation of Work Place Environment

Each hazardous industries is required to do initial appraisal of working environment by gathering following information:

- Identification of hazardous substances prevailing at the work site.
- Air borne contaminants of hazardous substances like toxic gas, dust, metal fumes etc.
- Quantification of specific hazards.
- Evaluation of the effectiveness of Engineering control installed to minimize exposure to the employees.
- Assessment of compliance status with respect to occupational health hazards.

Occupational Health Monitoring

It is necessary to have a periodical monitoring of health of the employees, who is behind the industrial machines, to ensure its suitability for the particular job, he is assigned.

Pre-employment Medical Examination

- To ascertain the employees ability to perform specific task.
- To determine the presence of medical conditions which could be potential source of injury and disease.
- To make relevant recommendation for elimination of the aforesaid risks and its effects and also enhance individual's performance.
- To establish a base line data of all findings for future requirement.

Periodical Medical Examination

At least once in a year and also frequency may vary upon the type and occupation and environmental factors.

Special Medical Examination

Some common occupational health hazards in Building and Other construction works like Hydropower project etc.

1. Occupational dermatitis.
2. Occupational cancer.
3. Asbestosis.
4. Silicosis.
5. Lead poisoning.
6. Benzene poisoning.
7. Occupational asthama.
8. Pesticide poisoning.
9. Carbon monoxide poisoning.
10. Toxic jaundice.
11. Toxic anaemia.
12. Compressed air illness (Caissons disease).
13. Noise induced hearing loss.
14. Isocyanates poisoning.
15. Toxic nephritis.

OHSAS 18001 - Consolidated OHSMS Standard

- Provides a common standard.
- The "New" standard may bring about a National and International standard in the future.
- Industries interest are very high for common standards.

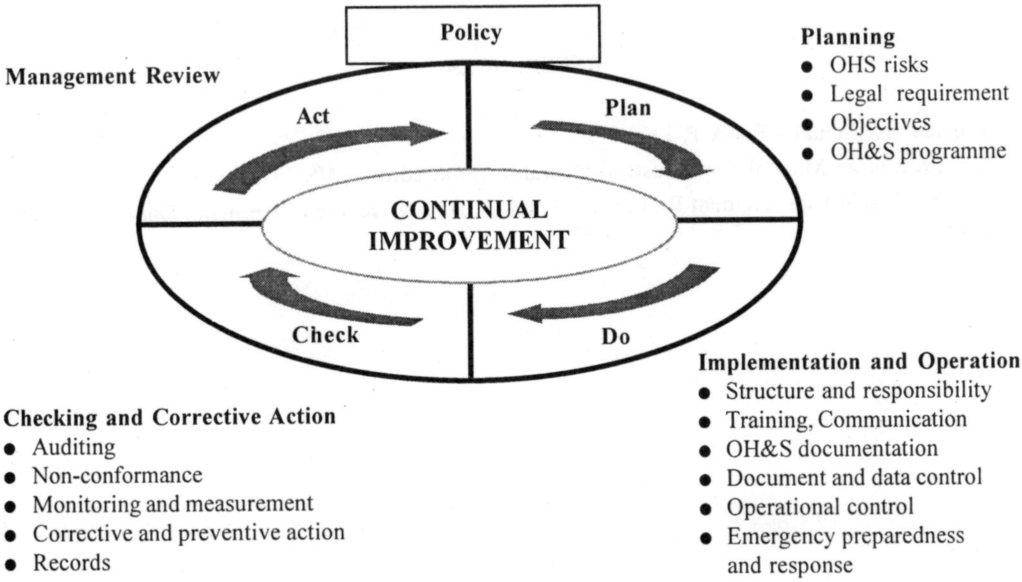

Important Safety Regulations in India

- Factories Act, 1948
- Building and Other Construction Worker's (Employment and Conditions of Service) Act, 1996
- Building and Other Construction Worker's (Employment and Conditions of Service) Central Rules, 1998

} Especially applicable for construction works like Hydro power etc.

- Indian Electricity Act, 1910
- Indian Electricity Rules, 1950
- Explosive Act, 1884
- MSIHC Rule, 1989
- Petroleum Act, 1934
- Gas Cylinder Rules, 1981
- U.P. Factories Rules, 1950 and Safety Officer's Rule

} Applicable for U.P. and Uttaranchal

Conclusion

To achieve high standards of Safety, minimize accidental losses and Business survival in present global scenario. The plant/processes/activities must be engineered from Occupational Health, Safety and Hygiene point of view also at planning stage itself. Since accidental losses mainly effect the ultimate objectives of any business directly or indirectly. However to control the accidental losses effective OH and Safety management/loss control programme must be there in the organization according to OHSAS-18001norms. Essence of OH and Safety Management is to avoid high risks, manage medium risks and live with low risks.

REFERENCES

Loss Prevention Manual - Frank P. Lee

Accident Prevention Manual for Industrial Operations, 6th edition, NSC Chicago

Heinrich, H.W. Industrial Accident Prevention, Frank E. Bird & George L. Germain, Damage Control

Internet

SAFETY CODE OF CONDUCT AND STANDARDS FOR CONTRACTORS AT JINDAL STAINLESS

A.K. Jain

Introduction

Safety code of conduct and standards for contractors at Jindal Stainless Limited (JSL) deals with the conditions/precautions to be taken for personnel safety, Equipment/material Safety, Environmental Safety, Process/Business Environment during contractual activity of Jindal Stainless Limited. This code is only for guidance of contractor. Contractor shall comply with the Environment, Health, Safety and Engineering Practices. Contractor shall also follow the guidelines for Environment, Health and Safety given during the course of contract.

Approach and Methodology

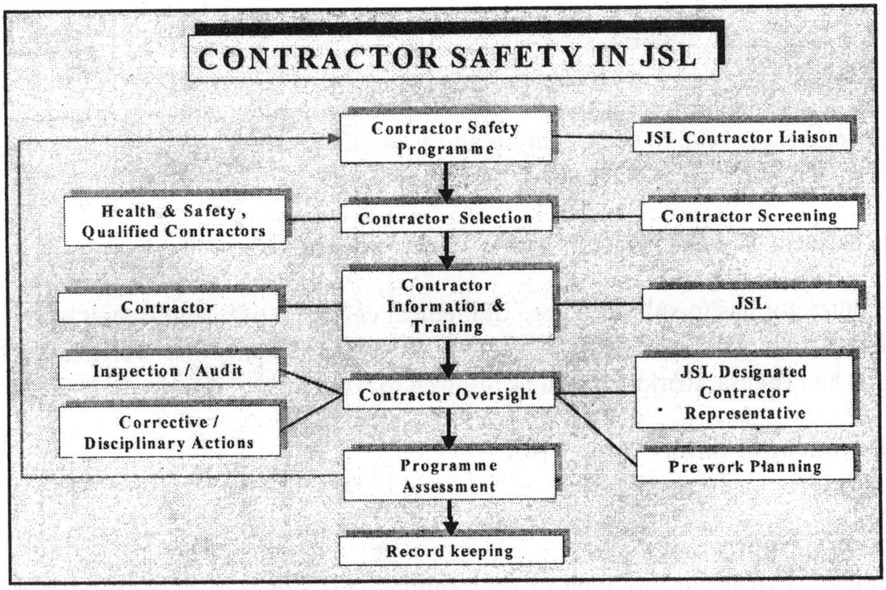

*Dy. G.M. (EHS), Jindal Stainless Ltd.

Q-SHE Policy

The contractor shall comply with the Quality, Occupational Health & Safety and, Environment policy of the JSL in letter and spirit.

Group Risk Assessment/Aspect and Impact Analysis

The contractor shall carry out the Group Risk Assessment/Aspect and Impact analysis of the job for which contract is to be awarded. The Group Risk assessment and aspect impact analysis will cover all hazards and environmental act and control measure to be observed by the contractor. The Group Risk Assessment and Aspect and Impact done by the contractor will be reviewed and approved by the concerned Department Head. A copy of the approved Group Risk Assessment/Aspect & Impact should be send to EHS Department and Quality System Department for monitoring and record for ISO -14001,OHSAS-18001 and ISO-9002.

Statutory Compliance

The contractor shall comply with all applicable Rules and Regulations framed by Central / State authorities pertaining to job such as:

Skill Competency

- The contractor has to submit the records of Job experience and past performance exclusively on Safety and Environmental matters along with the proposal.
- The contractor shall provide details of qualification, experience and skill of workforce being deployed by him at site as per the statutory norms and job requirements.

Lifting Tools and Tackles

The contractor has to submit Test Certificate by a competent agency of all lifting tools and tackles and cranes etc. to be used at his site as per the requirement of the Factories Act and the Haryana State Factories Rules.

Health Record and Physical Fitness of Workforce

The contractor has to submit record of medical check-up of his workmen by a competent Medical Officer (who should have a M.B.B.S degree) and only after verification of Health records and fitness, workmen will be allowed to work at site.

Appointment of Safety Supervisor

EHS Training

Contractor shall arrange to impart EHS training to his Supervisors and workmen at his end and maintain the record, which will be verified by JSL.

Safety Work Permit

The contractor shall comply the requirement of safety work permits of JSL and no work shall be carried out without safety work permit.

Personal Protective Equipment and Safety Appliances

Contractor has to provide standard quality safety appliances as per IS / European Standard such as Safety Helmet, Safety Shoes, full body harness safety belts, safety goggles etc.

Entry-Exit

The contractor shall ensure that all workers/supervisors engaged by him will carry Identity Card/Photo Pass. The same is to be displayed on their person during their presence inside.

Personal Conduct

Entering Plant premises while in the influence of alcohol/ narcotics is strictly prohibited.

Appointment of Sub-Contractor

The contractor will not appoint any sub-contractor without the written permission from JSL. EHS Code of JSL will also be applicable to the Sub-contractors. Details of the Sub-contractor is to be furnished to JSL.

Vehicular Safety

The contractors has to follow all norms of Central Motor Vehicle Act - 1988 and Rules made thereunder for the vehicles being deployed by them at site.

Civil Works

All civil works and activities undertaken therein should be carried out as per IS Code/ National Building code.

Electrical

Contractor should engage Licensed electrical supervisors for certifying their equipment and ensuring safety of their connection.

House Keeping

Good House keeping must be practiced by the contractor personnel at all times while within the plant.

Management of Accident

- In case of injury to persons, contractor shall first take the injured person to the JSL Health Centre/Hospital with the prescribed accident report form/in-plant injury form. In no case contractors are allowed to take the injured persons directly to their own doctors.
- Contractor shall ensure all his workmen against workmen's compensation Act.

Penalty

The penalties shall be imposed on the contractor in case of violation of safety & environmental norms and an accident (Personnel Injury, Damage to Plant Equipment/Structure, Business Interruption and Environmental Damage) take place at contractor's site on the recommendations of the Head of Department awarding/executing the work/Head of EHS Department.

Undertaking

I/We have gone through all the terms and conditions of EHS Code for contractors and undertake to strictly enforce the same during the execution of this work including mobilization period, if any. In case of violation of these terms and conditions, the owner is free to take action as deemed fit.

Result and Conclusion

- This safety code of conduct has developed a safety culture amongst contractors.
- The accident rate has gone to vanishing point.
- A partnership approach has developed between contractors and management.

REFERENCES

NSCIP Publications
WHO Web site
ILO Web site

Section–IV
Occupational Safety, Health and Environment

SAFETY MEASURES FOR COMMUNITY HEALTH

Lipi Mukhopadhyay*

ABSTRACT

Theme: Under occupational safety and health Occupational safety and health comes under the most urgent agenda of the state governance. Poor occupational safety and health costs to the State in multiple ways. For the long-term benefit of the community and the state it is necessary to observe standard health safety measures and legislation. Quality of road transport system, hygience and seats are hazards for health injury, backache, spine problems and asthma. These go unnoticed by the transport management, Emission is another hazard constantly affecting the commuters, pedestrians and the environment. The present paper intends to measure these aspects in road transport management system and suggest workable remedies.

Introduction

Getting healthy and safe work place environment is a human right. According to the declaration of Alma-Ata, which declared (1978) "Health for all by the year 2000"[1]. It was aimed that all the people of the world to have access to affordable, quality health care[2]. However, the overall increase in health facilities has not produced the expected result for the majority of the population.

There are many challenges to health care system with new concepts such as environmental legislation, ethical issues, new pathologies like cancer, stress, AIDS, new safety regulations, insurance and costly of heath care. Indian occupational health is a more complex issue, it includes child labour, poor industrial legislation, vast informal sector, less attention to industrial hygiene and poor surveillance of data[3]. In India there is always poor investment on industrial safety because labour is cheap and easily replaceable. Motivated by the desire to promote safety measures for community health and to prevent occupational hazards the objectives of this paper are to examine the existing occupational safety legislation in industries and road transport system in India compared to international constraints in promoting safety measures.

*Associate Professor, Indian Institute of Public Administration, I.P. Estate, New Delhi - 110002, India
E-mail: lipiiipa@yahoo.com

— The bottlenecks and constraints in promoting safety measures.
— Removing constraints with adaptable measures.

Occupational Safety Legislation

India also signed a bilateral collaborative agreement with United States in terms of emergency preparedness and response training , technology transfer and research. To meet local shortage of resources, to address all outstanding issues in the fields of environmental and occupational health, India should turn towards international community and World Health Organization (WHO).

A study by the National Safety Council States: "Statistical analyses indicate that companies promoting job safety have low workplace injury/illness rates"[4]. Another study published in the National Safety Council's Safety Health magazine shows that the higher a company's safety and health standards the higher its productivity and *vice versa*[5].

In most organizations, an off-the-job safety programme is probably the most overlooked way of improving on the safety and increasing income[6]. Every year more than seven million injuries, serious accidents and deaths[7]. Most of the victims are pedestrians and cyclists. About 2 per cent of GDP costs annually account for traffic accidents, which is exceptionally high. India has one of the highest vehicular fatality ratios. The rate of accidents among pedestrians, bicyclists, non-motorised transport users and motorised two-wheelers is even higher[8]. In Delhi bicyclists share 14 per cent of the road fatalities[9]. Despite such a large magnitude of loss of life and serious injury to survivors traffic safety sadly remains a low priority area for policy makers, law enforcement agencies and other associated executives. Most of the accidents occur due to poor traffic management system and infrastructure, lax traffic law enforcement, and poor public transport[10].

According to Delhi Traffic Police most accidents take place in Delhi due to high speed, non-observance of traffic rules and narrow space for pedestrians.

In India, the most vulnerable class who meet with road accidents are mostly from the middle and lower classes. The morbidity and disability following a traffic accident is also many a time the result of poor access to critical health facilities and services. It is found economically active people in the age group between 15-44 years[11] in most cases are the victims and are also the breadwinners in the family. Therefore, victim's families are badly affected by such accidents often resulting into denial of children's access to education and health, deteriorated living conditions and perennial mental health problems.

Bottleneck Constraints

India's poor traffic management and reduced priority to enforcement of occupational safety measures cause acute health problems among lower income groups. It is found that due to poor occupational health and safety systems and lack of awareness among factory workers about pulmonary malignancy related to asbestos is not detected at early stage. The Government should consider the ban of this material.

It is estimated that 10 million industrial or mine workers are exposed to asbestos or other dusts in India. Another crisis has emerged in West Bengal due to drinking water contaminated with arsenic. About 42 million people are exposed to arsenic till date according to an estimate[12].

Improper checks, negative attitudes and lack of implementation of rules influence some workers to consume alcohol during work. Alcoholism is often associated with road traffic accidents in case of Drivers. A study based on traffic policemen from six major towns of India attributed risk of exposure to benzene from fuel exhaust and also discussed how this inhaled benzene metabolized phenol, posed health consequences[13].

Another study reported that under the supervision of Inspector General of India, a small number of three hundred factory inspectors are responsible to check the industrial safety in this vast country, compared to 3000 factory inspectors in a small country like Japan[14].

UNICEF on emergency obstetric care reports that "communication and transport were far from adequate at the health centres. Only 26 per cent of health centres surveyed had radios.The road infrastructure is impassable in the rainy season and poses a great challenge to road users even in the dry season. Clients still have to cover great distances to reach a referral health facility"[15]. It is observed that the consequences of inadequate transport for the delivery of basic health care: "The most impoverished—usually rural—areas have few or no health care facilities, or the means to transport people for medical assistance. About 90 per cent of children dying from these diseases die and often without their families even seeking health care"[16]. At least 35 per cent of women in developing countries receive no antenatal care during pregnancy, almost 50 per cent give birth without a skilled attendant and 70 per cent of receive no post partum care in the six weeks following delivery reports UNICEF.

Need for Political Commitment

A significant number of people (seven million) are getting affected every year, among which a very large number are severely crippled or simply devastated—economically and emotionally. It surely calls for serious and strong willed measures. Such measures can be implemented when it is realized politically that traffic safety is a major issue, which needs immediate attention.

The whole emphasis during traffic safety campaigns is on the smooth flow of motorized traffic—especially cars—edging out the needs and safety of pedestrians, cyclists and non-motorized transport users, even though the highest fatalities are from amongst this class of road users. Traffic safety campaigns in India mostly voice the concerns of the powerful groups of motorized transport users while the safety interests of the pedestrians who are the majority, get squeezed out of the decision-making process.

Energy, Efficiency and Transport

Women walk and take public transport more frequently than men. In many countries, there are large differences between men and women in automobile ownership and access as well as in possession of driver's licences. Women tend to make a number of shorter and more complex daily trips for shopping, schools, part-time employment, and volunteer work. Current urban transport systems are not only energy-intensive, but can often restrict the mobility of those who do not use them (e.g., pedestrians, cyclists, and users of public transport).

Reducing women's drudgery is a frequently cited objective of rural electrification, and women have sometimes been cited as the prime beneficiaries of rural electrification, since they spend more time in the home and hence may use improved lighting and other appliances more. Rural electrification would end the health effects of kerosene fumes from lamps, since this is the main replacement effect.

Infrastructure and Office Furniture

It goes unnoticed that sitting arrangement and design of furniture do affect the spine and back. Continuous sitting on chair for long hours is not advisable for good health. But in most offices especially women perform jobs like telephone operators, Stenographers or manual or desk workers spending long hours sitting, which causes back pain. The design of chairs and height is a also important factor in causing back pain. Design of public transport, seats and backrest is crucial for promoting or affecting community health.

Conclusion

Based on above review and analyses it is clear that measures of community safety and health have low priority in India. Despite safety rules and legislation, enforcement of those rules in the interest of public health is minimal. The public in general is not aware of the safety rules and consequences of violations as public litigations are not given proper hearings.

It is necessary to enhance public awareness and build the desired capacity. It is also necessary to comply with the standardized laws relating to occupational safety. In order to have reduced accidents several factors in accordance with safety laws and regulations, infrastructure, technology and the human factor need to be taken into consideration. A joint committee for community safety and health may be introduced to facilitate training and awareness campaigning.

NOTES

1. Abdullahi, Siyad A. "Mobile health Services, and Health service delivery among nomadic population in northern Kenya" accessed from www.safetytimes.com/
2. Torfs, M.E. Mobile health care services in developing countries. *Public Health Reviews,* 1980; 9(1-2): 93-112.
3. Agnihotram, R.V.: An Overview of Occupational health research in India. *Indian Journal of Occupational and Environmental Medicine,* April 2005. Vol. 9, Issue-1.
4. National Safety Council (US). Access from www.safetytimes.com/
5. *Ibid.*
6. *Ibid.*
7. Sanjeev Shami, Road Traffic Safety, Cost of Government Neglect. *Economic and Political Weekly,* April 16, 2005.
8. *Ibid.*
9. *Ibid.*
10. *Ibid.*
11. *Ibid.*

12. Hassan, M.A., Khan, Z., Yunus, M., Bhargava, R. Health Profile of Lock factory workers in Aligarh, *Indian Journal of Public Health,* 2002, 46 : 39-45.
13. Verma, Y., Kumar, A., Rana, S.V., Biological Monitoring of Exposure to Benzene in Traffic Policemen of North India. *Japan Ind. Health 2003,* 41: 260-4.
14. Joshi, T.K., Smith, K.R., Occupational Health in India, *Occup. Med. 2002.* 17:371-89, iii-iv.
15. Cecelski, Elizabeth, *The Role of Women in Sustainable Energy Development,* National Renewable Energy Laboratory, US.
16. *Ibid.*

37

NOISE IN MILITARY AVIATION

Neeru Kapoor, A.P. Singh, M.R. Bhatia, R.K. Sharma and P.K. Banerjee*

ABSTRACT

The severe noise produced by aerospace systems greatly impairs the hearing and affects the health, safety and well-being of the personnel associated with or exposed to aerospace operations. A large body of persons is employed in the Indian Air Force and it is of paramount importance to preserve hearing and promote health in the work place. A survey of the prevalent noise level associated with Indian military aerospace activities was undertaken to evaluate the influence of aerospace noise on the hearing acuity and performance efficiency of the air personnel/crew and suggest remedial measures for controlling the incidence of hearing losses.

The measurement and analysis of the noise due to operation of high performance aircraft and helicopters at different locations within Air Force Stations was carried out with the help of sound level analyzer having snap shot and display facility. Effect of exposure to intense noise on the hearing acuity of the personnel was assessed by audiometric tests administered on 451 individuals comprising of pilots, ground duty personnel, families and school children by pure tone audiometry utilizing GSI 61 Audiometer having 1dB resolution. The linear and A-weighted sound pressure levels recorded during ground running, taxying, take-off were in the range of 110-145 dB for close range measurement while flying over head after take off the linear values increased from 80 to 120 dB and declined subsequently. The landing operations did not generate intense noise. In the similar conditions helicopter noise was relatively of lower intensity but more disturbing due to complex frequency variation and intermittent nature of noise from rotor and engine. The periodic noise due to blades and non-periodic aerodynamic noise pressure levels were more annoying. The sound pressure levels of noise measured in the hangers, tarmac, crew rest room and ATC complex were also high. In addition to work places, the sound pressure level monitored in the residential area, schools, play ground etc. during flying of air crafts were also higher compared to the accepted limit (90 dB). The audiometric tests conducted on 52 pilots indicated normal hearing in 28 of them (53.9%) and 20 (38.4%) with mild, 3 (5.8%) with moderate and only 1(1.9%) with severe hearing impairment. In comparison to this, the effect on ground crew personnel, their families and the school going children indicated that they were affected to a greater extent on account of their intermittent but frequent exposures to intense noise from short distances. Besides the high sound pressure level, the high frequency component of noise (3-4 kHz) makes the individuals more susceptible to hearing damages. The audiometric tests on pilots confirm the effect of protection rendered by the headgear and associated devices used by them.

*All at Defence Institute of Physiology and Allied Sciences, Lucknow Road, Timarpur, Delhi - 110 054

The study indicated that the noise levels associated with various mode of flying constitute auditory risk particularly to ground duty personnel. High levels of noise caused extra auditory effects like fatigue, irritation and annoyance in addition to decline in concentration. A definite requirement exists towards control of noise exposure by using barriers as well as effective ear defenders for the conservation of hearing of air force personnel. Sound barriers may be helpful in reducing the noise level. Their suitable design and commissioning may be used to absorb and prevent entry of noise in the crew rest room and service sick quarters etc. The utility of inhaling carbogen (a gas mixture of 5% CO_2 and 95% O_2) as a prophylactic intervention against noise induced hearing loss is discussed.

Introduction

Noise is unwanted and unpleasant sound that adversely interacts with the health and performance of man in his living and working environment. Intense noise leads to annoyance, irritability and fatigue, and long-term exposures are known to inflict high degree of hearing impairments and deterioration in several non-auditory functions of the body[1-4]. Several experimental investigations have demonstrated the influence of noise on cardiovascular, endocrine, metabolic, gastrointestinal and neurological systems of the body[5-7]. Studies conducted on industrial workers revealed that people working in noisy occupations often develop circulatory, digestive, metabolic and neurological problems[8-12]. Almost all extra-auditory effects are triggered as a result of noise interaction through autonomic nervous system[4,13]. Protecting the hearing from the assault of noise also helps in controlling the extra-auditory effects, and the efforts are now on to evolve abatement programmes to solve the problem[14].

The most important part of any work place hearing conservation and noise control programme involves the measurement of noise levels and the workers exposures. According to the place theory of hearing, noise entering the inner ear sets the basilar membrane into sympathetic vibration with the peak amplitude dependent on the frequency of the stimulating sound. Low frequency sounds excite vibrations of increasing amplitude that peak near the apical end while high frequency sounds cause vibration amplitudes that peak near the base of the basilar membrane. The mechanical disturbances so generated proceed like a travelling wave that travels from the stiffer base to the less stiff region, the apex15. Low frequency sounds cause a more extensive disturbance of the basilar membrane and its attached organ of corti than do high frequency sounds due to the travelling wave starting from the base and travelling right up to the apex of the membrane. Over-stimulation of the hearing system causes loss of hearing sensitivity known as temporary threshold shift (TTS) while intense and prolonged exposures lead to permanent hearing impairments (PTS). Noise Induced Hearing Loss (NIHL) results from the partial or complete loss of hair cells on the organ of corti. Another characteristic feature of NIHL is the fact that hearing loss begins at or above 4 kHz audiometric frequency irrespective of the stimulus frequency and proceeds towards lower frequencies with time. Yet another feature of the interaction of noise is that the threshold shift (TTS) or temporary hearing loss occurs at a frequency that is one and a half octave above the stimulus frequency.

The human auditory system is a tuned structure, with different parts of it demonstrating differential sensitivity to the different frequencies of sound and also functions as a filter. It has the ability to perceive sound frequencies from 20 Hz to 20 kHz and is very sensitive to frequencies in the range of 400 Hz to 4 kHz that makes it more prone to damages. For frequencies that are lower or higher, the sensitivity declines, enhancing the tolerance to the assaults of sounds. Further, when a sound of frequency below 2kHz enters the ear, it causes larger amplitude of vibration. To control the magnitude, the middle ear muscles tense altering the dynamic response of the middle ear and causing a reflex action[2]. This is a physiological defence mechanism. The aural reflex also helps to size down the passes of strong stimuli to the inner ear. In a large number of occupations, people are exposed to continuous noise with variable amplitudes. The noise in the industrial environment, traffic noise on the roads and the noise in the flying aircraft and engine room of naval ships constitute continuous noise whereas the noise due to hammering or striking of one object to another, weapon firing and explosions form impact or impulse noise[16]. The important physical parameters that contribute to hearing impairment are the sound pressure level (SPL, dB) frequency spectrum (Hz) and time period of exposure in the noisy environment. For impact and impulsive noise, Peak pressure level, rise time and the duration have to be estimated for the assessment of likely impact on hearing. Noise exposure standards and the Damage Risk Criteria (DRC) have been formulated on the basis of physical sound pressure level and the duration of exposure to noise with the objective to workout safe limits for noise exposure and prevention of injury to hearing[17-24].

In the context of military aviation, aircraft noise has been a major source of auditory and non-auditory reactions leading to discomfort, irritability and auditory damages due to intense, brief but repeated exposures. Development of high speed Jet aircrafts had resulted in a tremendous spurt in the noise emission and radiation into the outside environment. The main sources of noise from an aircraft are the propulsion system, generators and the aerodynamic interaction between the aircraft and the air through which it is flying. The noise generated by aerodynamic interactions is also called boundary layer noise. The acceleration of air masses affect propulsion in jet engines. Exhaust noise is the result of a series of rapid but small explosions within the jet system and it is broadly distributed in frequencies. The aircraft flying at or above the supersonic speed give rise to another very important phenomenon of the generation of sonic booms. To protect the hearing of the aircrew and other ground duty Air Force personnel from the assaults of intense aircraft noise is of paramount importance for their well being and performance. High noise levels are known to interfere with communication and disturb the passage of other auditory signals accurately and limit the accuracy, efficiency and personal safety of the individuals.

The application of safety measures like the use of ear defenders has been examined by several workers[25-27]. These included the use of earplugs and earmuffs and the use of some prophylactic agents like carbogen (a mixture of 5% CO_2 and 95%O_2) that has been extensively studied in this Institute[28-30].

The present study was undertaken with a view to study the noise problems associated with the ground and air operation of high performance aircrafts and helicopters at the Air Force Stations of the Indian Air Force, the impact of repeated exposure to brief but intense noise on the Air Force personnel and suggest measures to control the noise for better conservation of hearing and overall performance efficiency on the ground as well as in the air.

Paper Details

Measurement of sound pressure levels (SPL, dB) and frequency spectrum of noise produced by aerospace and ground operation of the aircrafts and helicopters was carried out at different sites and work and rest places at different Air Force stations using CEL-573 Sound Level Analyser and a Bruel and Kjaer Type 2230 Precision Integrated Sound Level Meter. The possible auditory impact of noise caused by aircraft flying overhead were also assessed by taking one third octave snap shot at quick successions to display the rise and decay of amplitude and frequency shifts in the process of take off, approach, flying overhead and going away. These measurements were taken till the noise level remained at 10 dB higher than background in order to assess the frequency shift that takes place during the processes at the instant of -15, -10, -5, 0, +5, +10 and +15 seconds of passages overhead.

The preliminary attempts made to explore the conditions of exposure of Air Force personnel to noise indicated that these took place under two environmental conditions; exposures to frequent aircraft flights at closer distances overhead and during maintenance and flight preparation. The aircraft flying at different speeds, altitudes and distances produce noise of varying amplitudes and frequency characteristics. Increasing speeds increase the aerodynamic interaction and therefore the noise levels but the increasing distances between the flying air craft and the individuals on the grounds tends to reduce the noise by atmospheric absorption due to damping, divergence and scattering of sound waves. Since the acoustic emissions of high-speed aircraft are sufficiently high, the attenuation by atmospheric absorbtions does not reduce the sound to a level acceptable to human beings from safety and work point of view. The exposures, therefore, constitute a definite risk of auditory hazards. Routine maintenance and preparation for the flights is still more deleterious because ground duty personnel are exposed to intense noise (> 130 dB) while working close to aircraft for several minutes. Such exposures without ear defenders are not permitted, even for brief periods, as per safety guidelines[31]. Since the exposure cannot be prevented there is every possibility for the service crew to run the high risk of auditory damages.

The sound pressure level of aircraft noise presented in Tables 37.1, 37.2 and 37.3 are higher as compared to the levels that appear in the permissible limit criterion for exposure to intense noise. Applying the exchange rules as suggested in DRC's the allowable exposure time to 120 dB 'A' noise level would be less than one minute. Exposure to higher noise levels above 120 dB 'A' without ear protection should not be allowed. The Occupational Safety and Health Administration (OSHA) has set up a ceiling limit of 115 dB 'A' for allowable exposures of 15 min or less and suggests that levels greater than 115 dB 'A' should not be permitted regardless of the duration of exposure. This is based on the assumption that a critical level for hearing losses exists and above which the human auditory responses to energy do not have a straight relation to the duration of exposure and instant structural damages may be expected for exposures above the critical level. The sound pressure levels recorded during the present investigations exceeded the 115 dB 'A' ceiling level at Crash Bay, Met Office terrace, ATC Officer's Cabin and the Hangars. The noise levels encountered by the maintenance personnel during ground operations are still higher (121.3 to 133.9 dB 'A') compared to the ceiling limit and there is every possibility for their developing high degree of hearing impairment. The data presented in Table 37.3 describes A-weighted noise levels measured around the aircraft during ground operation. These levels vary from

119.5 to 146.6 dB 'A' during ground operation of aircrafts with engine running at different rpm. This represents a situation of acute noise exposure capable of causing instantaneous damages to hearing system of the ground crew if ear-defenders are not in use.

Table 37.1: 'A' Weighted Sound Pressure Level (dB 'A') of Noise at Different Locations of Air Force Stations

Location	*SPL Range (dB 'A')*
CRASH BAY	72.0 - 122.5
CREW RESTROOM	86.8 - 100.8
MET OFFICE (Windows Closed)	62.0 - 102.0
MET OFFICE (Terrace)	68.0 - 115.9
ATC Control Room	60.4 - 99.0
ATC Officers Cabin	78.0 - 116.7
TARMAC	92.0 - 104.0
Hangar	99.7 - 120.0
Residential Area	64.0 - 109.7
RRecreation Area	70.0 - 94.0
School Class Room	60.0 - 88.0

Table 37.2: Sound Pressure Levels (dB 'A' & dB 'C') During Ground Operations of High Performance Air Crafts at 6 Meters

Air Craft	*'A' Weighted SPL (dB 'A')*	*'C' Weighted SPL (dB 'C')*
A	130.1 - 133.9	130 - 133.6
B	127.4 - 132.6	127.1 - 131.7
C	121.3 - 134.9	120.5 - 134.5
D	126.9 - 128.9	123.4 - 129.2

Table 37.3: 'A' Weighted Sound Pressure Level (dB 'A') During Different Stages of Ground Operations at Full Power at 6 Meters

Stage of Operation	*Air Craft A*	*Air Craft C*
Idling (40% rpm)	119.5	118.4
Warm up/Cool down (80% rpm)	135.0	126.6
Maximum rpm	139.2	134.8
After Burn up	146.6	139.5

The design of the modern day jet aircrafts is such that interior noise essentially is much lower as compared to that radiated out from the engine and aerodynamic interactions. This intense, brief and characteristically continuous noise with frequent intermittence interacts with the people on the ground. Though brief exposures and frequent intermittencies to noise can be beneficial[32-33], high intensity sounds can be increasingly damaging more so when tonal components are present[34]. Higher levels can be tolerated for shorter daily exposure but levels exceeding 135 dB without ear defenders are not permissible under any circumstances no matter how brief the exposure. The noise levels to which the ground duty personnel and also those working and living within the premises of the Air Force stations are intermittently exposed to for brief periods vary from 88dB 'A' to 122.5 dB 'A' (Table 37.1). The maintenance staff and repair team working at the Tarmac, Hangars and in areas close to the runway are exposed to still higher levels while attending to various jobs during ground run, preparation and repair of the aircrafts. The measurements taken during the present study indicate the noise level ranging from 120.5 to 134.9 as presented in Table 19.2 and still higher (146 dB) that surpasses the threshold of pain.

The helicopter noise during ground operations and hovering pose problems of different nature to that of aircrafts because of the characteristic differences in the nature of their operations. The maintenance staff is frequently subjected to noise levels between 100 to 120 dB 'A' (Table 37.4 and Table 37.5) while attending to check and maintenance jobs. The exposures in such conditions also constitute a potential risk of hearing damages due to the presence of strong tonal components, particularly in the speech frequency range, in helicopter noise.

Table 37.4: Sound Pressure Levels (dB 'A' & dB 'C') of Helicopters During Ground Operations at 10 Meter Distance

Angle	*H/C A SPL (dB 'A')*	*H/C A SPL (dB 'C')*	*H/C B SPL (dB 'A')*	*H/C BSPL (dB 'C')*
0°	108.9 - 111.0	108.4 - 110.0	104.5 - 109.8	106.0 - 109.1
45°	111.6 - 117.7	110.0 - 112.0	114.0 - 117.0	112.0 - 116.0
90°	114.5 - 116.5	113.4 - 116.4	114.6 - 118.0	114.8 - 118.7
135°	108.0 - 113.0	112.9 - 115.6	109.6 - 112.0	109.0 - 112.4
180°		–	100.0 - 101.0	103.0 - 107.0

Table 37.5: Sound Pressure Level of a Helicopter While Hovering at 6 Meters

Angle	*SPL (Lin.)*
0°	109.0 - 112.5
45°	115.9 - 120.0
90°	115.5 - 117.9
135°	115.0 - 117.0
180°	112.7 - 115.8

The one-third octave frequency spectrum of noise emitted by a jet aircraft in flight (Fig. 37.1) contains continuously distributed sound energy across the frequencies from 16 Hz to 16 kHz with uniformly high amplitudes from 200 Hz to over 4 kHz. The figure also shows the broadband linear sound pressure level (dB Lin) on the extreme left which is over 130 dB. The high acoustic energy concentration in the speech frequency range makes the sound radiated by aircrafts in flight increasingly hazardous for hearing system of the individuals. The propulsion at high speed acoustic emission from engines transmission systems and jets along with aerodynamic interaction of the fast moving aircraft with the moving air may be responsible for the characteristic broad band noise of this nature. In modern high-speed jet aircrafts the aerodynamic noise is the principal contributor to uniformly distributed intense noise predominantly in the speech frequencies.

One-third-octave band frequency spectrum of helicopter noise in static mode of operation is presented in Fig. 37.2. Fluctuating amplitudes are seen prominently in the lower and higher frequencies of auditory interest. This characteristic variation in the amplitude has been recorded in the frequency spectrum of almost all the helicopters that were examined for their noise emission. Although the helicopters do not cruise with high speed as the aircrafts, the sound radiated is more annoying because of the characteristic differences in the manner the noise is produced during helicopter flight.

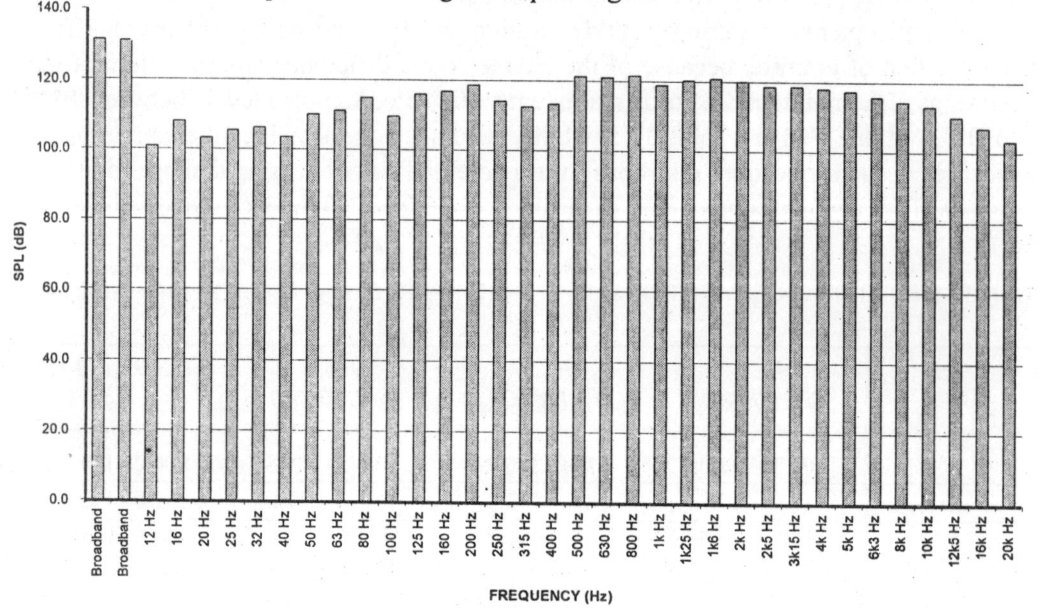

Fig. 37.1: One-third Octave Band Frequency Spectrum-Aircraft 'A'

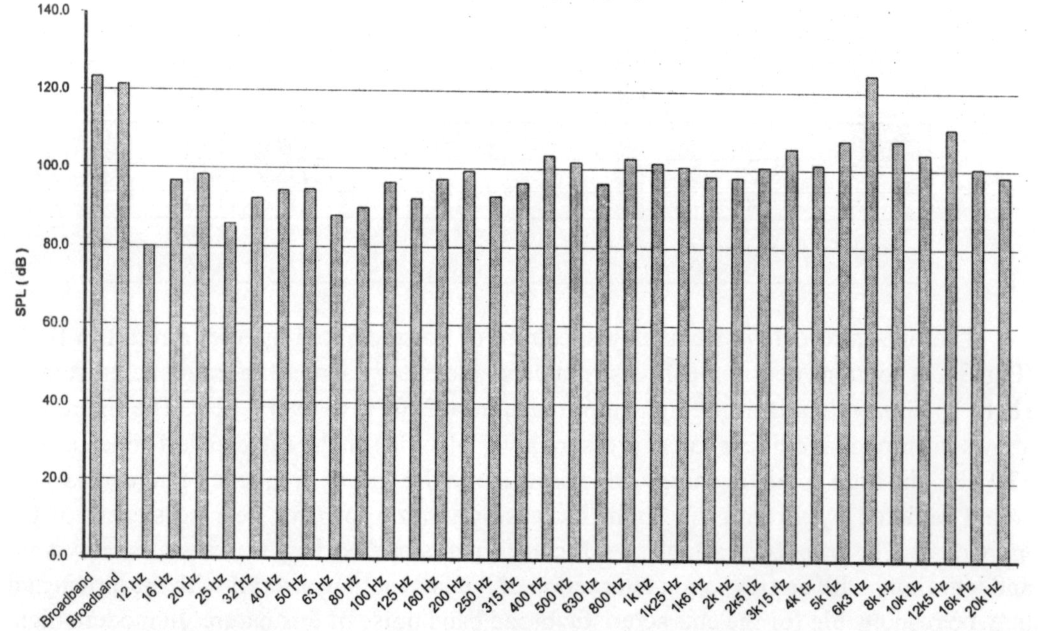

Fig. 37.2: One-third Octave Band Frequency Spectrum-Helicopter 'B' Static Running

The audiometric tests conducted on fliers, ground duty personnel, families and children living in the close vicinity of the runway indicated widespread hearing impairment amongst ground crew and their families compared to the pilots. The audiometric evaluation was carried out using GSI 61 Audiometer. This study was undertaken in a quiet room. Audiometric tests conducted concerned the frequency range of 125 Hz to 12 kHz for both ears. On the basis of the hearing levels, the subjects were categorized as those having normal hearing (Hearing Level up to 25 dB), mild loss of hearing (Hearing Level between 25-40 dB), moderate hearing impairment (Hearing level between 40-60 dB). Hearing levels higher than 60 dB were adjudged as severe hearing impairments.

Table 37.6: Percent Distribution of Hearing Losses in Different Categories of Exposed Population

Category	*Total Nos. (n)*	*Normal*	*Mild*	*Moderate*	*Sever*
Pilots	52	53.9%	38.4%	5.8%	1.9%
Ground Crew	278	29.1%	45.0%	18.0%	7.9%
School Children	59	64.4%	30.5%	1.7%	3.4%
Families	62	9.7%	64.5%	16.1%	9.7%

As seen from Table 37.6, 53.9 per cent of pilots presented normal hearing as against 29.1 per cent seen in ground crew. The incidences of moderate and severe hearing losses in pilots were 5.8 and 1.9 per cent and in ground crew these losses were 18.0 per cent and 7.9 per cent respectively. The differences were statistically significant ($p< 0.001$). The difference between the number of ground duty personnel having normal and impaired hearing was also statistically significant. Severe hearing losses amongst the families and school going children were 9.7 per cent and 3.4 per cent respectively. The differences between those having normal and impaired hearing in these two groups were statistically significant ($p < 0.001$).

The frequency wise distribution of hearing losses among the pilots and the ground duty personnel indicated that the most affected frequencies were 6 and 8 kHz. Hearing losses in the lower frequencies were not seen in pilots but many of the ground duty personnel presented hearing impairments in the lower frequencies below 1 kHz indicative of their prolonged repeated exposure to brief but intense noise.

The level of fatigue, irritability and annoyance were evaluated with the help of questionnaires about which the subjects were explained and about which the subjects gave their frank and free subjective impression. Majority of the ground duty personnel, families and school going children residing and studying in the close vicinity of the runway experienced, during the flying operations, symptoms of irritation and annoyance leading to loss of concentration and fatigue besides interfering with oral communication.

Conclusions

The study has demonstrated that the aerodynamic noise due to high performance aircraft and helicopter flights is fairly high compared to the safe limits for human exposure and it constitutes a potential risk of noise induced hearing loss for ground duty Air Force personnel and others working and living in the close vicinity of the runway at Air Force Stations. Noise encountered by the maintenance personnel during routine maintenance and preparation for flights far exceeds the recommended norms and even surpasses the threshold of pain. Brief but repeated exposure to intense noise has inflicted concernable degree of hearing impairment that is more prominently seen in ground duty and maintenance personnel. Mild tc moderate level of hearing loss was also noticed among children and families of Air Force personnel living in the close proximity of the flight path.

In view of the high levels of noise and resulting hearing losses in Air Force personnel, it is felt that use of sound barriers, hearing protective devices or any kind of safety aspect based on ingestion of prophylactic agent or breathing of Carbogen, a mixture of 5 per cent CO_2 and 95 per cent O_2, may be of benefit for the better health and performance and efficiency of the Air Force personnel.

NOTES

1. Kryter K.D. (1970) : *The effects of noise on man*. Academic Press, New York pp. 139.
2. Kryter K.D. (1985) : *The effects of noise on man* - 2nd Edition. Academic Press Inc. Orlando FL.
3. Burns W. and Robinson D.W. (1970) : *Hearing and noise in industry.* London. Her Majesty's Stationery Office.
4. Jonsson A. and Hansson L. (1977) : *Prolonged exposure to a stressful stimulus (noise) as a cause of raised blood pressure in man.* The Lancet 1 : 86-87.
5. Arguelles A. E.. Martinez M. A., Pucciarelli E., Disito M.V. (1970) : Endocrine and metabolic effects of noise in normal, hypersensitive and psychotic subjects. In Welch B. L., Welch A. S. (eds.). *Psychological effects of noise.* Plenum Press, New York, 43-55.
6. Davis R. C., Berry T. (1964) : Gastrointestinal reactions to response-contingent stimulation. *Psychol. Rep.* 15 : 95 - 113.
7. Jansen G. (1961) : Adverse effects of noise on iron and steel workers. *Stahl Eisen* 81 : 217.
8. Zhao Y., Zhang S., Selvin S., Spear R.C. (1993) : A dose response relationship for occupational noise induced hypertension. *Schriftenr . Ver. Wasser. Boden. Lufthyg.* 88:189 - 207.
9. Wu C.C., Chen S.J. and Yen M.H. (1992) : Effect of noise on blood pressure and vascular reactivities. *Clin. Exp. Pharmecol. Physiol.* 19 (112):833 to 838.
10 Tomei F., Tomao E., Papaleo B., Baccolon T.P., Alf P. (1991) : Study of some cardiovascular parameter after chronic exposure to noise. *Int. J. Cardiol.* 33(3):393-399.
11. Verbeek JHAM, van Dijk FJH and de Vries FF (1987) : Non-auditory effects of noise in industry. IV. A field study on industrial noise and blood pressure. *Int. Arch. Occup. Environ. Health* 59 : 51-54.
12. Anticagalia J.R., Cohen A. (1970) : Extra auditory effects of noise as a health hazard. *Am. Ind. Hyg. Assoc. J.*, 31, 277-281
13. Singh A.P., Rai R.M., Bhatia, M.R. and Nayar H.S. (1982) : Effect of chronic and acute exposure to noise on Physiological functions in man. *Int. Arch. Occup. Environ. Health* 50 : 169-174.
14. Ogle C.M., Lockette M.M. (1966) : The release of neurohypophyseal hormone by sound. *J. Endocrinol.* 36 : 281-290.

15. Bekesy, G von (1960) : *Experiments in Hearing.* New York : McGraw-Hill Co.

16. Hamernik R. P., Henderson D., Salvi R. (1981) : *Potential for interactions of low level impulses and continuous noise.* Air Force Aerospace Research Laboratories Wright Patterson Air Force Base, Ohio. : Report No AFAMRL 80-68.

17. ISO 1981 : Acoustic-Determination of sound pressure levels of noise sources - Engineering methods for free field conditions over a reflecting plane. ISO 3744, CH-1211, Geneva 20, Switzerland.

18. ISO 1984 : Acoustics-Pure tone air conduction threshold audiometry for hearing conservation purposes. ISO/DIS 6189.2. Int. Orgn. for Stand. Ch-1211, Geneva 20, Switzerland.

19. ISO 1999 : Acoustics-Assessment of occupational noise exposure for hearing conservation purposes. ISO 1999-1975 (E), Int. Orgn. Stand.

20. CHABA (1968) : Proposed Damage-Risk Criterion for Impulse Noise (gunfire). Report of Working Group 57, ed. W.D. Ward. *National Academy of Sciences,* National Research Council Committee on Hearing, Bioacoustics & Biomechanics,Washington, D.C.

21. Pfander F., Bongartz H., Brinkmann H., Kietz H. (1980) : Danger of auditory impairment from impulse noise: A comparitive study of the CHABA damage risk criterion and those of the Federal Republic of Germany. *J. Acoust. Soc. Am.* 67 : 628 - 633.

22. Price G.R. (1983) : "Relative hazard of weapons impulses". *J. Acoust. Soc. Am.* 73 : 556 - 566.

23. OSHA (1983) : CPL 2-2.35A-29CFR 1910.95(b)(1), Guidelines for noise enforcement; Appendix A, Washington DC : US Dept. of Labor, Occup. Safety & Health Administration OSHA Directive No. CPL 2-2.35A (Dec 19,1983)

24. EPA (1974) : Information on levels of environmental noise requisite to protect public health and welfare with an adequate margin of safety. Washington, DC : US Environ Protection Agency. EPA Report 550/9-74-004.

25. Forrest M.R., Coles R.R.A. (1970) : Problems of communication and ear protection in the royal marines. *J.R. Nav. Med. Serv.* 56 : 162

26. Murthy V.S.N., Pranesh Rao B.N. (1973) : Testing of ear defenders. Institute of Aviation Medicine, Bangalore, Report No. 22.

27. N. Kapoor, R.C. Chaturvedi, A.P. Singh, R.K. Sharma, K.K. Srivastava, W. Selvamurthy 2001: A Critical Analysis of The Noise Attenuation And Speech Intelligibility Performance of Hearing Protective Devices. *J. Acous. Soc. Ind.* 29 (1 - 4) : 364 - 374

28. Chaturvedi R. C., Sharma R. K., Lakhers S. C., Tiwari R. S. , Rai R. M. 1984 : Role of Carbogen in protection against noise induced hearing loss in Man. *Indian J. Med. Res.* 80 : 583.

29. Chaturvedi R.C., Rai R.M., and Sharma R.K. (1988) : Prophylactic action of carbogen against noise induced hearing loss. Ind. J. Med. Res. 88 : 60 - 63.

30. Chaturvedi R.C., Rai R.M., and Sharma R.K. (1990) : Therapeutic role of carbogen in impaired hearing. *Ind. J. Med. Res.* 92 : 420 - 423.

31. Guignard J.C., King P.F. (1971) : North Atlantic Treaty Organization Advisory Group for Aerospace Research and Development : AGARDograph No. 151. Aeromedical Aspects of Vibration and Noise : 115-257.

32. Bohne B.A., Zahn S.J. and Bozzay D.G. (1985) : Damage to the Cochlea following interrupted exposure to low frequency noise. *Ann. Otol., Rhinol. & Laryngol* 94 : 122-128

33. Clark W.W., Bohen B.A. and Boettcher F.A. (1987) : Effect of periodic rest on hearing loss and cochlear damage following exposure to noise. *J. Acous. Soc. Am.* 82 (4) : 1253-1264.

34. Acton W.I. 1967 : A review of Hearing Damage Risk Criteria. *Ann. Occup. Hyg.*, 10, 143-153.

EFFECTIVE AIR QUALITY MANAGEMENT IN MEGA CITIES, INDIA

Suresh Jain* and Prapti Mittal**

ABSTRACT

Air pollution in mega cities and industrial centres is one of the major environmental problems in India, which results in health impacts that affect urban residents, especially the poor. The movements of people into urban areas together with the increase in consumption patterns and unplanned urban and industrial development have led to the problem of air pollution. By doing economic valuation, it is possible to draw the attention of peoples, policy-makers and, of course, researchers to the extent of damage and the value of it. Further, if the urbanization and rapid industrialization foreseen in the next decade materialize, the technology and environment nexus will only assume greater importance. Thus it is now very clear that enhancing the capability of key players to improve the air quality of the megacities of India must take place in the context of an integrated and comprehensive policy framework.

Key Words: Air quality; pollution; management; mega cities, policy.

Introduction

Air quality management (AQM) is the monitoring and control of air emissions to eliminate or limit the impact on the environment. Air pollution consistently ranks as one of the main environmental concerns for the general community and it is also one of the primary causes of adverse health effects on human beings (Hall, 1996; WHO, 2000; Elsom and Longhurst, 2004; Marshall *et al*., 2005). In India, estimated annual health damage of pre-Euro vehicle emissions for 25 Indian cities were estimated between US $14 million and US $191.6 million per city (Mashelkar Committee, 2002). In mega cities, such as Bombay, Calcutta, Delhi, Dhaka, and Karachi in South Asia, and Bangkok, Beijing, Shanghai, Jakarta and Manila in

*Research Scholar, Department of Civil Engineering, Indian Institute of Technology Delhi, Hauz Khas, New Delhi-110016, E-mail: sureshjain_in@yahoo.com

**Student, Department of Civil Engineering, Punjab Engineering College, Chandigarh, E-mail: praptimittal@gmail.com

East Asia, the pollution levels often exceed the WHO air quality guidelines by a factor of 3 or 4 (Faiz and Sturm, 2000; World Resources Institute 1992, 1998). It is shown that the pollution in Mumbai can lead to high health costs. In general, the average cost amounts 0.26 per cent of income due to highly subsidized treatment and poverty. However, 5 per cent of patients who suffer severe attacks may pay as much as 19 per cent of their income (Parikh and Hadker, 2003). Further, if the urbanization and rapid industrialization foreseen in the next decade materialize, the technology and environment nexus will only assume greater importance. As such, all industries need to effectively manage their air discharges to protect the quality of their local environment, maintain good relations with neighbouring land users and to ensure compliance with air quality legislation and license limits (Weatherleya and Timmisb, 2001). Thus it is now very clear that enhancing the capability of key players to improve the air quality of the megacities of India must take place in the context of an integrated and comprehensive policy framework. Implementation of the policies, strategies and plans are by way of a variety of instruments like, regulatory, voluntary and involves public awareness and action. Only such a framework will make it be possible to achieve the desired environmental and related outcomes related to effective AQM system.

Air Pollution in Megacities

The severity of the air pollution problems reflects the level and speed of development in megacities. As cities undergo economic and industrial development, air pollution becomes an increasing problem. In the past, the major causes of environmental degradation occurred sequentially rather than simultaneously. Presently megacities are suffering the pressure of a combination of different driving forces (e.g. motorization, industrialization and an increase in urban population density), each with a greater intensity than has occurred elsewhere or in the past and without a sufficiently well developed institutional capacity or the financial resources to control them. As a consequence, the ability of many cities in the region to cope with the combined pressures is often exceeded thereby leading to deterioration of environmental quality.

Improving Urban Air Quality

Although several attempts have been made at regional cooperation to address air pollution control in India, much progress needs to be made to deal with the issue of urban air quality in megacities like Delhi, India. Delhi has worsened due to the cumulative effects of population growth, industrialization and increased vehicular use. Therefore, need a well coordinate, integrated approach to address the fundamental problem of urban air pollution and provide the basis for future regional cooperation. AQM can be improved by different management practices (Fig. 22.1).

Environmental Policy Framework

(i) Environmental policy should be supported by all responsible ministries in the government, can lead to sound and rational air quality management. The objective of air quality management should be sustainable development.

(ii) For efficient policy framework, areas like land-use planning, energy, transport and industrial development will reduce the adverse impacts of air pollution on human health and the environment.

(iii) Increased commitment to AQM and its enforcement from all stakeholders (industry, local authorities, non-governmental organizations, international agencies, media and the general public), strengthening the legal basis of AQM in national laws and regulations.

(iv) An integrated AQM process can inform, educate and train all stakeholders and strengthen stakeholder participation in all aspects related to air quality, e.g. adverse health and environmental impacts, prevention and reduction of air pollution.

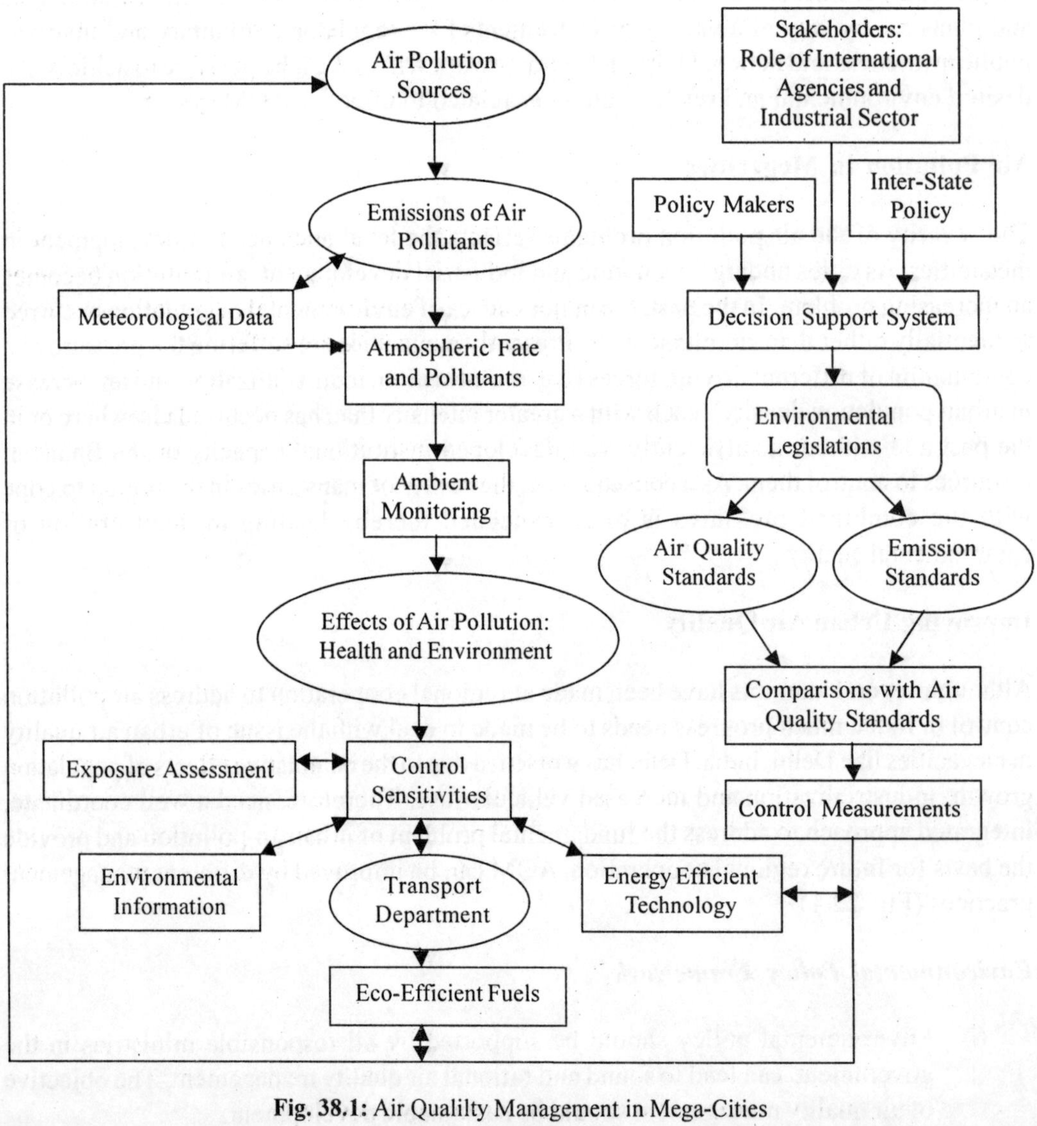

Fig. 38.1: Air Qualilty Management in Mega-Cities

Emissions Control Strategies

AQM can be improved by effective measures to assess and reduce emissions using different control strategies (Fig. 38.2):

(i) Alternative fuels and energy efficient technologies are less expensive as compare to end of pipe reduction of pollution.
(ii) By adopting long-term strategies for emission prevention and reductions instead of short-term strategies.
(iii) Develop low-cost and effective alternative technologies.
(iv) Emissions from mobile sources can only be reduced through a combination of measures like, cleaner vehicle technology, tighter emission standards, improved inspection programmes, public transport system (e.g. Metro Train).

Emission Inventory System

Prior to developing any options for emission reductions or mitigation there is a need to establish an integrated emission inventory system. This system should account not just for the 'base year' but also for future year stationary and mobile source emissions that can be used for forecasting ambient air quality through the exercise of appropriate air quality models. Emission inventories for each source of emissions and each pollutant of interest must be collected for the 'base year', and predicted for the subsequent years up to the target year when air quality targets will have to be met (Fig. 38.2). Predictions of emissions will be linked to various socio-economic scenarios that are used for simulating levels of economic and social activity.

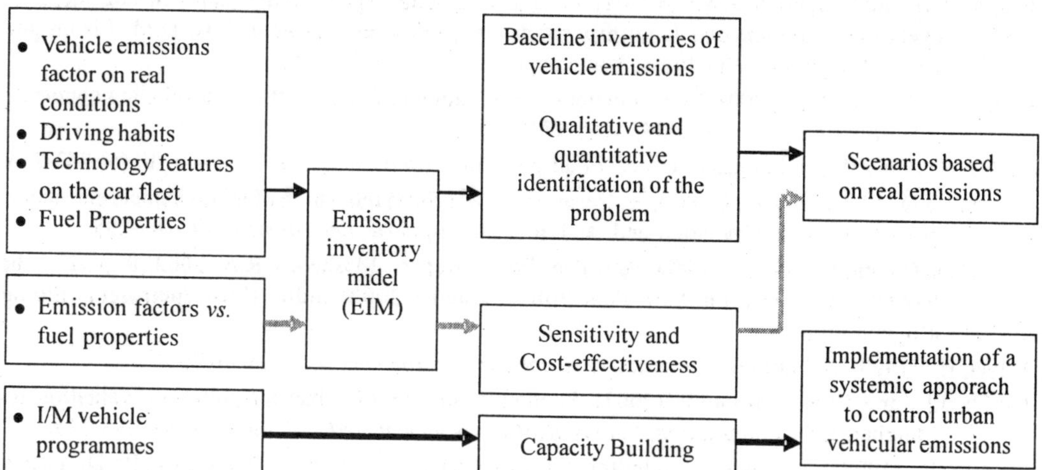

Fig. 38.2: Determining emissions control strategies—a schematic example based on the general approach taken by the ARPEL/Clean Air Initiative project in Latin America (*Source*: APPEL)

Environmental Efficient Technology

The technologies that support implementation of AQM plans should be an environmentally sound, efficient and sustainable. Environmental Efficient Technologies protect the environment, are less polluting, use all resources in a more sustainable manner, recycle more of their wastes and products, and handle residual wastes in a more acceptable manner than the technologies for which they are substitutes. Moreover, environmentally sound technologies should be compatible with nationally determined socio-economic, cultural and environmental priorities.

Conclusions

Effective air quality management programmes are those that seek to integrate the different contributors into a working partnership—a partnership that is involved and committed to the definition of air quality goals, and to the elaboration and implementation of the programmes through locally acceptable solutions aimed at meeting the predefined air quality goals. All of the partners are individually committed, from the outset, to implementing the least-cost, prioritized recommendations that the assessment will show to be necessary to achieve future air quality goals. Without those prior commitments the value of this structured approach to AQM will not be realized.

REFERENCES

ARPEL, (2001). Systemic Approach to Vehicular Emission Control in Latin America and the Caribbean, ARPEL, Montevideo, Uruguay, May 2001. Available on line from the ARPEL library at: http://domino.arpel.org/aplicaciones/biblioteca/ML_lib.nsf

Elsom, D.M. and Loughurst, J.W.S., (2004). Local and regional aspects of air quality management: an overview, *Regional and local aspects of air quality management,* Eds. D.M. Elsom and J.S.M. Loughurst, WIT Press: Southampton, UK, pp. 1-13.

Faiz, A. and Sturm, P.J., (2000). New directions: air pollution and road traffic in developing countries. *Atmospheric Environment* 34, 4745-4746.

Hall, J.V. (1996). Assessing health effects of air pollution, *Atmospheric Environment,* 30 (5) 743-746.

Marshall, J.D., McKone, T.E., Deakin, E., Nazaroff, W.W. (2005) Inhalation of motor vehicle emissions: effects of urban population and land area. *Atmospheric Environment* 39, 283-295.

Mashelkar Committee, (2002). India Auto Fuel Policy Report, Mashelkar R.A. 2002. Report of the Expert Committee on Auto Fuel Policy, August 2002. India. (http://petroleum.nic.in/afp_con.htm.

Mayer, H., (1999). Air pollution in cities. *Atmospheric Environment* 33, 4029-4037.

Parikh, J.K. and Hadker, Nandini, (2003). Economic impacts of urban air pollution: valuation for Mumbai, India. *International Journal of Environment and Pollution* 19 (5), 498-515.

Sharma, P., (1998). Air quality modelling for an urban intersection of Delhi City. Ph.D. Thesis, Department of Civil Engineering, Indian Institute of Technology, New Delhi.

Sharma, P., and Khare, M., (2001). Modelling of vehicular exhaust—a review: *Transportation Research* D 6, 179-198.

Weatherleya, N.S. and Timmisb, R.J., (2001). The atmosphere in England and Wales: an environmental management review. *Atmospheric Environment* 35 (2001) 5567-5580

WHO (2000). *Guidelines for Air Quality.* WHO/SDE/OEH/00.02. World Health Organization, Geneva. Website: http:/www.who.int/peh/

World Resources Institute WRI, (1992). *Urban Air Pollution in Mega Cities of the World.* Blackwell reference, Oxford, UK, 39pp. In collaboration with World Health Organization and United Nations Environment Programme.

World Resources Institute WRI, (1998). *A Guide to the Global Environment.* Oxford University Press, New York. In collaboration with United Nations Environment Programme (UNEP), United Nations Development Programme (UNDP) and World Bank (IBRD), World Resources 1998-99.

39

NOISE POLLUTION
Sources and Effects

R.K. Gupta* and Subhash Chander**

ABSTRACT

Noise is a necessary evil because no one on earth can escape the unwanted, disturbing sound that causes a nuisance. It is on such a rapid and alarming increase especially in urban areas that it has become a major threat to the quality of human life. Noise has a number of adverse effects on the human environment. We seem to have accepted noise and the accompanying physiological and psychological deterioration. We have set standards but unfortunately not able to monitor them. We sometimes accept and enjoy some sounds from loud music to the sound of racing motorcycles even after knowing fully well that it can damage hearing. It is the high time to adopt the measures of abatement of noise for our ownselves and also for the generations to come. This paper is aimed at presenting causes and effects of noise pollution.

Webster's New World Dictionary defines noise as any loud, discordant or disagreeable sound. Nobody is inescapable from noise. Some of the noise we are confronted with is either ignored or else we get used to it. It is, however, becoming more and more difficult to do so as our convenience oriented society races around in singly occupied cars with our stereos turned on. On coming back home, we turn to our stereo systems and television sets to seek solace. Not only this even while we burn off our calories by walking or jogging, our heads are wedged between the head phones whose thumping bass help us to keep in stride. All these actions on our parts show that we don't seem to mind it to some extent.

Sources of Noise Pollution

The commonly faced sources of noise pollution are:

1. Street Traffic
2. Aircrafts

*ME (Public Health Engg.), MIE, Honorary Secretary, The Institution of Engineers (India), Delhi State Centre, Delhi

**MIE, Council Member, The Institution of Engineers (India)

3. Railways
4. Industry
5. Construction
6. Consumer Products
7. Other Sources

Street Traffic

Of all the known sources of noise pollution, the most prevalent and most damaging is the street traffic. Experiments have shown that more people are exposed to noise from motor vehicles than any other source. Noise emitted by the street traffic is generated by engines, horns and the tyres interacting with the road surface. The noise produced during acceleration can be as much as 20 dBA greater than that produced at cruising speed.

The noise produced by heavy truck or bus might sometimes be equivalent to that from 10 to 15 private cars together, the simple reason for this being the use of diesel engines.

In this category the motorcycles are the worst offenders not only for others but for the users themselves because the rider is not protected by an enclosed compartment. The motorcycles can generally produce noise levels in the range of 72-85 dBA but can reach as high as 120 dBA immediately behind the cycle as against the noise level range of 65-75 dBA in case of a car.

In addition to above there are other factors which influence the amount of noise produced. These factors are

(i) *Traffic parameters* - Speed, density, fluidity and driver behaviour.
(ii) *Road Parameters* - Road surface conditions, gradient, width of road and propagation of traffic.
(iii) *Environmental parameters* - distance and height from the road the receiver is located, presence or absence of natural or artificial screens and the amount of noise absorption due to the condition of ground between road and receiver.
(iv) *Weather parameters* - precipitation conditions, wind direction.
(v) *Design of buildings* - insulations.

Aircrafts

The increased amount of air traffic has resulted in an increased number of people who are affected by aircraft noise, both as passengers and as members of communities located near airports. The noise derives from the propulsion systems and from aerodynamic noise. Under normal cruising conditions, the aerodynamic noise predominates, especially among higher frequencies. However, propulsion noise tends to predominate at the lower takeoff and landing speeds.

In subsonic flight, aircraft noise peaks when the aircraft is approximately overhead and then gradually diminishes to the ambient noise level. Noise that is heard as the aircraft approaches tends to be dominated by higher frequency sound, whereas the one that is heard after the aircraft has passed tends to include more lower-frequency sound. In helicopters, noise includes both periodic noise produced by the "blade slaps" and a non-periodic broadband noise resulting from air interacting aerodynamically with the rotor.

Railways

The majority of noise emitted by trains is produced by the engines and by the interaction of wheels with the track. The whistles and horns, hawkers, porters and passengers and equipment used in construction and maintenance add to the woes.

The railways noise is though confined to areas near tracks, it still represents a significant danger to human health. However, it is much less pervasive than street traffic. Noise levels experienced by passengers in trains generally increase by about 10 dBA in tunnels.

Industry

The following industrial activities mainly result in noise generation occurring at the lower frequency:

(i) *Product fabrication :* The industrial noise affects both the people around it and within it, the latter category generally bear the brunt. Product fabrication is a highly noisy operation especially in metal fabrication because of cutting, shearing, pressing and riveting. It may be interesting to note that riveting a large steel structure can produce noise levels greater than 130 dBA. Moulding can also be highly noisy with the use of high pressure air in the operation, pneumatic control and cooling of machinery (100 dBA).

(ii) *Product assembly :* Product assembly also produces dangerous noise levels. Such activities produce broadband noise including high frequency noise of high levels due to operation of electric and pneumatic tools such as grinders etc.

(iii) *Power generation :* The power generation emits noise produced by turbine generation and compressors and also from blowers and fans.

(iv) *Processing :* The processing actively involves emission of noise due to furnaces, heat exchangers, pumps and compressors etc.

Construction

This a major source of noise pollution. The noise is emitted by construction equipment utilized at construction sites for the fabrication, erection, modification, demolition, removal of any structure or facility including all related activities such as clearing of land, site preparations, excavation etc. The construction equipments tend to produce more noise in the lower end of the frequency spectrum. Construction equipment is since used outdoors, affects many other people besides the site workers.

The earthmoving equipment, materials handling equipment, stationary equipment and impact equipment are some of the categories of construction equipment. In addition vibrators and saws are also used invariably at site. The noise levels emitted by some of the construction equipments at a distance of 15m are as under :

Sl. No.	*Category of Equipment*	*Noise level range dBA*
1.	Earthmoving, e.g. Compactors (Rollers), tractors, front loaders, back hoes, powers, graders, trucks etc.	70-95
2.	Material Handling, e.g. Concrete mixers, concrete pumps, Cranes, etc.	75-95
3.	Stationary, e.g. Pumps, Generators and Compressors	65-85
4.	Impact Equipment, e.g. Jack hammers, Rock drills, Pile drivers	80-105
5.	Other equipments, e.g. Vibrator and saws	65-85

Consumer Products

They represent a wide range of noise-producing gadgets. The noise levels emitted by some of the consumer products are as under:

Category	*Items*	*Noise Level Range dBA*
Recreational	Guns, model airplanes, motorcycles, go carts, all-terrain vehicles, video arcades etc.	80-120
Hobbies/Workshop	Lawn mowers	80-115
Household	Food blenders, vacuum cleaners, washers, dryers, air conditioners and air coolers, refrigerators	45-85
Music	Personal stereos, rock concerts, home stereos	80-120

Other Sources

In addition to the above sources of noise pollution, there can be several other sources, as well, such as sirens, military and noise generated by humans. Since the sirens are intended to convey urgent messages, these have to be loud enough to attract attention. Sirens are used by the police vehicles, ambulances and fire tenders. Military activities can generate some of the loudest man-made noise. Military noise sources included explosions, airplanes and rockets, engines and other equipments.

A final sources of noise to consider is humans. While activities such as commentaries on cricket matches and the excitements during the matches, voices, dropping of objects, children jumping can be irritating especially in poorly insulated houses. Our pets can be a significant source of the damaging effects of noise. Above all the music played by the bands at the marriage processions and parties and loud speakers engaged at the Religious functions, e.g. *Jagrans* are also the human created sources of noise.

Effects of Noise Pollution

Noise is more than just a pain in the ear. Various studies show that noise pollution has many effects on humans, e.g. hearing loss, non-auditory physiological effects, sleep disruption, annoyance, communication interference.

Noise-Induced Hearing Loss

Exposure to sufficiently intense noise for a long enough duration results in damage to the inner ear and thus decreases one's ability to hear. In addition, the quality and clarity of auditory perception also can be affected. While these effects are generally temporary, there may be some residual permanent damage to persist for the sufferer's remainder life.

In the noise-induced hearing loss, the organ of corti in the inner ear is worst affected especially when the noise persists over a significant period of time. The sound induced motion of the fluid in the cochlea induces shearing and bending movements of the hair cells in the organ of corti, which in turn, result in electrical stimuli transmitted by the auditory nerve. Excessive noise exposure over a significant period produces deterioration and finally, destruction of hair cells and thus disrupts the mechanism of sound transmission.

The effect of noise exposure on hearing can be divided into three categories:

(i) Acoustic trauma
(ii) Noise-induced temporary threshold shift (NITTS)
(iii) Noise-induced permanent threshold shift (NIPTS)

Acoustic trauma is defined as immediate organic damage to the ear from excessive sound energy. It results from a single or relatively few exposures. If the noise is intense enough, other structures outside the inner ear such as ear drum may also be affected. Such damage is, however, rare but causes some degree of permanent damage to the auditory system.

NITTS and *NIPTS* involve a decrease in a person's auditory sensitivity. The difference between these two affects, as the name suggests is that the threshold eventually returns to its original level, i.e. temporary in *NITTS* whereas in *NIPTS* the threshold shift is permanent. NIPTS results from long periods of repeated exposure to noise. This exposure is often the result of one's occupation.

Non-Auditory Physiological Effects

Non-auditory physiological effects include cardiovascular, autonomic and gastric effects. In an experiment carried out on rhesus monkeys, it showed that their exposure to elevated noise levels that did not decrease their auditory sensitivity, raised their blood pressure. An increase in antihypertensive medication consumption in a village near an airport was correlated with an increase in aircraft traffic. Loud noise is purported to cause vaso-constriction with a consequent increase in blood presure. This leads in turn to smooth muscle hypertrophy, narrower lumen in small vessels and increased resistance to blood flow. The end result is hypertension.

Autonomic Effects

Autonomic effects have also been implicated in addition to Cardio Vascular effects. Urinary adrenaline and non-adrenaline levels were higher in subjects who were exposed to short duration of noise.

Gastric Effects

A study has shown that exposure to 80 dBA noise levels resulted in a reduction in stomach contraction strength. Since the gastric changes are related to ulcers, noise may be related to ulcer development.

Effects on Sleep

Noise disturbs sleep. Increased level of noise are associated with an increased use of sedatives by people in a village as per a study conducted. The people are awakened by 70 dBA truck noises (Considered a modest intensity). Increased levels of noise reassociated with sleep disturbances, including shorter sleep duration, more frequent awakenings, downward shifts in sleep stages and increased sleep latency (i.e. difficulty in falling asleep).

Several factors such as type of noise, repetition, duration, intensity, stage of sleep at which the disturbance occurs and state of health of individual affect sleep disturbance. It also varies from one individual to another. Some people feel disturbed by levels as low as 35 dBA, others being able to sleep through 90 dBA levels. It requires greater intensity stimuli to awaken people in the deeper stages of sleep and REM sleep appears to be particularly easy to disturb. Lower intensities of noise are generally required to awaken people as they grow older. Certain disorders, e.g. depression are known to affect sleep behaviour.

Annoyance

It is one of the most salient effects of noise on human beings. A noise is said to be annoying if an exposed individual or a group of individuals would reduce the noise, avoid or leave the noisy area, if possible. It is a response to noise rather than a dimension of auditory experience. Annoyance due to noise depends on many factors including several parameters of the noise itself, e.g. louder noises are more annoying than quieter noises, though two sounds with equal intensity (loudness) may result in different levels of annoyance. Patterned sounds appear to be less annoying than those which are randomly produced.

Noise that are higher than pitch are more annoying than lower frequency noise. Annoyance is also dependent on regularity of the noise. The noises that remain constant in pitch and intensity are generally rated as less annoying than those that change in pitch and intensity. The noise of a given intensity is more annoying during the night than during the day. Similarly, noise appear to be more annoying in the summer than in the winter.

Continued exposure to noise does not appear to decrease annoyance. Rather in some cases noise increases annoyance.

Communication Interference

Noise pollution has a considerable effect on the communication. There is no doubt that noise masks speech. The background noise results in greater pains on the part of the talker as well as the listener than otherwise would be needed. Noise that has a similar frequency to speech will mask it better than noise at other frequencies, especially higher ones. Noise-induced speech interference is affected by many other factors. Speech content is very important, since a person sharing personal information is less likely to raise his voice to compensate for background noise. Culture is another factor, which influences the noise-induced speech interference. People in cultures that emphasize personal space are more likely to encounter communication difficulties in noisy situations.

Age of the individual is another factor influencing communication interference. Children have poorer articulation skills than adults. Noisy conditions are more likely to interfere with the speech of the children than with that of adults. The ability to understand partially masked or distorted speech appears to begin deteriorating at around the age of 30. The older the listener, the lower the background noise must be for practical or satisfactory communication.

Noises produced in areas containing highly reverberant materials become less localized, resulting in greater communication interference. Forms of non-verbal communication such as lip-reading or bodily gestures are often utilized to compensate for noisy environments.

Noise can be very hazardous, in that it can preclude the conveyance of vital life-saving information. Everyday conversation disruptions can lead to increased annoyance and anxiety and may indirectly contribute to physiological complications such as the non-auditory physiological effects.

Other Effects

Other than the effects discussed above, some other effects involve cognitive performance, occupational performance, psychological functioning, and effects on pro-social behaviour and effects on aggression. Recall is detrimentally affected by background noise to some extent. Vigilance and attention are also impaired by noise. Even if a person performs well on a cognitive task, there may be other costs, including a reduced psychological or physiological capacity to react to additional demands and increased fatigue after completion of the task.

Occupational performance is also hindered by noise. Noise impairs occupational performance mainly when it is unexpected and unfamiliar to the worker. In situations involving noise levels over 90 dBA there are increases in errors and in failures to notice unexpected events in tasks requiring continuous attention.

Noise can have other psychological effects also. There are many cases of noise apparently driving a person to insanity. Other psychological effects of noise are increased irritability and problems of adjustment to hearing loss. Individuals who suffer from less than severe hearing loss have less severe psychological ramification than individuals with severe hearing loss.

Noise can cause property and environmental damage, especially when sonic booms are involved. The people who are located near airports not only complain about sleep disruption but also report damage to their homes.

Conclusions

Noise is produced by almost everything we humans do. It makes sense, then, that we would get used to it. But unfortunately this attitude also leads us to accept excessive and damaging amounts of noise. Exposure to noise not louder than people shouting for eight hours a day, five days a week can lead to substantial hearing loss in a matter of a few years, yet we continue to work and allow others to work amidst heavy machinery, impact tools, street and railway traffic. And we live along approach paths to airports. We accept hearing loss as a part of the normal aging process, even though in many cases it doesn't have to be.

Many of the problems discussed here are the result of ignorance. Ignorance may be bliss, but it is a lonely, frustrating bliss when you can't hear a word spoken to you unless it is shouted. Educating oneself may not guarantee a lifetime of good hearing, but it certainly gives a person better odds. And the people around them too. A parent educated in safe auditory practices has kids who will grow up in a safe environment and who will learn safe auditory practices themselves. And may be with enough educated parents, teachers, co-workers, and policy-makers our children and grandchildren won't have to grow up in an annoying world where it is difficult to communicate, a world where noise-induced hearing loss is accepted as part of the normal aging process.

REFERENCES

Bragdon, C.R., *Noise Pollution, the unquiet crisis.*

Broadbent and *Karl Kryter, Effects of Noise on man.*

Donand Broadbent, *Effects of Noise on Behaviour.*

Gupta, R.K., Dissertation of Noise Pollution Monitoring and its effects on hearing organs.

Michal Paul, *Noise in the News.*

Michael Rodda, *Noise in Society.*

Noise Pollution (Regulation and Control) Rules, 2000.

Webster's New World Dictionary.

40

ROLE OF CERTIFICATION IN ENHANCING ENVIRONMENTAL AND WORKPLACE SAFETY

R.K. Singh*

ABSTRACT

This paper seeks to examine the role of certification in enhancing environmental and workplace safety, particularly in Indian context. Taking stock of framework of Bureau of Indian Standard (BIS) for such certification including organizational structure, reference certifiable standard and competent certification personnel, it probes how BIS plays a major role in guiding industry and other stakeholders in the field of EMS/OHSMS by way of laying down national standards on Code of Practices, guidelines, method of test and specifications.

Introduction

India is one of the richest countries in terms of natural health, housing 10 bio-geographical zones and 25 biotic provinces incorporating an exceptional range of biodiversity; in fact, it is one of the 12 countries of the world with a mega bio-diversity. In this scenario, India's development priority envisages doubling the per capita income by 2012, reducing the poverty level by 10 per cent, providing gainful employment and ensuring food, energy and economic security for the country. This pace of development requires 8 per cent GDP growth per annum.

In order to achieve this priority, a lot of efforts in diverse sectors require to be taken. However, unplanned and indiscriminate use of resources may play havoc on environment and health of the mankind. The results of such usage are not far to seek; global warming and rise in surface air temperature, change in pattern of rainfall, climate changes causing intensities of extreme heat and cold and of floods and draughts, changes in infectious disease occurrences, vector borne diseases, aeroallergens, etc., are few to name. If we look into the pollution caused by Volatile Organic Compounds (VOCs) contributed mostly by transport sector and solvent industry, it will be clear that a lot of work still remains to be done in this area. VOCs (including Non-methane VOCs) are responsible for the formation of photo-oxidants, such as ground level ozone, which in turn, cause depletion of stratospheric ozone

*Director, Management Systems Certification Deptt., Bureau of Indian Standards, New Delhi

as well as enhance greenhouse gas effect. If we relate these to the following statements made by Jukka Takala, Director of ILO's Safety Work Programme during the opening session of 17th World Congress on Occupational Safety and Health (held during 18-22 September 2005), the issue becomes critical:

> 'The sad truth is that in some parts of the world, many workers will probably die for lack of an adequate safety culture. This is a heavy price to pay for uncontrolled development. We must act swiftly to recover this trend.'

According to the latest report of ILO, each year about 2.2 million people die of work related accidents and diseases. It further states that coverage of occupational health and safety in many developing countries, is poor and in some cases it is deteriorating. The situation becomes further complex when many cases of incidents and accidents are under reported, particularly in developing countries, including India.

Therefore, it is clear that though developmental activities has to take place but not at the cost of everything; development has to be made in a safe and sustainable way.

Certification—a BIS Perspective

Certification, especially management systems certification in the field of environment and occupational health and safety, provides first layer of protection to stakeholders. The second layer of protection could be provided through monitoring mechanism such as sampling and testing, surveillance, internal audits, management reviews etc. It is a fact that without certification also protection to environment and health could be achieved by implementing an appropriate management system. However, when it comes to demonstration and evidence for compliance of such a system, certification is the only answer. The framework for such certification includes an organizational structure, a reference certifiable standard and competent certification personnel. Bureau of Indian Standards (BIS) has all the three components in place.

BIS, as national standards body of India, is operating environmental management systems (EMS) and occupational health and safety management systems (OHSMS) certification under the Bureau of Indian Standards Act, 1986 and Rules framed thereunder. To provide prompt and better service to its customers, BIS has de-centralized its certification activities to our five regional offices located in Delhi, Chennai, Mumbai, Chandigarh and Kolkata. Each regional office, headed by a Deputy Director General, deal such cases starting from receipt of application up to grant of certification. The regional distribution of EMS/OHSMS certification (by BIS) is as follows:

REGION	*EMS*	*OHSMS*
Central	30	3
Eastern	12	3
Northern	15	4
Southern	25	—
Western	21	4
Total	103	14

BIS has played a major role in guiding industry and other stakeholders, in the field of EMS/OHSMS, by way of laying down national standards on Code of Practices, Guidelines, Method of Tests, Specifications, etc. Consequent to publication of an International Standard on Environmental Management Systems by ISO/TC 207 (in 1996), of which India (BIS) is a participating member since its inception. BIS has adopted this standard as IS/ISO 14001 and has started certification from 1997, after effecting necessary amendments to its Regulations. On revision of this standard in 2004, BIS has also adopted the revised standard and has started certification/renewals as per revised version.

In the field of occupational health and safety, the Industrial Safety Sectional Committee (CHD 25) of BIS took a lead in formulating a certifiable management systems standard on occupational health and safety to meet the growing demand from industry and related sectors. The efforts of the technical committee resulted into publication of an Indian Standard 'Occupational Health and Safety Management Systems—Specification with Guidance for Use (IS 18001:2000)'. This standard intends to assist the organizations to develop a systematic approach to management of occupational health and safety issues in such a way as to protect their employees and others whose health and safety may be affected by the organizations' activities. It is to be noted that this is the first certifiable standard on occupational health and safety management systems developed by a national standards body. Another specification which is equally popular and available in the market, is OHSAS 18001 (Occupational health and safety management systems—specification) developed by a consortium of certification bodies operating in occupational health and safety sector. However, it is to be noted that this specification is neither a national standard nor a international standard.

It is a fact that success of any certification scheme of an organization depends upon the competency of its personnel (auditors). In this regard, BIS has abundant reservoir of qualified, trained and competent personnel in all major scope sectors; regional distribution of such personnel is shown below:

Stream	Team Leaders	Team Leaders-in-training	Auditors	Auditors in-training
EMS	9	2	14	27
OHSMS	6	—	5	13

There is no other Certification Body in India having such auditor resource and/or network of offices to provide service anywhere in India at minimum cost.

Control of Environmental Risks and Occupational Hazards

Some time ago there was a news item in a national daily highlighting a peculiar situation, industrial units emitting acidic or other corrosive fumes were eating away the railways' signal equipment, electrified overhead wires and even the track; so corrosive and concentrated are the emissions that overhead wires used to power electric trains, have begun to snap. Apart from this, many illegal hazardous waste sites still exists throughout the country which pose a serious threat to the environment and public health, particularly those living near such sites. The most internationally published case was that of Love Canal, situated in a suburb of Niagra Falls in New York State, which ultimately caused it to be declared as local emergency by the President of United States. We do not know how many such Love Canals are hidden in India. Living in such hazardous situation, amid filth and garbage, among all chaos, day and night, has now become an art in itself.

Many methodologies exists and are available for controlling risks to environment and occupation related hazards; however, to reduce occupational and safety hazards, following principles has to be kept in mind:

— Keep person away from hazards,
— Guard the hazard and keep person free,
— Guard the person, when hazard cannot be guarded, and
— Guard the person and guard the hazard.

In present day situation, when both public and governments are aware about these issues, it would be prudent to deploy adequate resources and pay attention to total safety and environmental management practices and particularly to more systematic, periodic and structured audits. EMS standards (like IS/ISO 14001:2004) and OHSMS standards (like IS 18001:2000) provide such a management tool to have an assurance about its environmental performance and safety.

Auditor's Perspective

An EMS/OHSMS auditor requires to look into those activities which have potential to cause environmental risk and occupational hazard. A safe and environmentally sound culture, followed by robust operational control procedures and work instructions go a long way in strengthening the system. As a thumb rule, existence of an efficient storm water drain in a process industry, is a good benchmark for EMS. It is felt that if organizations follow strictly and honestly all applicable Acts, Rules, Regulations and Notifications, the system would automatically be effective. However, practical experience show that apart from these sometimes non-availability of following programmes causes hurdle in smooth certification process.

Non-consideration of Safety and Environmental Aspects at Design Level

Consideration and incorporation of these at the initial design stage makes the system robust. This is particularly true for multi-functional organizations like hospitals, power plants, etc .having different utilities and cross frontier impacts/hazards. General safety deficient features may include:

— unprotected floor openings
— sleeping/defective floors
— stairway surfaces
— cross-overs over pipings
— platforms/handrails

— inadequate illumination
— ventilation
— walk ways in tank farm area

Lack of Operational Control Procedures for Significant Activities

Organizations are required to identify those activities associated with significant aspects/hazards and make such plans, the absence of which could lead to deviations from the policy and objectives. It is vital to remember that before finalising such procedures, appropriate consultation among all stakeholders including specialist may be required. A few examples of such activities might include:

— Working in confined spaces (oxygen deficit atmosphere)
— Procedure of work permit for hazard prone activities
— Working in atmosphere having higher concentration of organic vapours
— Fire prevention procedures during welding/cutting and other hot work processes
— Culture of machine guarding, particularly moving parts
— Procedures for emergency along with contingency plans
— Critical incident recovery plan
— Instructions for use of personal protective equipments
— House keeping and storage of materials
— Maintenance of tools and tackles
— Material handling (mechanized/manual)
— Material balance (based on mass balance technique)

Conclusion

It can be safely stated that today good management practices, guidelines, code of practices, technologies, etc., are available but what is lacking is the poor implementation and dissemination of these documents to those who matter. With growth of Indian economy coupled with growth of population, a lot of stress has been put on energy sector. The real challenge facing the Indian economy in general and Indian industry in particular is now to reduce energy intensive technologies while keeping a sustained increase in GDP growth rates. We would need to adopt energy efficient methods which could be safe and sustainable. Cement sector has shown a way by converting them from low capacity energy inefficient wet plants to high capacity energy efficient dry processes. It is felt that certification process, particularly for EMS and/or OHSMS, in such organizations will provide the desired impetus in this direction.

Section–V

Safety in Construction

THE ROLE OF EXPERIENCE MODIFICATION RATING AND WORKER'S COMPENSATION
Insurance for Safety in Construction Industry

G.C.S. Reddy* and V.S.S. Kumar**

ABSTRACT

Construction industry is one of the most dangerous industries in the nation. It reported high incidents of injuries causing many human tragedies. A marginal reduction or prevention of accidents in construction industry is an important aspect for the construction industry. Various methodologies have been developed to reduce the accidents and OSHA provides many safety measures. Based upon the past knowledge future accidents can be predicted and can be reduced marginally. The impact of accidents can be reduced by insuring the project and workforce can work with assurance. Workers Compensation Insurance (WCI) plays a vital role to provide incentives for employers to improve safety measures. In addition experience modification rating provides the guidelines for employer's WCI costs to reduce the accidents in the industry. The cost of insuring against occupational injuries, illnesses, and fatalities has increased dramatically in recent years. Thus, there is a need to develop a methodology to reduce the accidents in construction industry. A case study has been considered to illustrate the methodology to reduce construction accidents.

Key Words: WCI, EMR, Safety, Construction Accidents.

Introduction

Construction industry is employing approximately 5 per cent of the industrial workforce and accounted for 14 per cent of all workplace deaths and 9 per cent of disabling injuries. The combined injury and illness rate for the construction industry is higher than for all other industries except for agriculture. The accidents rate in construction industry is at least four times higher than any other industry. The construction industry continues to be a leader in

*Research Assistant, Department of Civil Engineering, University College of Engineering (A), Osmania University, Hyderabad-500 007

**Associate Professor, Department of Civil Engineering, University College of Engineering (A), Osmania University, Hyderabad-500 007, Email: vsskumar1958@hotmail.com

the frequency of fatalities and disabling injuries. The major causes of accidents are related to the unique nature of industry, human behaviour, difficult work site conditions, and poor safety management practices. The prevention and even marginal reduction in their cost will have a significant human and financial impact. Prevention of construction accidents usually requires predicting future accidents and their nature under given circumstances. Making such predictions must be based on knowledge about past accidents, and can be conducted using various decision support tools, including those utilizing machine learning.

Accidents in the construction industry tend to be costly in both human and financial terms. These expenses are concentrated in the areas of health care, litigations, management time, and worker's compensation. Safety is not a luxury and may be considered an important function to be used against unnecessary loss of property, injury or death. Preventing occupational injuries and illness should be a primary concern of all employers. Especially in developing countries, there must be an effort to raise the level of awareness among both employees and employers of the importance of health and safety at work-sites.

Much research effort has been directed at examining accident records. Despite the importance of such study, construction accident investigation stopped at a premature level to identify the root cause of accidents. Construction organizations regard workers' compensation insurance (WCI) and Experience modification rating (EMR) are important factors to compare an individual employer's actual loss experience to its expected record. Many contractors believe that WCI spreads the costs of their losses throughout the industry. Few contractors understand how their WCI premiums are determined and the degree to which their WCI costs can be controlled. The objective is to expose the true impact of an employer's loss experience on that employer's WCI costs and the firm's bottom line.

Background

Ninety-eight per cent of all accidents are caused by unsafe behaviour. Workers injured on the job have probably committed the same unsafe act many times before getting injured (Watson, 1986, Heinrich, 1980). In the construction industry it is a common, but unsafe practice to accommodate workers, heavy mobile equipment, and trucks in confined areas. Often equipment operators work with very limited visibility and must depend on backup alarms to warn workers on the ground. The general construction environment tends to encourage unsafe behaviour. Conventional backup alarms intended to warn workers on the ground are no more than non-directional area wide announcements that a vehicle has changed into reverse gear. These alarms add to the noise level. The alarms are not directed towards any specific group of workers. Workers must use senses other than their auditory senses to interpret the environment and clues from co-worker to assess potential hazards from moving equipment. However, the auditory sense is still the primary sense used to detect such hazards. The California Department of Industrial Relations study noted that over a period of 3 years, 15 deaths were caused by backing trucks, graders, loaders, and other heavy equipment (Helander, 1991). Four major factors were involved in the incidents (1) inoperative backup alarm; (2) construction noise that masked the sound of the backup alarm; (3) poor visibility and (4) worker's in-attention. It appears from this literature review that there are inherent limitations associated with each safety performance measure and a combination of measures. This paper makes an attempt to reduce hazards at work using new construction technology.

Safety in Construction

Construction in India is more labour intensive than in developed countries, involving 2.5 to 10 times as many workers per activity. Typically, workers tend to be unskilled and migrate in a group, with or without their families, throughout the country in search of employment. For example, in India, productive labourers in the excavation and earthwork are often come from the state of Andhra Pradesh, concrete labourers from Karnataka, carpenters from Rajasthan, masons from Uttar Pradesh, and fitters from Bihar. Communication problems related to differences in language, religion, and culture tend to inhibit safety on work-site.

Unsafe conditions exist on many sites, both large and small, and labourers are subjected to numerous hazards. Even large national firms who execute mega projects in different parts of India with a typical work force of more than 100 technical staff, 500 skilled labours and 2000 unskilled labours do not have safety programmes. On many sites no training programmes for the staff and workers exists. No orientation for new staff or workers is conducted, hazards are not pointed out and no safety meetings are held. Employees are required to learn from their own mistakes and experiences. In addition, lack of medical facilities, shanty housing and substandard sanitation tend to exist on the remote projects. Construction is one of the most dangerous occupations. There is a growing interest in improving safety for humanitarian reasons and because of rising costs of worker's compensation and Occupational Safety and Health Administration fines. Falls both from an elevation and from the same level accounted for 34.6 per cent of the injuries and is shown in Fig. 41.1.

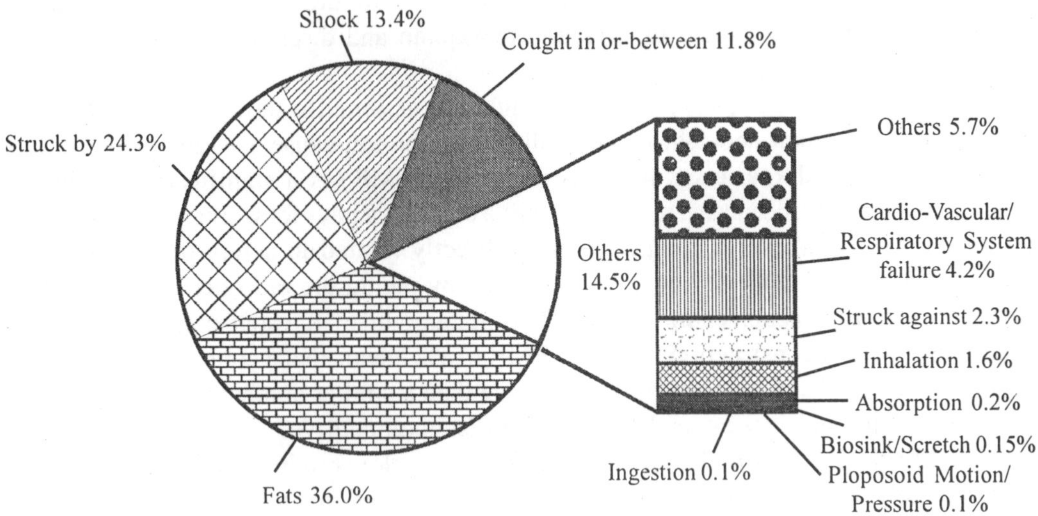

Fig. 41.1: Causes of Construction fall Accidents Investigated by OHSA (*Source:* Huang and Hinze, 2002)

By quantifying these inputs the project manger can allocate limited resources is a more cost-effective manner to attain improved safety performance and to improve the company safety record. Combinations of qualitative and quantitative safety measures are used to evaluate the safety performance of a construction company.

WCI Premium Calculation

Workers' compensation laws are governed by individual countries and may vary from country to country, but all employers are required by law to provide some type of WCI for their employees. Some employers are self-insured and others are covered by insurance carried by clients and project owners. However, the vast majority (more than 90% in all industries) of WCI premium come from experience rated plans. For experience rated plans the WCI standard premium is based on the following formula

$$Standard\ Premium = Manual\ Rate \times Payroll\ Units \times EMR$$

To get the WCI premium, number of other factors are applied to the standard premium. A premium discount accounts for certain costs that do not vary with the size of the policy, and giving large insured credit for economies of scale in overhead expenses. The application of these other factors varies widely from insurer to insurer within the states. Generally, these are marketing options that give insurers an opportunity to reward employers with good safety records and penalize.

Experience Modification Rating

Experience Modification Rating (EMR) is the ratio used to compare a contractor's insurance claims experience to other companies in the contracting industry. EMR is defined as the ratio of actual losses to expected losses over a rolling three-year period. A contractor typically needs an EMR of less than 1.0, preferably between 0.80 and 0.85. If the contractor is above these rates, they should be prepared to explain and discuss steps being taken to reduce job site accidents.

On the other hand EMR is criticized because it is assumed to show bias towards larger contractors. One small accident for a small contractor can cause a large change in that contractor's EMR. Experience modification rating (EMR) is a significant indicator in construction industry, because it reflects the cost a contractor pays for workers' compensation insurance (WCI). Worker's compensation is directly related to safety performance via claims paid due to accidents. EMR is calculated by taking the ratio between the rupees amounts of adjusted actual claims filed to the rupee amount of adjusted expected claims for a particular type of construction. Thus, an EMR of 1.2 means that a contractor pays 20 per cent more for workers' compensation insurance than a similar company with an EMR of 1.00. This additional cost can make a contractor less competitive.

Finally, the EMR provides an incentive for the employer to improve its safety record. Unfortunately, the incentive often gets lost in the complex method used to calculate the EMR.

Research Methodology

The research methodology for the safety in construction is as follows:

Perform a Literature Review to Identify Significant Factors

Collect the literature from various journals, text books and construction firms. Identifying attributes associated with enhanced construction safety performance provides a brief over view of key safety research. Summarize the significant factors associated with successful safety performance. Both EMR and OSHA reportable incidence rates are useful in evaluating company safety performance over a number of years.

Develop a Questionnaire Related to Project Safety Programmes

Prepare a questionnaire for the information pertaining to the project safety programs. The questionnaire should contain data related to type of organization, number of skilled personnel, number of years in construction, information about the company safety programme, and percentage of time devoted to safety by field safety managers. Data are also to be collected on number of days away from work, and the number of fatalities. In addition, the total amount of fines charged and paid out are also to be collected.

Collect Data

Different types and sizes of construction companies are requested to collect the data. Reasonable number of questionnaires are to be mailed to potential respondents. The questionnaires with insufficient data could not be used in the analysis.

Analyze the Data

The data analysis approach involves statistical comparison between safety inputs for projects with differing levels of safety performance. At the company level, EMR will be used as measures of safety performance. Insufficient data collected on the lost time and severity rates, make it difficult to obtain meaningful results using these measures. Companies are to be divided into three categories: (1) EMR > 1.00; (2) EMR between 1.00 and 0.75 (inclusive); and (3) EMR < 0.75. This categorization would provide more meaningful information compared to simply having two categories (i.e., EMR > 1 and EMR < 1).

Summarize the Result

Based on the company characteristic data summarize the findings related to significant variables for company. This information can be used as a guideline to establish a base level for a viable safety programme.

Provide Recommendation for Improving Construction Safety Performance

Recommendations will be made for improving company safety performance. This relates to lowering EMR from >1.00 to between 1.00 and 0.75, and lowering EMR from between 1.00 and 0.75 to < 0.75. Recommendations include providing a more detailed written safety programme, providing better indoctrination to new foremen related to company policies and guidelines, increasing the number of meetings to discuss safety performance with field supervisors, and increasing the number of informal inspections on each project.

The above research methodology is demonstrated with the following case study to achieve the excellence for providing safety measures in the construction industry.

Case Study

A construction company is constructing a multistoried commercial complex. The total cost of the project is Rs. 3 crores with duration of 24 months. The resources required are 75 skilled and 250 unskilled workers. Lift crane, material hoist etc. have been used in the construction work. It has a multilevel parking provided in the cellar and mezzanine floor. The hard disintegrated rock is encountered and is excavated by blasting and manual chiseling. Due to the drainage pipe flowing in the neighborhood, the area has a lot of seepage. To overcome this problem, water is pumped out at regular intervals during excavation and construction of footings.

Literature Review to Identify Significant Factors

Literature has been collected from various journals, text books and construction firms. Identified the relevant attributes associated with Construction Company in terms of project cost, manpower employed, machinery utilized and duration of the project.

Preparation of the Questionnaire

A questionnaire has been prepared to collect the information pertaining to the project safety programs. The questionnaire includes the data related to company characteristics in terms of the number of skilled personnel, and number of years in construction, information about the company safety program, and safety performance measures.

Collection of the Data

Different types and sizes of construction companies are requested to respond for the questionnaire for collecting the data. Ten questionnaires are mailed to potential respondents to collect the relevant information. The filled in questionnaires with insufficient data are ignored in the analysis.

Case Illustration

To set up a case library, a set of 5 cases is studied. The accidents occurred and the preventive measures taken are recorded for all the cases along with project details such as cost, duration, and equipment used. The accuracy of working of this system depends on the number of these input parameters that are considered for each case and the number of cases considered. The accuracy increases as the number of cases in the case library increases.

Analyze the Data

The data collected from the above questionnaires and field survey have been compiled for statistical comparison between safety inputs for projects with differing levels of safety performance.

Summarize the Results

Based on the characteristics of the company the information obtained from the above analysis has been used as guidelines to establish a base level for a viable safety programme. The data obtained has been used to prevent the accidents for the project.

The following recommendations are suggested:

- temporary wall with hollow blocks may be constructed across the entrance of the lift pit to prevent accidental slipping,
- safety belts to be provided as shown in Fig. 41.2.
- steel scaffolding may be used in place of that of timber,
- gloves to be provided during fixing of glass sheets, and.
- workers are to be suitably trained to act carefully while involved in a risky job.

Fig. 41.2: Worker wearing safety belt

Discussions

In a particular construction work, the site engineer has not provided any safety barricades all-round the lift hole. On a no work day, some kids were playing on the slab and have fallen down from second floor to ground floor through the lift gap as the boy steps into it without looking down. If the site engineer had followed simple safety measures such as, providing temporary safety barricades all-round the lift gap, by not allowing kids to play on rooftop or by putting danger signals near the lift gaps this accident would have not been occurred. The above accident would not have occurred, if the company had adopted some safety rules. The accidents can be prevented by providing additional training to part-time safety managers, providing better indoctrination to new foremen related to company policies and guidelines, increasing the number of meetings to discuss safety performance with field supervisors, hiring project managers with more experience with an average of approximately 20 years of experience in the construction industry. Increasing the number of formal safety meetings with supervisors to approximately three times per month also has a positive impact on the reduction of accidents. These recommendations provide quantitative guidelines to help companies establish effective safety programmes to achieve their goals in the most efficient manner. The costs associated with implementing these recommendations may be minimal, especially for companies with well-established and formal safety programmes. This research can help these companies by providing them with direction in terms of what is truly important in achieving desired safety results. It is believed that the benefits of an effective safety programme should far exceed the costs, increased productivity, lower project costs, and improved reputation with project owners.

Conclusions

Construction industry is implementing variety of programmes to improve safety performance. More importantly, these programs have been associated with good safety performance, measured in injury incident rates, lost time injury incident rates, experience modification rates etc. While there is some concern regarding the merits of safety incentives there is general agreement that the safety techniques play a vital role in safety performance.

The following conclusions are made from the above research work:

- The poor safety records adversely affect contractors' ability to operate a profitable operation.
- To improve the safety record, company management must take lead and empower every employee to make meaningful safety related decisions.
- Implementation of a successful safety programme requires planning and improved communication.
- Temporary wall with hollow blocks may be constructed across the entrance of the lift pit to prevent accidental slipping.
- Steel scaffolding may be used in place of that of timber.
- Gloves to be provide during fixing of glass sheets.
- Workers are to be suitably trained to act carefully while involved in a risky job.
- Provide the better quantitative strategies for achieving better construction safety

performance at the company and project level. Numerical profiles of companies and projects with varying levels of safety performance are to be considered for improving safety performance based on the record able incidence rate and EMR.

- Upper management support, number of informal safety inspections made by the company safety manager, meetings with the field safety representatives and skilled workers reduces the accidents.
- At the project level, project manager experience, increased time devoted to safety for the project safety representative, reduced skilled worker penalties, and increased budget allocation to safety awards influence for the prevention of accidents.

Acknowledgements

The Authors express their sincere thanks to the All India Council for Technical Education for funding this research work through permit No. 8017/RDII/MON-606 (98-99)/2001. Authors also wish to express their gratitude to the Department of Civil Engineering, OU for providing facilities to carry out this research work.

REFERENCES

Browm, C.B. (1988). "Errors in Structural Engineering." *Journal of Structural Engineering, ASCE,* 114(11), 2580-2582.

Galvan, J.R., and Tucker, R.L. (1991). "Forecasting design related problems - Case study." *Jl. Constr. Engrg. and Mgmt.,* ASCE, 117(1), 47.

Huang, X., and Hinze, J. (2002). "Analysis of Construction Worker Fall Accidents." *J. Constr. Engrg. and Mgmt.,* ASCE, 129(3), 262-271.

Jaselskis, E.J., Anderson, S.D. and Russel, J.S. (1996). "Strategies for Achieving Excellence in Construction Safety Performance." *J. Constr. Engrg. and Mgmt.,* ASCE, 122(1), 61-70.

Lonergan, B. (1992). *Collected works of Bernard Lonergan; insight.* University of Toronto Press, Toronto, Ontario, Canada, 301.

Puri, S.P.S. (1994). "Failure as source of engineering judgment: Case of John Roebling." *Journal of Performance Construction Facilities,* ASCE, 8(2), 158.

Watson, C.E. (1986). "Does behavior based safety management work." *Professional safety,* 31(9), 20-25.

42

RISK SHARING THROUGH INTEGRATION OF TRANSPORT INFRASTRUCTURES

Pardeep Kumar Gupta* and R.K. Khitoliya**

ABSTRACT

Integrated economic development and infrastructure policies are interdependent. Successful policy must include access, scale and nature of infrastructure. Infrastructure generally involves significant initial outlays and continual on-going expenditures on maintenance and management. There is an increased amount of international traffic and more international personal mobility. Integrated economic development allows for a more efficient use of resources and the potential for the realization of higher living standards. There should be adequate 'inter-operability', 'inter-connectivity' and 'inter-modality'.

Introduction

If there is to be integrated economic development across a number of states, the concern has to be with looking at the types of infrastructure policies that are important. In doing this it looks not only at the more transparent issues of infrastructure investment, but also at the use made of infrastructure. No one wants infrastructure for its own sake. It is a means to an end. Consequently, successful policy must invariable embrace the terms on which people have access to infrastructure as well as the scale and nature of the infrastructure itself. From the perspective of integrated economic development, effective transportation infrastructure policies are vital. Wealthier counties already have substantial infrastructure systems. Nevertheless, however difficult it is for poorer nations, there is a need for effective infrastructure policies if the longer terms benefits of integrated economic development are to be reaped.

*Lecturer, Civil Engineering Department, Punjab Engineering College, Chandigarh-160012
**Professor, Civil Engineering Department, Punjab Engineering College, Chandigarh-160012

Financing

Infrastructure generally involves significant initial outlays and continual on-going expenditures on maintenance and management.

More recently there have been attempts to better structure the ways in which local resources can be used. These have taken the forms of trying to attract more private sector finance and in trying to improve the management of finances by public agencies. Such changes will be vital to ensure adequate infrastructure capacity within an integrated economic structure and to enable the system to be efficiently maintained and operated. The World Bank guidelines in the context of transportation infrastructure financing include:

- Contracting out more design and implementation work to the private sector or expose in-house work to competition from outside contractors.
- Increase revenue mobilization by simplifying road user taxes and charges, restructuring them and improving revenue administration to reduce avoidance, evasion and leakage.
- Allocating additional revenues from the government's consolidated budget. Financial plans need to identify where the additional revenues might come from and at what cost—whether by taking funds away from other sectors and/or raising clearly identified taxes and charges.

Attempts at Creating Integrated Transport Systems

The globalisation and internationalisation of production is inevitably leading to a growth in the demand for international transportation in all continents. This is compounded by a rapid increase in tourism in many areas as the industrialized world enjoys rising personal incomes; better health, longer retirements and people are more adventurous in their leisure pursuits. The result is increased amounts of international traffic and more international personal mobility.

Borders have traditionally proved to be bottlenecks in the international transportation system. In generally there are few viable border-crossing points and traffic must funnel through them. At the very least they provide relatively easy locations to control trade and collect revenues. They can also serve as a means of meeting other non-economic objectives such as the seeking of illegal drugs or the prevention of disease or harmful insects entering a country. In many cases the bottlenecks are, therefore, deliberate and deemed to be an effective way of meeting explicit non-transportation objectives.

Infrastructure and Economic Development

Integrated economic development allows for a more efficient use of resources and the potential for the realization of higher living standards. Political agreement between states to facilitate such integration is important in fostering the creation of larger markers and opening up the supplies of materials and the human ingenuity that allow economic development. But there is also the requirement of appropriate and well-managed infrastructure. The issue if often one of the roles of this infrastructure and the form it should take.

There are also issues of definition and considerations of how infrastructure is maintained and the ways in which it is utilized. Simple calculus seldom offers much assistance in examining what are often qualitative issues of judgment.

There are also the much larger issues of sustainable infrastructure. Providing investment is not of itself enough to ensure the long-term social and economic viability of infrastructure.

Further, infrastructure has no single definition. It is most commonly discussed in terms of its characteristics—longevity, scale, inflexibility, and higher investment costs. Increasingly in the economically advanced parts of the world its meaning has also been shifting from one focusing on physical fixed assets such as roads, airports, sea ports, telecommunications systems, water distribution systems and sanitation. It now often embodies notions of softer types of infrastructure such as information systems and knowledge bases.

Where there does seem to be something of a consensus is that infrastructure, and transportation and communications in particular, is important to lubricate the wheels of trade and to allow the benefits of these economic gains to be distributed across those living in the states involved in the integration process. While there is a tendency for emphasis to remain on the pure efficiency advantages of having a coordinated infrastructure policy, notions of sustainable social development also requires these gains to be spread in a socially acceptable manner.

Transportation is very important in the economic development process acting as a lubricant for both domestic and international transactions. Investment in transportation infrastructure also constitutes a large part of multilateral and bilateral foreign aid contributed to economically developing countries. When it comes to developing integrated infrastructure policies, transportation also poses some of the most difficult challenges.

The Facets of Effective Coordinated Infrastructure Policy

Each type of infrastructure has its own features and each poses a particular set of challenges. Although transportation infrastructure is the focus here, there are parallel considerations relating to other types of infrastructure. Here the focus is on some of the important technical and institutional policy issues.

There should be adequate 'inter-operability', 'inter-connectivity' and 'inter-modality'. These are interrelated, multi-dimensional concepts that embody institutional as well as technical considerations.

Inter-operability means that the operating equipment (trucks, trains, ships, etc.) can operate on either side of the border equally efficiently. This means common technical specifications, or at least sufficient flexibility in specifications. It also means common institutions such as licences, insurance, computer and information systems, safety standards, and labour laws and practices. In other words, it means equity of access, on comparable terms, to the entire integrated transportation infrastructure network.

Inter-connectivity is largely, but not exclusively, a technical matter in its relationship to infrastructure. Railways require the same gauge on either side of a border to be efficient and, with electric locomotion, the same power system. Roads must be of comparable engineering quality to carry heavy trucks.

The idea of *inter-modality* is not strictly only a trans-border concern but involves the more generic issue of be[illegible] to switch between transportation modes at minimal generalized cost (i.e., the full costs of movement including delay costs, transhipment, etc.). It concerns efficient inter-change between modes. Confronting these necessary features on an integrated transportation infrastructure requires institutional as well as technical coordination. Further, while in some cases it is possible to isolate those elements of transportation infrastructure that are immediately important for economic integration, in the longer term the ability of this diminishes as economic growth takes place. Essentially, the production process itself becomes more integrated as populations become more deeply involved in a larger money economy.

The focus on infrastructure policy should also be tempered by considerations of how these matters of inter-connectivity, inter-operability and inter-modality within transportation fit with other transportation policies. In practice there is a need for crosscutting, or at least policy coordination and recognition, policies that also embrace energy and communications. Developing an integrated infrastructure policy in one sphere can often be wasteful of resources without cognisance being taken of what is happening elsewhere.

Conclusions

One of the potential advantages of integrating economies is to enjoy the synergies of risk sharing and lower overheads that can be generated. Integrated development almost inevitable requires coordinated and possible shared infrastructure policies. There are problems of creating institutions that can achieve this. One involves deciding on an efficient form of coordination.

REFERENCES

World Bank. 1993. *The East Asian Miracle: Economic Growth and Public Policy.* New York: Oxford University Press.

Newbery, D. 1994. "The Case for a Public Road Authority", *Journal of Transport Economics and Policy* 28 (3): 235-253.

Allport, R.J. 1994. "Defining the Role for Private Finance In Transport Infrastructure Provision—the Asian Experience." Paper presented to the PTRC Summer Annual Meeting, London, England.

SAFETY IN INFRASTRURE DEVELOPMENT

K.M. Soni*

ABSTRACT

Infrastructure is the basic requirement for the development. In a developing nation, construction plays a vital role in the development of the country. Since demand is always more than the supply in a developing nation, quality and safety aspects become secondary issues compared to the availability of the infrastructure which becomes a primary issue. But such a concept ultimately proves to be uneconomic. A developing nation cannot sustain recurring expenditure on recreating such infrastructure facilities. Also such a concept provides a back seat to construction companies in an era of globalization. Hence, quality and safety must be considered as a primary issue in construction and maintenance. Hence, safety aspects must be incorporated in planning, execution and maintenance stages for the structure as well as for the users of the structures. Like education has become a routine feature for road users, similar education needs to be imparted to professionals and users of other structures also. In the present paper, all such issues have been discussed.

Introduction

Safety in infrastructure development like in buildings, roads, water supply and sewerage projects, waste management projects, irrigation structures, railway projects, nuclear projects etc. is of paramount importance. Therefore, safety aspects needs to be considered from the concept stage of a project to the post construction stage, i.e. maintenance stage.

Agencies involved in various stages of a project are architects, town planners, designers, engineers who include civil, electrical, plumbing and other specialized engineers based on the type of the project. The execution is done mainly by the contractors or builders; hence they are also very important part of the team responsible for the safety of the structures. Apart from these, users are required to be aware of the safety aspects of the structures and should be involved in the process of the safety management.

*Superintending Engineer, CPWD, DCCIX, B-108, IP Bhawan, IP Estate, New Delhi

Various Stages of a Project

Various stages thus can be categorized as,

(i) Conceptual/preliminary stage
(ii) Planning stage
(iii) Execution stage
(iv) Maintenance stage

Conceptual/Preliminary Stage

Whenever a project is conceived, concept designs are prepared for the approval of the client. Preliminary drawings are submitted to local bodies along with the layout plan for approval. Various local bodies from which the approval is sought may involve,

(i) Municipal corporations/municipalities
(ii) Fire authorities
(iii) Environmental authorities
(v) Highway authorities
(v) Aviation authorities

Therefore, it may be construed that the safety of the structure is to be seen by the corporation, and other agencies will examine the project based on the guidelines. In such circumstances there should be no problem of the safety of the structure once this has been approved by the local bodies. Still in many cases of the calamities, all the structures have not been found safe. Hence, there is a need to review the existing system. Let us consider a metropolitan city in which a number of projects are in pipeline. Architect or the town planner can superficially examine the drawings based on the following considerations as number of projects may be put up to him,

(i) Form of the building
(ii) Basic plan/elevation/usage of the building
(iii) Area of the building, green area, plot area, FAR etc.

Still many issues like safety from earthquake considerations, structural safety of the building, plumbing designs etc. need to involve specialized professionals to just assess the basic concepts followed in the preparation of the designs and not for proof checking. Hence either these professionals should be available in the corporations or drawings should be prepared in consultation with them. Considering the volume of the work, basic responsibility should rest with the architect/builder/engineer submitting the drawings to ensure the safety, which should just be examined by the local bodies.

Other important factors which are to be considered along with a project are traffic pattern to avoid accident after the completion of the project, road network, site particulars and other infrastructure facilities in the master plan of the area.

In few cases, it has been reported that the foundation of the structures has failed as neither the foundation was designed nor any geotechnical investigation carried out. Therefore, local bodies should also ensure that proper geotechnical investigation report is available before approval particularly for multi storeyed structures. Local bodies should also ensure availability of the safe and potable water for the projects approved by them.

Planning Stage

Planning is very important in any project. It is said that bad planning results only in bad work. Planning involves all disciplines such surveying, architectural, quantity surveying, designing, electrical services, plumbing, landscaping and even financing. From safety considerations each is required to provide a proper input. Starting from the beginning, site selection and even geotechnical investigation is to be based on the project, site parameters and soil available at the site. Architect is to prepare designs so that structure is earthquake resistant and less susceptible to other damages. Many buildings failed in Gujarat in the last earthquake were aesthetically very good but lacking from the earthquake safety considerations. Provision of soft storeys without proper design, too many cantilevers and imbalanced structures were such architectural features, which contributed to their failures. A lot of literature and codes are already available in this regard for designing earthquake resistant designs. Similarly, architect is to include fire safety requirements in a building. Environmental safety can be adopted by adopting environment friendly materials and technologies. User's safety is also to be given prime importance when preparing the architectural drawings. The architect is to consider the durability of the components of the building so that they are safe without much maintenance during its life span.

Quantity surveyors play an important role from safety considerations in framing the conditions and the items for the bidding documents and even for selecting the credible contractors. Normally the following conditions for safety of life and property are to be considered,

(i) Proper specifications of the scaffolding
(ii) Safety code for the workers
(iii) Workmen's Compensation
(iv) Insurance clause for loss of life and property

Apart from this, environment friendly materials and technologies can be stipulated in the bidding documents. Various quality control measures stipulated in the bidding documents also ensure safety of the structure during its life.

Structural designer has a very important role for safety of structures during normal but particularly during critical conditions. The structures are to be designed based on latest codes and drawings issued with proper detailing. Badly designed or structures constructed based on thumb rules failed at many places during critical conditions like earthquakes.

Electrical engineers have to ensure the safety of the structures from the fire as well as have to provide fire alarm and detection measures. Safety considerations are to be considered in determining the properties of electric cable, gadgets and electrical load.

Sometimes it has been observed that wrong plantation of trees and even wrong adoption of species of plants lead to accidents particularly when they affect vision and the eye sight distance. Plantation of trees on slopes sometimes help in landslides as the roots attract water and sometimes plantation on the banks of the rivers/nallah itself blocks the river in case trees get uprooted and leads to erosion of banks and ultimately floods. Hence, landscaping and horticulture officers also need to consider the safety aspects during plantation.

Execution

A good planning needs to be supplemented with good execution to make a structure safe. During execution the following safety factors are adopted. These may not be exhaustive and may vary from project to project,

(i) Proper shoring/strutting/bracing during deep excavations.
(ii) Proper safety arrangements during working in sewers, under water conditions, in caissons, nuclear stations etc.
(iii) Proper scaffolding.
(iv) Proper safety measures for workers.
(v) Abiding with all structural design and detailing.
(vi) Proper supervision and quality control.
(vii) Proper arrangements to check pollution.
(viii) Insurance for loss of life and property during execution.

Maintenance Stage

Maintenance in a structure is very important. Poor maintenance may shorten the life of the components of the structure and ultimately the life of the structure. In such conditions, safety of the users as well as the structures directly gets affected as in many cases; the users occupy structures. Also such structures fail during disasters easily.

Poor maintenance of roads, irrigation structures, buildings, water supply/sewerage projects or any other project not only poses safety problems of structures, but also to users. They also lead to inconvenience and lessen the life of structures which directly affect the economy of the country.

Education

In general, well-designed government buildings remain safe during earthquakes. Also during fire incidents, such buildings pose a minimal causality. But un-designed and under-designed buildings cause havoc during calamities. Local bodies can take steps for the safety of the buildings during the approval of drawings of such buildings as started at few places but risk involved in the buildings in rural/semi-urban areas is tremendous whether these areas are in rural areas of Delhi or in a remote area in Himalayan belt. Even measures like plinth band, lintel band and corner reinforcements are not provided in load bearing structures.

Unauthorized structures constructed in mega cities without any approval, which are regularized by the government at a later date, also pose a major threat during natural calamities.

Apart from this, an infectious disease is spreading in many cities particularly in Delhi regarding addition/alteration in the flats constructed by the government organizations. At many places, walls are even removed and columns are provided on ground floor in the construction of four storeyed buildings. Non-uniform blocks and cantilevers, which are not even anchored properly, are added without any technical consideration. These unsafe structures during earthquakes are waiting to the moment for disaster to come and kill many people. No agency is taking any action at present, however many agencies will come forward and raise the issue once disaster would occur. Otherwise also we hardly learn anything from the past and even forget disaster events easily and wait for news of the other disaster.

RCC structures are becoming part of modern construction in northeastern states at a very fast speed both at government level and private level without working out the consequences. In Uttaranchal region RCC roofs are being provided on traditional structures having stonewalls sometimes constructed even without mortar. Probably cheapest commodity in the country has become life of a human being and due to this probably thus required consideration is not given to the safety. Thus education regarding safety provisions in the structures need to be taught to the public and artisans in case disasters and loss of life and property is to be prevented. These can be done through posters, banners, conducting programmes at panchayat and municipal levels and by including basic safety requirements of structures during natural disaster like earthquakes in school curriculum.

Conclusions

Road accidents kill few people but failure of structures kills many people in few moments. Therefore, structures must be planned, designed, executed and maintained with safety considerations. In an era of information technology and engineering excellence, it would be disasters if available knowledge is not utilized and loss of life and property remains a continuous tragedy. Safety aspects should be included in pre-construction, execution and in maintenance stage. Dissemination of available knowledge of safety aspects of the structures is also essential not only among professionals but also to the public.

REFERENCE

CPWD Specifications 1996, Nabhi Publications, N.Delhi.

44

SAFETY PROVISIONS IN NATIONAL BUILDING CODE OF INDIA 2005

Sanjay Pant* and Aman Deep**

ABSTRACT

In the field of Civil Engineering, 'safety' has immense importance, which needs to be kept in mind and built-in at all stages such as planning, designing, construction, operation and maintenance stages of civil engineering projects. During the above stages, aspects like orientation, lighting and ventilation; fire safety; safety through structural design; safety during construction and in operation and maintenance; safety in electrical/other services related aspects; etc., needs to be taken care of.

A series of Indian Standards have been developed to cover the whole gamut of safety considerations and necessary recommendations have been provided therein for guidance and implementation by all concerned. The aspects like orientation, lighting and ventilation and functionality have a bearing on hygiene and health of people including workers in industrial units, for which a number of Indian Standards have been formulated covering aspects like day lighting, ventilation, noise reduction, facilities for physically challenged, etc. as applicable to various building occupancy types. Fire safety is assuming an increasing importance in view of complexities now encountered in buildings and installations built to meet the challenging requirements of today. Indian Standards have, therefore, been developed in the vast area of fire safety covering provisions and guidelines on fire safety of buildings of various types including residential, educational, industrial, etc. and their further sub-categories. Structural design of buildings and other structures takes care of various possible loads, forces and effects likely to be encountered during lifecycle of building/structure, affecting strength, stability and safety. Loading Codes and Design Codes have, therefore, been developed providing comprehensive guidance to design a structure taking cognizance of dead load, imposed load, wind load, snow load, seismic forces, cyclone resistance and landslide control. Such comprehensive Design Codes are made available for design of stable structures using various material streams like concrete, steel, timber, bamboo, masonry and even using prefabricated construction techniques. Safety during construction is another area which has been given the attention it deserves and a number of Indian Standards covering aspects like safety during excavation, blasting work, tunneling work, deep foundation work, while working at height, handling of materials likely to cause harm, etc., have been formulated. Safety from electrical and allied installations has also been dealt with in number of standards developed in the field. Similarly the standards on

*Joint Director, (Civil Engg), Bureau of Indian Standards, New Delhi

**Asst Director (Civil Engg), Bureau of Indian Standards, New Delhi

drainage and sanitation have an important role to play as regards health and hygiene of occupants/ users of buildings. Equally important are the provisions laid down regarding gas supply for ensuring safety.

Specific to buildings, all the above and many other aspects and details have also been in-built and interwoven in the National Building Code of India 2005 (NBC 2005), which can be effectively used as an instrument for regulating the building activity in the country and thereby ensuring implementation of safety aspects dealt with in India Standards. Particularly, NBC 2005 provides administrative provisions for implementing the various provisions specially the safety related provisions.

Introduction

The National Building Code of India is a single document in which, like a network, the information contained in various Indian Standards is woven into a pattern of continuity and cogency with the interdependent requirements of its different chapters carefully analyzed and fitted in to make the whole document a cogent continuous volume. A continuous thread of 'preplanning' is woven which, in itself, contributes considerably to the economies and safety in the construction of building and thereafter in safety of building and its occupants.

NBC 2005 contains 11 Parts with some parts further subdivided into sections (thus making it a total of 26 chapters) covering almost all aspects of building planning, design, safety, execution, maintenance, etc. A complete list of various Parts/Sections under NBC 2005 is given at Annex A. The safety aspects have been covered mainly in Part 4 (which comprehensively deals in fire safety aspects for various types of buildings), Part 6 (which deals in safe structural design of buildings), Part 7 (which deals in constructional practices and safety) and Part 8/Section 2 (deals with the safety issues related to electrical and allied services). Part 9 deals with the public safety by way of provisions for water supply, drainage and sanitation (including solid waste management) and gas supply. Part 3 of NBC provides the development control rules and general building requirements which have immediate bearing on health, hygiene and safety. Part 2 covers the provisions for the administration and implementation of the provisions of all the parts of NBC 2005. Some of the basic safety provisions are given hereunder.

Fire Safety Provisions

(a) For the purpose of fire safety provisions/requirements, the buildings have been classified as follows, which have been further subclassified:
   (i) Residential (Group A)
   (ii) Educational (Group B)
   (iii) Institutional (Group C)
   (iv) Assembly (Group D)
   (v) Business (Group E)
   (vi) Mercantile (Group F)
   (vii) Industrial (Group G)
   (viii) Storage (Group H)
   (ix) Hazardous (Group J)

(b) As per NBC 2005, a building of height 15 m or above has been classified as high rise building.

(c) For high rise buildings non-combustible materials should be used for construction and the internal walls of staircase enclosures should be with minimum of 2 h rating.

(d) A building or portion of the building may be occupied during construction, repairs, alterations or additions only if all means of exit and fire protection measures are in place and continuously maintained for the occupied part of the building.

(e) A high rise building during construction shall be provided with the following fire protection measures, which shall be maintained in good working condition at all the times:

  (i) Dry riser of minimum 100 mm diameter pipe with hydrant outlets on the floors constructed with a fire service inlet;

  (ii) Drums filled with water of 2000 litres capacity with two fire buckets on each floor.

(f) Smoke venting facilities for safe use of exits in windowless buildings, underground structures, large area factories, hotels and assembly buildings (including cinema halls) shall be automatic in action with manual controls in addition.

(g) Natural draft smoke venting shall utilize roof vents or vents in walls at or near the ceiling level; such vents shall be normally open, or, if closed, shall be designed for automatic opening in case of fire, by release of smoke sensitive devices.

(h) Where smoke venting facilities are installed for purposes of exit safety, these shall be adequate to prevent dangerous accumulation of smoke during the period of time necessary to evacuate the area served, using available exit facilities with a margin of safety to allow for unforeseen contingencies.

(i) An exit may be a doorway; corridor; passageway(s) to an internal staircase, or external staircase, or to a verandah or terrace(s), which have access to the street, or to the roof of a building or a refuge area. An exit may also include a horizontal exit leading to an adjoining building at the same level.

(j) Every exit, exit access or exit discharge shall be continuously maintained free of all obstructions or impediments to full use in the case of fire or other emergency.

(k) Every building meant for human occupancy shall be provided with exits sufficient to permit safe escape of occupants, in case of fire or other emergency.

(l) Exits shall be clearly visible and the route to reach the exits shall be clearly marked and signs posted to guide the occupants of the floor concerned. Signs shall be illuminated and wired to an independent electrical circuit on an alternative source of supply.

(m) To prevent spread of fire and smoke, fire doors with 2 h fire resistance shall be provided at appropriate places along the escape route and particularly at the entrance to lift lobby and stair well where a 'funnel or flue effect' may be created, inducing an upward spread of fire.

(n) All exits shall provide continuous means of egress to the exterior of a building or to an exterior open space leading to a street.

(o) Exits shall be so arranged that they may be reached without passing through another occupied unit.

(p) Notwithstanding the detailed provision for exits, the following minimum width shall be provided for staircases -

(i) Residential buildings (dwellings) 1.0 m
(ii) Residential hotel buildings 1.5 m
(iii) Assembly buildings like auditorium, theatres and cinemas 2.0 m
(iv) Educational buildings up to 30 m in height 1.5 m
(v) Institutional buildings like hospitals 2.0 m
(vi) All other buildings 1.5 m

(q) The minimum width of tread without nosing shall be 250 mm for internal staircase of residential buildings. This shall be 300 mm for assembly, hotels, educational, institutional, business and other buildings. The treads shall be constructed and maintained in a manner to prevent slipping.

(r) The maximum height of riser shall be 190 mm for residential buildings and 150 mm for other buildings and the number shall be limited to 15 per flight.

(s) The design of staircase shall also take into account the following :

(i) The minimum headroom in a passage under the landing of a staircase and under the staircase shall be 2.2 m.

(ii) For building 15 m in height or more, access to main staircase shall be through a fire/smoke check door of a minimum 2-hour fire resistance rating. Fire resistance rating may be reduced to 1 h for residential buildings (except hotels and starred hotels).

(iii) No living space, store or other fire risk shall open directly into the staircase or staircases.

(iv) External exit door of staircase enclosure at ground level shall open directly to the open spaces or through a large lobby, if necessary.

(v) The main and external staircases shall be continuous from ground floor to the terrace level.

(vi) No electrical shafts/AC ducts or gas pipes, etc, shall pass through or open in the staircases. Lifts shall not open in staircase.

(vii) No combustible material shall be used for decoration/wall panelling in the staircase.

(viii) Beams/columns and other building features shall not reduce the head room/ width of the staircase.

(ix) The exit sign with arrow indicating the way to the escape route shall be provided at a suitable height from the floor level on the wall and shall be illuminated by electric light connected to corridor circuits. All exit way marking signs should be flush with the wall and so designed that no mechanical damage shall occur to them due to moving of furniture or other heavy equipment. Further, all landings of floor shall have floor indicating boards prominently indicating the number of floor as per byelaws.

The floor indication board shall be placed on the wall immediately facing the flight of stairs and nearest to the landing. It shall be of size not less than 0.5 m x 0.5 m.

(x) Individual floors shall be prominently indicated on the wall facing the staircases.

(xi) In case of single staircase it shall terminate at the ground floor level and the access to the basement shall be by a separate staircase. The second staircase may lead to basement levels provided the same is separate at ground level by ventilated lobby with discharge points to two different ends through enclosures.

(t) *Pressurization of Staircases (Protected Escape Routes)* : Though in normal building design and compartmentation plays a vital part in limiting the spread of fire, smoke will readily spread to adjacent spaces through the various leakage openings in the compartment enclosure, such as cracks, openings around pipes ducts, airflow grills and doors, as perfect sealing of all these openings is not possible. It is smoke and toxic gases, rather than flame, that will initially obstruct the free movement of occupants of the building through the means of escape (escape routes). Hence the exclusion of smoke and toxic gases from the protected routes is of great importance. Pressurization is a method adopted for protected escape routes against ingress of smoke, especially in high-rise buildings. In pressurization, air is injected into the staircases, lobbies or corridors, to raise their pressure slightly above the pressure in adjacent parts of the building. As a result, ingress of smoke or toxic gases into the escape routes is prevented. The pressurization of staircases shall be adopted for high rise buildings and building having mixed occupancy/multiplexes having covered area more than 500 m^2.

(u) The staircase of basements shall be of enclosed type having fire resistance of not less than 2 h and shall be situated at the periphery of the basement to be entered at ground level only from the open air and in such positions that smoke from any fire in the basement shall not obstruct any exit serving the ground and upper floors of the building.

(v) In multi-storey basements, intake ducts may serve all basement levels, but each basement levels and basement compartment shall have separate smoke outlet duct or ducts.

(w) Use of basements for kitchens working on gas fuel shall not be permitted, unless air conditioned.

(x) All floors shall be compartmented with area not exceeding 750 m^2 by a separation wall with 2 h fire rating, for floors with sprinklers the area may be increased by 50 per cent. In long building, the fire separation walls shall be at distances not exceeding 40 m. For departmental stores, shopping centers and basements, the area may be reduced to 500 m^2 for compartmentation. Where this is not possible, the spacing of the sprinklers shall be suitably reduced. When reducing the spacing of sprinklers, care should be taken to prevent spray from one sprinkler impeding the performance of an adjacent sprinkler head.

(y) From fire safety point of view, electrical services shall conform to the following:

(i) The electric distribution cables/wiring shall be laid in a separate duct. The duct shall be sealed at every floor with non-combustible materials.

(ii) Separate circuits for fire fighting pumps, lifts, staircases and corridor lighting and blowers for pressurizing system shall be provided directly from the main switch gear panel and these circuits shall be laid in separate conduit pipes, so that any fire in one circuit does not affect the others.

(iii) The inspection panel doors and any other opening in the shaft shall be provided with air-tight fire doors having fire resistance of not less than 2h.

(iv) Medium and low voltage wiring running in shafts, and within false ceiling shall run in metal conduit. Any 230 V wiring for lighting or other services, above false ceiling, shall have 660 V grade insulation.

(v) Suitable circuit breakers shall be provided at the appropriate points.

(z) From fire safety point of view, gas supply shall conform to the following:

(i) *Town Gas/L.P.Gas Supply Pipes* : Where gas pipes are run in buildings, the same shall be run in separate shafts exclusively for this purpose and these shall be on external walls, away from the staircases. There shall no interconnection of this shaft with the rest of the floors.

(ii) All wiring in fume hoods shall be duly insulated. Thermal detectors shall be installed into fume hoods of large kitchens for hotels, hospitals, and similar areas located in high rise buildings. Arrangements shall be made for automatic tripping of the exhaust fan in case of fire; if LPG is used, the same shall be shut off. The voltage shall be 24 V or 100 V dc operated with external rectifier. The valve shall be of the hand re-set type and shall be located in an area segregated from cooking ranges. Valves shall be easily accessible. The hood shall have manual facility for steam or carbon dioxide gas injection, depending on duty condition.

(aa) All high rise buildings shall be equipped with manually operated electrical fire alarm (MOEFA) system and automatic fire alarm system.

Safety Through Structural Design

Safety through structural design lies in the proper planning and design for the following condition and the combination thereof:

(a) Soil liquefaction study and stabilization;

(b) Soil shear strength and safe bearing capacity;

(c) Load during construction and during the service life. The various loads which need to be considered as applicable and with appropriate load factors for the safety shall be

(i) Dead load

(ii) Imposed load

(iii) Earthquake load

(iv) Wind load

(v) Snow load

(vi) Cyclone load

(vii) Landslides

(viii) Loads and effects due to equipments and machinery;

(d) Adequate selection of building materials and quality control;

(e) Adequate construction and erection techniques; and

(f) Quality construction and supervision.

Taking the above into cognizance, the Part 6 of the NBC 2005, through its seven sections provides provisions and good practices for structural adequacy of buildings to deal with both internal and external environment, and provide guidance to engineers/structural engineers for varied usage of material/technology types for building design.

Construction Safety

The safety of personnel engaged in building construction should be ensured through a well planned and well organized mechanism. For this, depending on the size and complexity of building construction project, as per Part 7 of NBC 2005, safety committee shall be constituted to efficiently manage all safety related affairs. The site in-charge or his nominee of a senior rank shall head the committee and a safety officer shall act as member-secretary. The meetings of the safety committee shall be organized regularly depending on the nature of the project, however, emergency meetings shall be called as and when required. The safety committees shall deal with all the safety related issues through well structured agenda, in the meetings and all safety related measures installed at the site and implementation thereof shall be periodically reviewed. Following safety provisions need to be taken care of:

(a) Safety during excavation,
(b) Safety during drilling and blasting,
(c) Safety during piling and deep foundations,
(d) Safety during tunneling,
(e) Safety during road making,
(f) Safety during site transport,
(g) Safety during floor and wall opening,
(h) Safety during demolition,
(i) Safety during structural steel erection,
(j) Safety during concrete framed structure,
(k) Safety during material handling,
(l) Safety during compressed air,
(m) Safety during tools,
(n) Safety during construction machinery, and
(o) Safety during common hazards.

Particularly persons working at heights may use safety belts and harnesses. Provision of cat-walks, wire mesh, railings reduces chances of fall-ladder and scaffoldings, staging etc. should be anchored on firm footing and should be secured and railing should be provided as far as possible. All accesses should be barricaded to prevent accidental fall. Also, during demolition operation, it should be ensured that:

(a) On every demolition job, danger signs shall be conspicuously posted all around the structure and all doors and openings giving access to the structure shall be kept barricaded or manned except during the actual passage of workmen or equipment. However, provisions shall be made for at least two independent exits for escape of workmen during any emergency.

(b) During nights, red lights shall be placed on or about all the barricades.
(c) All the necessary safety appliances shall be issued to the workers and their use explained. It shall be ensured that the workers are using all the safety appliances while at work.
(d) The power on all electrical service lines shall be shut off and all such lines cut or disconnected at or outside the property line, before the demolition work is started. Prior to cutting of such lines, the necessary approval shall be obtained from the electrical authorities concerned. The only exception will be any power lines required for demolition work itself.
(e) Adequate natural or artificial lighting and ventilation shall be provided for the workmen.

Notwithstanding the individual standards provisions in NBC 2005 on all the above aspects, Special Publication, namely SP 70 *'Handbook on Construction Safety Practices'* has been prepared for providing guidance in this area.

Safety Provisions for Electrical Installations

The safety provisions fo electrical installations have been dealt with in Section 2 of Part 8 Electrical and Allied Installations. Some of the provisions are given hereunder:

Location and Requirements of Substation

(i) The substation should preferably be located in separate building and could be adjacent to the generator room, if any. Location of substation in the basement floors should be avoided, as far as possible.
(ii) The ideal location for an electrical substation for a group of buildings would be at the electrical load centre on the ground floor.
(iii) The floor level of the substation or switch room shall be above the highest flood level of the locality.
(iv) Substations with oil filled equipment will require great consideration for the fire detection, protection and suppression. Oil cooled transformers require a suitable soak pit with gravity flow to contain the oil in the event of the possibility of oil spillage from the transformer on its failure. Substations with oil filled equipment shall not be located in any floor other than the ground floor or a semi-basement. Such substations with high oil content may be housed in a separate service building or a substation building, which is not the part of a multistoreyed building.
(v) In case electric substation has to be located within the main multistoreyed building itself for unavoidable reasons, then it should be located on the floor close to ground level, but shall have direct access from the street for operation of the equipment.
(vi) Substations located within a multi-storeyed building shall not have oil filled transformers, even if it is at the ground level. Substations with very little combustible material, such as a Dry type transformer, with Vacuum (or SF6) HT switchgear and ACB or MCCB for MV can be located in the basement as well as upper floors in a building with high load density in the upper floors.

(vii) The power supply control to any such substation or transformer (located at basement levels or upper floors) shall be from a location on ground floor/first basement level having direct access from outside so that in case of fire, the electrical supply can be easily disconnected.

(viii) Oil filled transformers may be used only in substations located in separate single or two storeyed service buildings outside the main building structure and there shall at least 6 metre clear distance between the adjoining buildings and substation such that fire tender is able to pass between the two structures.

Requirement for Overhead Lines and Cables

(i) Any person responsible for erecting an overhead line will keep informed the authority(s) responsible for services in that area for telecommunication, gas distribution, water and sewage network and roads so as to have proper coordination to ensure safety. He shall also publish the testing, energizing programmes for the line in the interest of safety.

(ii) Any part of an overhead line which is not connected with earth and which is not ordinarily accessible shall be supported on insulators or surrounded by insulation.

(iii) Any part of an overhead line which is not connected with earth and which is ordinarily accessible shall be:

  (1) made dead; or

  (2) so insulated that it is protected, so far it is reasonably practicable, against mechanical damage or interference; or

  (3) adequately protected to prevent danger.

(iv) Any bare conductor not connected with earth, which is part of a low voltage overhead line, shall be situated throughout its length directly above a bare conductor which is connected with earth.

(v) No overhead line shall, so far as is reasonably practicable, come so close to any building, tree or structure as to cause danger.

(vi) Every support carrying a high voltage overhead line, and every support carrying a low voltage overhead line incorporating bare phase conductors, shall have attached to it sufficient safety signs and placed in such positions as are necessary to give due warning of such danger as is reasonably foreseeable in the circumstances.

(vii) Poles supporting overhead lines near the road junctions and turnings shall be protected by a masonry or earth fill structure or metal barricade, to prevent a vehicle from directly hitting the pole, so that the vehicle, if out of control, is restrained from causing total damage to the live conductor system, likely to lead to a hazardous condition on the road or foot path or building.

Water Supply, Drainage, Sanitation and Gas Supply

The water supply provisions covered in NBC 2005 encompass the requirements of water supply, plumbing connected to public water supply, design of water supply systems, principles of conveyance and distribution of water within the premises, hot water supply system, inspection and maintenance of water supply systems. It also covers design of water supply systems in high altitudes and/or sub-zero temperature regions.

The drainage and sanitation provisions covered in NBC 2005 encompass the drainage and sanitation requirements of buildings, design, construction and maintenance of drains inside buildings and from the buildings up to the connection to the public sewer, private sewer, individual sewage disposal system, cesspool or to other approved point of disposal/ treatment work. It also covers drainage systems peculiar to high altitudes and/or sub-zero temperature regions of the country.

A minimum set of safety provisions have also been laid down to safeguard the gas piping installation and the mode of operation in the interest of public safety.

Conclusion

Notwithstanding a large number of Indian Standards formulated in the field of safety covering structural safety, fire safety, electrical safety and safety in construction, the success hinges upon administratively implementing the same. NBC 2005, while giving comprehensive treatment to the above aspects in tandem with the Indian Standards concerned, provides administrative provisions in Part 2 Administration also providing the necessary guidelines and proformae, to ensure that these are followed by construction and other agencies and fully implemented by the concerned authorities. Apart from this, Part 3 dwells on the development control rules and general building requirements giving necessary minimum provision regarding open spaces and setbacks/safe distances, road widths, etc., all leading to a habitat in an environment of safety, health and hygiene. Safety has to be centre of attention on a continual basis. NBC 2005, therefore, provides for period examination for approval of completed/occupied buildings with respect to structural, fire and electrical safety.

ANNEX A

The NBC 2005 consists of the following Parts and Sections

Part 0	Integrated Approach—Prerequisite for Applying Provisions of the Code	
Part 1	Definitions	
Part 2	Administration	
Part 3	Development Control Rules and General Building Requirements	
Part 4	Fire and Life Safety	
Part 5	Building Materials	
Part 6	Structural Design	
	Section 1	Loads, Forces and Effects
	Section 2	Soils and Foundations
	Section 3	Timber and Bamboo
	Section 3A	Timber
	Section 3B	Bamboo
	Section 4	Masonry
	Section 5	Concrete
	Section 5A	Plain and Reinforced Concrete
	Section 5B	Prestressed Concrete
	Section 6	Steel
	Section 7	Prefabrication, Systems Building and Mixed/Composite Construction
	Section 7A	Prefabricated Concrete
	Section 7B	Systems Building and Mixed/Composite Construction
Part 7	Constructional Practices and Safety	
Part 8	Building Services	
	Section 1	Lighting and Ventilation
	Section 2	Electrical and Allied Installations
	Section 3	Air conditioning, Heating and Mechanical Ventilation
	Section 4	Acoustics, Sound Insulation and Noise Control
	Section 5	Installation of Lifts and Escalators
Part 9	Plumbing Services	
	Section 1	Water Supply, Drainage and Sanitation (including Solid Waste Management)
	Section 2	Gas Supply
Part 10	Landscaping, Signs and Outdoor Display Structures	
	Section 1	Landscape Planning and Design
	Section 2	Signs and Outdoor Display Structures

SAFETY CONCERN IN CONSTRUCTION
A Need of the Hour for Developing Countries

Shashank Mehta*

ABSTRACT

Construction industry is one of the most hazardous industries to work in. This fact is well established. However despite the large number of fatalities, safety in construction remains on a low priority. Only recently has this issue received some attention in the developed countries. But the scenario in the fast growing construction sector in developing countries is far from satisfactory. This paper highlights the importance of safety in construction industry by looking at trends of safety and accidents in the construction sector in many countries. Further the main reasons for accidents in construction industry of developing countries are identified. Along with accidents the possible health hazards effecting workers in these countries are also discussed.

Key Words: safety, construction, health hazards

Introduction

The construction industry has traditionally been considered as a hazardous occupation, due to the high incidence of occupational accidents and, specially of fatal injuries, as is shown in the data from some countries with statistics on the issue. In the United States (US), in 2001, the construction sector employed 7.1 per cent (9,581,000 workers) of the country work-force. Nevertheless, it accounts for 9.7 per cent of all occupational injuries and 20.7 per cent of the fatalities. In France, in 2000, construction employed 5.6 per cent (1,215,000 workers) of the total paid employment, but it accounted for 19 per cent of all occupational injuries and 26 per cent of the fatalities. In Spain, in 2001, the construction sector employed 11.6 per cent (1,850,200 workers) of the total workforce. However it accounted for 26.4 per cent of all work related injuries and 26.1 per cent of the fatalities. In Japan, in 2000, construction employed 10 per cent (5,690,000 workers) of the total workforce, but it accounted for 25.1 per cent of all occupational accidents and 38.7 per cent of the fatalities. From the above data, it is easy to see the significant contribution of the construction industry to

*Department of Civil Engineering, Punjab Engineering College, Chandigarh-160012, U.T.

employment creation. Another common feature is the high proportion of the work-related accidents of a particular country that occurs in the construction sector, which makes construction safety a compelling issue.

However, the most striking aspect of the above data is the enormous proportion of the occupational fatalities of a particular country that occurs in the construction sector, which makes construction safety one of the highest priorities for national policies and programmes in the field of occupational safety and health (see Fig. 45.1). Fig. 45.2 shows the fatality rates in the construction industry in a number of countries.

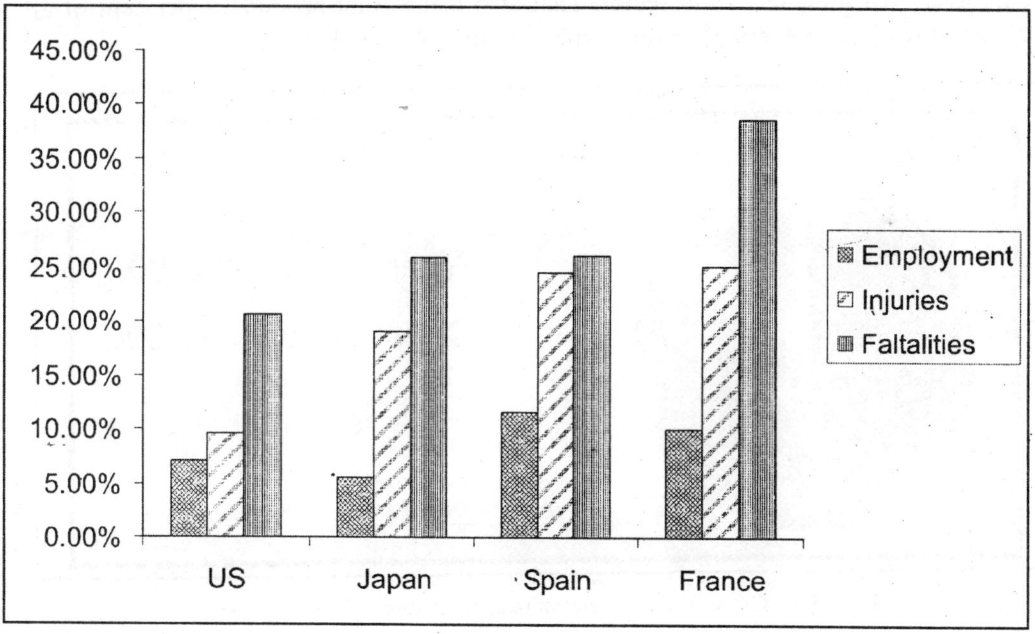

Fig. 45.1 : Employment and occupational accidents in construction as a percentage of all economic activities

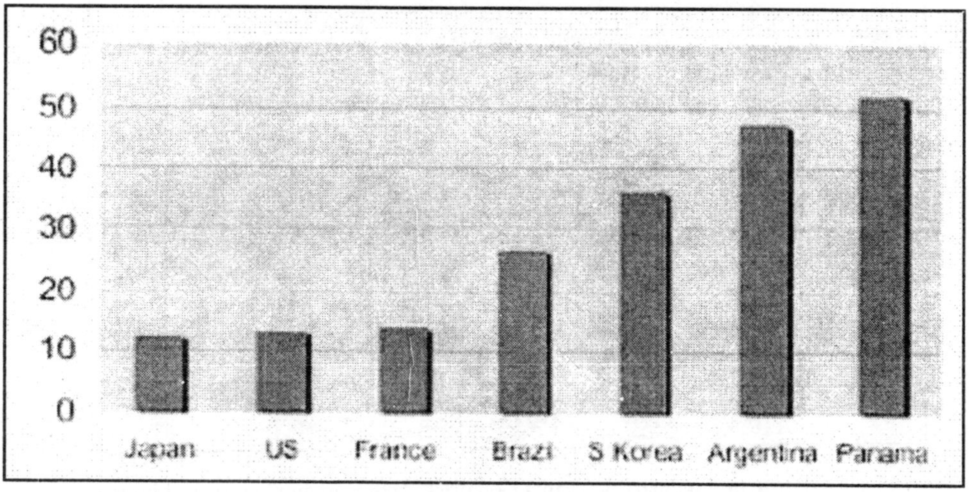

Fig. 45.2 : Fatality rates in construction industry. Different countries. Number of fatalities per 100,000 workers

It should also be noted that in most of the industrialized countries the fatalities involved in construction work is now stabilized at below 20 fatalities for every 100,000 workers.

The case of developing countries is different. Their situation is far from being uniform. Some developing countries have already brought down their fatality rates in the sector below the level of 40 (per 100,000). Yet, in most of the developing countries these rates are believed to be still well above this level. The number of fatal occupational accidents in construction all over the world is not easy to quantify, as information on this issue is not available for most countries. Nevertheless, the ILO estimates that at least 60,000 fatalities occur at construction sites, around the world, every year. This means that one fatal accident occurs every ten minutes in the sector. That also means that around 17 per cent of all fatalities at work (1 in every 6) are construction fatalities (see Fig. 44.3).

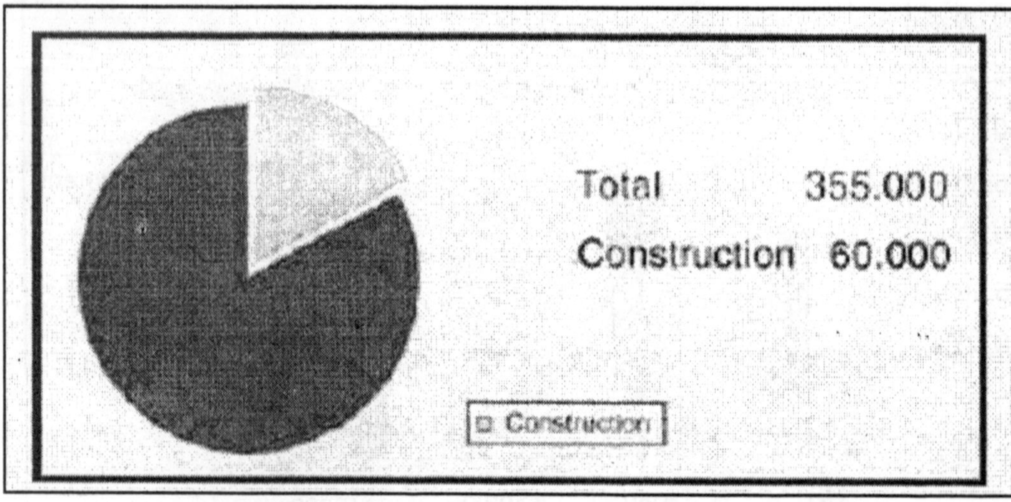

Fig. 45.3 : Global ILO estimates of fatal occupational injuries, 2003.

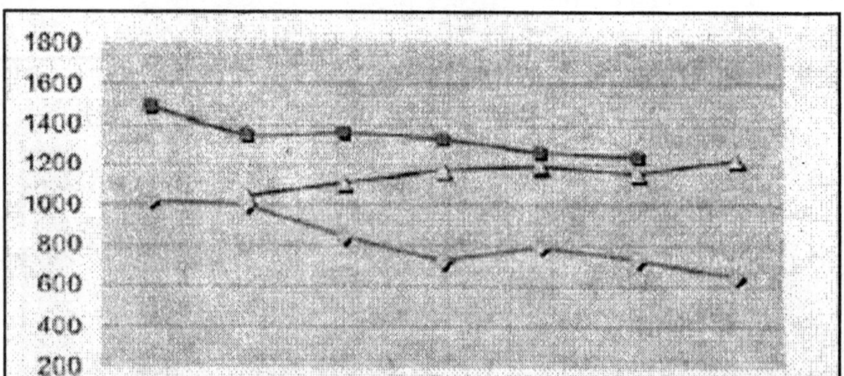

Fig. 45.4 : Construction industry. Different Countries. Trends in number of fatal injuries

The Case of Developing Countries

There is little doubt that many developing countries like India and China have seen a dramatic increase in output and employment in the construction industry in the past 30 years. For example for China, in the year 2002, the output of the construction industry was around US$ 100 billion. Numerous studies have shown that construction output is growing particularly rapidly, often exceeding the overall economic growth rate. The industry employed more than 36 million people, doubling its share of the total workforce from 2.3 per cent to 5.2 per cent. The construction industry plays a major role in combating the high levels of unemployment and in absorbing surplus labour from the rural areas.

In India the construction sector has been steadily growing since Independence. It is the prime mover for the growth of the Nation's infrastructure and produces goods and services worth Rs. 2.1 trillion. The industry also is the second largest employer, as it employs 31 million persons round the year (Source: Construction Industry Development Council, New Delhi).

The construction workers are one of the most vulnerable segments of the unorganised labour in our country. They are exposed to a wide variety of serious Occupational Safety and Health (OSH) hazards. The rate of fatal accidents in this industry is 4 to 5 times that of the manufacturing sector. The workers are also exposed to a host of hazardous substances, which have a potential to cause serious occupational diseases such as asbestosis, silicosis, lead poisoning, etc. There is also a serious potential for fires due to the storage and use of flammable substances and a potential for disasters due to collapse of the structures and subsidence of the soil on which the construction activity is being carried.

In 1996 the Building and Other Construction Workers (Regulation of Employment and Conditions of Service) Act, 1996 was promulgated. The Central Rules under this Act were notified in November, 1998. Since the Rule making powers and enforcement rest with both the Central and State Governments depending upon which one is the "Appropriate Government" under the Act, notification of Rules by different States is essential. Except Kerala, other States have not yet notified their Rules.

Table 45.1: Break-up of Construction Costs

	Materials %	*Construction Equipment %*	*Labour %*	*Finance %*	*Enabling Expenses %*	*Admin. Expenses %*	*Surplus %*
Building	58-60	4.5	11-13	7-8	5.5-6.5	3.5-4.5	5-6
Roads	42-45	21-23	10-12	7-8	5.5-6.5	3.5-4.5	5-6
Bridges	46-48	16-18	11-13	7-8	5.5-6.5	3.5-4.5	5-6
Dams, etc	42-46	21-23	10-12	7-8	5.5-6.5	3.5-4.5	5-6
Power	41-43	21-24	10-12	7-8	5.5-6.5	3.5-4.5	5-6
Railway	51-53	6-8	16-18	7-8	5.5-6.5	3.5-4.5	5-6
Mineral Plant	41-44	20-22	12-14	7-8	5.5-6.5	3.5-4.5	5-6
Medium Industry	50-52	7-9	16-18	7-8	5.5-6.5	3.5-4.5	5-6
Transmission	49-51	5-7	19-21	7-8	5.5-6.5	3.5-4.5	5-6

Source: Construction Industry Development Council Survey

Considering the significance of the construction sector, it is necessary to identify the major issues affecting the efficiency of the sector and take corrective action.

The Main Occupational Health Hazards

Experience from developing countries shows the following to be the major occupational health hazards:

Social Issues

The stresses inherent in site work: low pay, long working hours, poor hygienic conditions, poor quality of food and water, and the abuses that an all-male workforce living away from home seems to indulge in, such as heavy smoking, alcohol consumption, and sexual-psychological issues, including even sexually transmitted diseases.

Noise

Many studies have shown that exposure to high levels of noise is widespread in the construction industry. The extensive use of heavy equipment and hand tools within small, sometimes enclosed, areas produces personal exposure to noise. The enclosed nature of tunnels and cut and cover areas increases the problem, because sound reverberates off the concrete walls and ceilings. Hearing loss due to noise exposure is a prevalent occupational disease in developing

Lifting and Handling Heavy Objects

These are caused by manual handling of heavy loads, sometimes over long distances—for example, bricks, cement blocks and cement bags weighing 50 kilos. Confined spaces, awkward postures, heavy tasks and productivity demands and long hours can aggravate the problem. Lower back injuries, sciatica, hernias and slipped discs can put people out of the labour market for good.

Vibration Syndromes

Hand-arm vibration can cause damage to blood vessels and nerves. This leads to a lack of sensitivity in the fingers. Called Raynauds Syndrome, this condition is particularly due to the use of pneumatic tools. Whole body vibration is caused by operating heavy machinery and vehicles, and can damage the spine.

Other Musculo-Skeletal Disorders

These are injuries to muscles, nerves, tendons and joints caused by physically demanding work. Risk factors include: uncomfortable postures, forceful and repetitive movements, awkward tools and sustained effort. In many developing countries, construction work is really labour-intensive. There is little mechanization and tools are rudimentary, recycled and improvised. Typical injuries include: *bursitis*, from kneeling, for example floor laying;

tenosinovitis—the inflammation of the tendon sheaths due to overuse and repetitive and forceful movements; *tendonitis*—inflammation of the tendons, especially in the shoulder, often due to working with the arms reaching above shoulder level; neck problems—for the same reason; epicondilitis ("tennis elbow"), caused by the impact absorbed when making repeated blows

Some Facts and Figures

The construction industry consistently ranks as one of the leading industries with regard to the occurrence of occupational fatalities in most developing countries. For example in China in the year 2002, construction work had the highest number of fatal work injuries to date, with over 1,948 accidents leading to 2,042 deaths. Compared with the figures for the year 2001, these represented increases of 16.4 per cent and 24 per cent. In all, 80 per cent of all fatal occupational injuries are attributable to six causes: motor vehicle incidents, machine-related incidents, being struck by falling objects, collapse of structures, falls, and electrocution. Of these accident types, falls (47.2%), collapse of structures (19%), being struck by falling objects (8.8%), electrocutions (9%), and machine-related incidents (6%) were found to be the leading causes of death. Injuries resulting from falling are usually classified as occurring from an elevated work surface (e.g. ladders, scaffolds, roofs, buildings, stairs, vehicles, etc.) or from the same level. The magnitude of these fatal occupational injuries resulting from falling is a major public concern.

With respect to India except for few pilot surveys in some of the segments of the unorganized (agriculture, construction etc.) sector, no authentic statistics at the national level are available on accidents and occupational diseases in these segments.With regard to the construction sector the rate of fatal accidents is 4 to 5 times that in the manufacturing sector. The workers are also exposed to many types of hazardous substances, which have a potential to cause serious occupational diseases such as asbestosis, silicosis, lead poisoning, etc.

Possible Reasons

The reasons for the high accident rate in construction Industry especially in the developing countries are numerous and not difficult to ascertain. Some of the probable causes are listed here below:

1. Construction is labour-intensive, and the workforce tends to be itinerant, unskilled, foreign, and, above all, subcontracted.
2. Many of the construction workers come from outside urban areas. Their educational level is relatively low and their notions of health and safety are less than adequate. Furthermore, most of them usually receive little training before beginning the job, especially in some small companies.
3. In addition, their work environment changes frequently. Construction site workers operate moving machinery and vehicles, work underground or at elevated levels, and use sharp tools and heavy equipment. This is also a major cause of the high accident rate in the local construction field.

4. Yet another main reason is that the existing legislation does not address all of the problems. The legislation does not provide a sufficient prescriptive umbrella that defines accountabilities, supplemented by succinct regulations and industry-led codes of practice or guidelines.
5. Another important reason is inadequate management and organization. For example, even though this country has some safety and health legislation, compliance and enforcement are sometimes lacking or minimal. This situation needs to be rectified.

Recommendations and Corrective Measures

There is an urgent need to identify and declare established technically competent authority under Ministry of Labour for enforcement of the technical provisions and also development of standards based on field conditions. State Government has to expedite the notification of their Rules under the Building and Other Construction Workers (Regulation of Employment and Conditions of Service) Act, 1996 and identify the enforcement agencies (Annexure I). There should be a Central Plan Scheme for covering the following activities:

1. Developing and conducting appropriate training of the Inspection Officials of both the Central and State enforcement agencies.
2. Developing Model Guidelines and terms and conditions for identification and use of the services of experts and technical agencies for the purpose of carrying out technical inspections or enquiries into accidents or dangerous occurrences as required under the statute.
3. Developing a suitable enforcement strategy for the purpose of achieving best results in terms of the compliance with the key statutory provisions for the construction sector.
4. Developing and conducting training courses for key categories of workers (such as scaffolding erectors, operators of lifting machines, etc.) and the supervisory and middle management staff from the industry.
5. Development of Do's/Don'ts, Codes of Practice/Training Manuals/Guidelines/ Check-list are important for safe performance by those employed in this sector is essential.

For concurrently and effectively implementing the activities under the above Scheme, it is suggested that different measures be contracted out on project funding basis to independent/autonomous institutions having competence in the above fields.

At present Chief Labour Commissioner (CLC) enforces the 'Building and Other Construction Workers (Regulation of Employment & Conditions of Service) Act, 1996 and the central rules framed there under. Directorate General Factory Advice Service & Labour Institutes (DGFASLI) provides technical support in drafting model rules, carrying out surveys and conducting training programmes in construction sector. The National Safety Council (NSC) undertakes training and awareness creation activities in this sector.

Further improvement and a new high in safety standards in the construction sector can be achieved by implementing following suggested corrective measures in a phased manner:

- The scope of the Building & Other Construction Workers (Regulation of Employment and Conditions of Service) Act be enlarged by suitably reducing the existing eligibility criteria to 5 workers and making appointment of safety officers by the employers of establishments employing 100 workers and above as a statutory obligation.
- The Act applicable to the construction sector is presently overseen by non technical department. In order to supplement their efforts in inspection, the existing professional bodies/organizations, may be associated to oversee technical inspections and investigation of accidents.
- DGFASLI and professional independent bodies/organizations should organize trainers' training programme for education officers Central Board of Workers Education and Supervisory and Management personnel from industry.
- The Central Board of Workers Education under the Ministry of Labour may be given the task of training of construction workers in collaboration with professional independent institutes involved in construction safety.

The Path Ahead

Because of the inherently hazardous nature of the construction activity, it may never be possible to completely prevent accidents and serious injury in this field. The stress should be on safety and accidents prevention to try to reduce the number of accidents. Safety is everybody's job. All member of the construction industry must work together to prevent accidents. In construction industry risk can be reduced to an acceptable level by investing in human capital to improve the quality of labour. This must be done through more training and certification institutes to augment the supply of skilled labour; associating ITIs with training of construction workers; stipulating the widespread use of trained labour as a prequalification condition for bidding. The percentages stipulated could be increased over time. Safety in construction must be given high priority or else the industry will continue to be one of the most hazardous industries to work in.

REFERENCES

BLS (Bureau of Labour Statistics) Workplace Injury, Illness and Fatality Statistics 2001, USA http:/ www.osha.gov/oshstats/work.html

Khitoliya, R.K., and Sen, A.K. Various aspects to make construction safe and hazard free, *Journal of Institution of Engineers*, Vol 75, August 1994, pp 83-86.

LABORSTA. ILO on-line database on labour statistics http://laborsta.ilo.org/

CNAMTS (Caisse National de l'Assurance maladie des Travailleurs Salariés).Statistiques nationales 2000 des accidents du travail, France (*Travail & Securité*, July - August 2002, No. 620)

INSHT (Instituto Nacional de Seguridad y Salud en el Trabajo). Estadísticas de Accidentes de Trabajo 2001: http://www.mtas.es/insht/statistics/mtas.htm

JCSHA (Japan Construction Safety and Health Association). Visual Statistics of Industrial Accidents in Construction Industry 2001: http://www.jicosh. gr.jp/english/statistics/jcsha/2001/2001.pdf

ILO, 1992. ILO Code of Practice Safety and Health in Construction. (ILO, Geneva) http://www.ilo.org/ public/english/protection/safework/cops/english/ download/e920894.pdf

Report of Working Group on Occupational Safety and Health, Tenth Five Year Plan (2002-2007) Government of India, Planning Commision-2001

Xia, Z., LU, G.L, Liang, Y. 2004. Challenges and strategies of occupational health and safety in the construction industry in China, *Asian-Pacific Newslett on Occup Health and Safety* 2004;11pp.:8-9

ANNEXURE I

Legislations Relating to Occupational Safety and Health in India

1. The Factories Act, 1948 and the State Rules notified thereunder
2. The Dock Workers (Safety, Health & Welfare) Act, 1986 and the Regulations framed thereunder
3. The Mines Act, 1952 and the Rules framed thereunder
4. The Plantation Labour Act, 1951
5. The Shop and Establishments Act
6. The Explosives Act, 1884 and the Rules framed thereunder
7. The Petroleum Act, 1934 and the Rules framed thereunder
8. The Insecticides Act, 1968 and the Rules framed thereunder
9. The Indian Electricity Act, 1910 and the Indian Electricity Rules, 1956
10. The Indian Boilers Act, 1923 and the Indian Boilers Regulations
11. The Dangerous Machines (Regulation) Act, 1983
12. The Environment (Protection) Act, 1986 and the Manufacture, Storage and Import of Hazardous Chemicals Rules, 1989 and other Rules framed thereunder
13. The Indian Atomic Energy Act, 1962 and the Factories Rules framed thereunder.

Section–VI

Innovative Safety Measure by the Award Winners

SAFETY MANAGEMNT AT NTPC, RAMAGUNDAM

The safety, health and environmental protection are major concerns of NTPC Ramagundam management and have been given high priority. As a part of it, Safety department, exerting all its efforts to spread its activities, not only in Industrial Safety, but also Safety at Home and on road to make the life of NTPC family is more safer. Safety Department at NTPC Ramagundam is headed by Manager (Safety) and supported by twc Sr. Engineers, two supervisors and a stenographer. Head of Safety directly reports to the Head of the project and shall advise and monitor the implementation of safety procedures.

Hazard Identification

To identify the unsafe acts and unsafe conditions, safety department conducts daily site inspections, periodical special inspections, surprise checks, joint site inspections. The unsafe acts and unsafe conditions thus found are brought to the notice of concerned Engineer-in-charges with suitable remedial measures.

Special Safety inspections are conducted on daily basis during Unit overhaul works and other critical works. Safety Assurance Plans are prepared by Safety department for overhaul works and critical works identifying various works and precautions to be taken during the specified works. Responsibilities for the implementation of same are fixed on the concerned working agencies including Engineer in-charges. The Safety Assurance plans are written document signed by the Engineer in-charge, Concerned Contractor and Safety Department.

Safety Department conducts special inspections/Surveillance Checks during overhaul works to monitor the above for effective implementation. The points identified are discussed in overhaul review meeting for immediate implemaentation.

Safety at Overhaul Works

Testing of Sky climbers: Sky climbers are tested by Competent authority before put them into use during overhauls. Apart from the above, joint inspection by Safety, Quality and Concerned EIC is done before actual use in respective boilers to ensure safe erection of sky climber, healthiness of sky lock, hoist motors and wire ropes and protocols are signed and maintained.

Inspection of Scaffoldings : Scaffoldings erected for working at heights are tested jointly by Safety, Quality, concerend EIC and Protocols are signed and maintained.

Inspection of portable electrical tools: Portable electrical tools like drilling machines, grinding machines etc., are tested jointly by Safety, Electrical and protocols are signed before put them into use.

Usage of ELCBs : Usage of Earth Leakage Circuit Breakers is ensured while tapping temporary lighting supply, welding machines and all other portable drilling machines etc.

Daily inspections are conducted during overhauls by the members of 24×7 cross functional team members apart from safety department staff. Exceptions are discussed in the daily overhaul review meetings.

24×7 Cross Functional Safety Task Force Team

Exclusive 24×7 cross functional safety teams on weekly rotational basis have been constituted to ensure compliance of safety requirements and safe working environment in the O&M area.

The team is collecting the list of works identified for each day from the Maintenance Planning department and inspecting the work areas accordingly. The members of the above cross-functional safety teams are inspecting all areas of O&M including overhaul works. One of the members is coming in B shift everyday during their tenure and in case any critical activities are planned during night shift one member is also coming in the Night shift for close monitoring of safety precautions.

The team leader discusses the observations on daily basis in the planning meetings and also in overhaul review meeting for immediate implementation. Any unsafe practices are noticed by the team, thc work is stopped till the safety precautions are implemented. They are also giving pep talks to all the working group on the spot about the unsafe practices being followed by them and safe procedure to be adopted.

Head of Operation and Maintenance is reviewing the observations in daily planning meetings and exceptions are reviewed by General Manager once in a week in the daily planning meeting.

The observations are prepared by the respective teams on weekly basis and submitted to all concerned with copies to Head of O&M, General Manager and Head of Safety and the team leader of succeeding week's group. Exception report is submitted to Regional Headoffice once in a month.

Suggestions for Safety

The hazards identified are analysed and apprised to the concerned Engineer-in-charges with suggestions for their rectification.

Safety work orders are also issued for implementation of unsafe condtions identified and the same are being monitored.

Implemetation of Safety Suggestions

The identified unsafe conditions/actions through above means were regularly followed up and reviewed with concerned section in-charges, HODs and site incharges of contracting agencies by safety department for implementation. The pending observations and critical points are discussed in daily planning meetings of Main plant and Coal Handling Plant for their rectification.

Various Safety audits recommendations and other major safety issues/problems which needed the attention of General Manager are discussed every Wednesday in the planning meeting chaired by the General Manager and also in special safety review meetings with HOD's/DGM's for their compliance.

Fire and Safety Review Meeting

These meetings are conducted once in every month and the issues related to the Fire and Safety are being discussed and reviewed for implementation. All heads of departments/ sections, Incharge of CISF Fire wing, Head of Safety will attend these meetings. The meeting is chaired by Head of O&M. Minutes of meetings are recorded.

Employee Participation in Safety Management

Station Safety Committee

Safety Committee is constituted as per the statutory requirement with equal participation of workers and management representatives. Safety Committee meetings are conducted once in every two months.This meeting is participated by all the HODs, Employees representatives nominated by recognized Unions of Workers and Supervisors. This meeting is chaired by Additional General Manager (O&M).

Twelve safety committee meetings were held during the year 2004 in which 64 points were discussed. Out of 64 points, 41 points were implemented and rest of the points action plan made for implementation.

Project Safety Committee

Exclusive Safety Committee was constituted for Unit VII Project area to discuss and review safety aspects of Construction and Erection. This meeting used to be conducted on a fixed date of 13th of every month. All the HODs of Project (C&E), all Site In-charges of agencies, Safety Officers NTPC and chairman of Safety Task Force and DGM (Commissioning) were the members of the above committee. These meetings were chaired by General Manager.

Safety Stewards Committee

To involve supervisory employees in safety promotional activities, Safety stewards scheme is introduced and is in operation in NTPC Ramagundam. These meetings are conducted once in every 2 months. The safety problems pertaining to the concerned stewards department are being discussed and remedial measures are suggested. This meeting is chaired by Head of Safety.

Six safety steward meetings were conducted during the year wherein 16 points discussed and 13 points were implemented.

Safety Circles

Safety circles are formed with voluntary membership of employees irrespective of cadre within a department. The circles meet in their department and select the problems related to safety and put efforts to solve for improvement in safety standards. 4 Safety circles from OS-M, AHP, Operation, BM-PP1 and BM-PP2 are functioning.

Safety Challan System

To caution the contracting agencies for strict adherence to the safety systems/proceedures and for immediate rectification of unsafe conditions/acts safety challan system is in pratice. On repetative violations by any contracting agency, safety challan is issued through Engineer -in-Charge with a recommendation for imposing penalty.

Permit To Work (PTW) System : A systematic and comprehensive Permit To Work (PTW) system is in vouge. Whenever, any maintenance work or repair work is undertaken, the concerned maintenance department will take the PTW prior to the start of the work. Before issuing PTW by Operation department isolation of all the respective equipments is ensured and locking is also done by the operation dept to prevent inadvertent start of the equipment during maintenance or repair.

Testing of Lifting Tools and Tackles: All the lifting tools and tackles being used by NTPC and contracting agencies and pressure vessesls are tested by a competent person as per statutory norms and records are maintained.

Safety Clauses in Conditions of Contracts: Safety clauses were incorporated in Conditions of Contract while awarding contracts and safety conditions were monitored by Engineer Incharges and Safety Officers while executing the works.

Safety Audits

Third Party Safety audit is conducted every year by reputed organisations like M/s.National Safety Council, M/s. Disaster Management Institute etc.

HAZOP study and Risk analysis of crtical locations are conducted through reputed organisations like M/s National Safety Council to ensure system safety.

Internal safety audit is conducted by the safety officers of other projects once in every year.

Permit to Work system audits are conducted jointly by Safety and Maitenance Planning departments once in every month and the exceptions are discussed in the daily planning meetings for implementation of the recommendations.

Occupational Medical Checkup

Occupational Health checkup for employees working at hazard prone areas is conducted annually as per Statutory and Corporate guidelines. In the year 2004-05, 659 employess were covered out of 667 employees identified.

Use of Information Technology for Safety

Safety website is created and the following information is kept on the web site of Godavari Intranet for the benefit of all the employees.

- Recommendations of Accident Enquiry Committee Reports
- Reporting of Incidents/Unsafe Acts/Unsafe Conditions
- Accident Reporting System/Formats
- Material Safety Data Sheets
- Safety Assuracne Plans
- Disaster Management Action Plans
- Emergency Prepardeness and Response Plan
- Accident analysis
- Monthly reports
- Safety Committee minutes
- Safety stewards minutes

Inspection/Testing of Fire Detection/Protection Systems

Inspection/testing of Fire detection/protection systems are conducted as per the schedules involving Safety, Maintenance Planning, Operation, Maintenance and CISF- Fire wing departments. Protocols are maintained. Exceptions are discussed in daily planning meetings and also reviewed in Fire and Safety review meetings.

Safety Education and Training

Demonstrations

To create awareness in usage of fire extinguishers theory-cum-practical demonstrations are conducted at various work locations periodically in coordination with CISF(Fire Wing). 11 theory-cum-fire fighting demonstrations were conducted covering 535 NTPC and Contractors' employees during the year 2004.

Practical demonstrations are also conducted periodically at worksites for acquaintance in the usage of self-contained breathing apparatus (SCBA) sets and Chlorine leak arresting kits. During the year 2004 five such programmes were conducted.

Electrical Safety and First aid demonstrations at work locations are organized with the help of company doctors; 05 Programmes were organised among all the employees working in Electrical systems during the year 2004.

Four progrmmes on use of modern safety gadgets like Fall arrestors, safety nets etc., were organised at Unit-VII Project activities through a supplier and live demo also was organised for the benefit of all the employees during the year 2004.

Three programmes on the "Usage of flash back arrestors for gas cutting sets" were organised at Unit-VII project area and two programmes at O&M area including live Demo for highlighting the importance of flash back arrestors in gas cutting sets.

Pep Talks

Pep talks are conducted at work locations wherein job specific safety precautions are explained to the working group regularly by Safety and 24×7 cross-functional teams at various work locations.

Safety Training Programmes

To educate the employees for safety in-depth various safety training programmes are conducted for NTPC employees and contractors employees on various safety subjects. During the year 2004, 25 Safety training programmes covering 640 NTPC employees and 8 safety training programmes for contractors' employees covering 153 persons were conducted. Employees were also sponsored for outstation safety training programmes organised by reputed safety organisations.

To educate the school children on road safety, 7 road safety training programmes were organised at different schools covering 1475 school children during the year 2004.

Home safety training programme was also conducted for ladies of NTPC family covering 30 participants through M/s. Loss Prevention Association Ltd., Secunderabad during the year 2004.

Safety Promotional and Motivational Measures

National Safety Day celebrations: National Safety Day is celebrated every year on 4th, March in a befitting manner. To involve and motivate all the NTPC employees, Contractor employees and their families towards safety, competitions are organised in Safety Quiz for employees, Painting competition for School children, Usage of SCBA set for employees and contractors workmen. Prizes are distributed to the winners as party of the celebrations.

Safety Awards: To encourage healthy completion among various departments of the Plant, "Good Safety Record" and "Good housekeeping" award schemes are in practice. The departments adjudged as best in safety record and Housekeeping standards are given Rolling shields and commendation certificates on National Safety Day.

To motivate the contracting agencies to implement safety norms, Suraksha Jyothi Safety Award in the form of Rolling Shield along with a certificate are given away to the agency based on their safety performance. Special commendation certificates are also given to the agencies who achieved ZERO accidents during the year.

Suraksha gynapika award in the form of rolling shield and a certificate is given to the agency working for project activities and maintaining best safety standards.

Display of Safety Messages: Safety hoardings are displayed with safety messages and pictures by the side roads from Township to Main Plant. Safety pictorial messages are displayed on hoardings and on the walls inside the plant area.

Safety propagation: To propagate safety message among employees booklets, pamphlets on various safety topics are prepared and distributed. For ensuring safety, while doing any maintenance/repair works or during operation of the plant, a set of standard safety procedures have been developed and brought out in a booklet form under the title "safety document" for all the systems and critical equipment. It encompasses work place hazards, statutory requirements, manufacturers recommendations, emergency procedures and safety requirements for safe operation of all the systems. The safety documents are developed for various operations like Electrical maintenance, turbine operation, boiler operation, operation of the main plant, Ash Handling, Pressure Parts, Rotating Machinary etc.

Selection, Purchase and Distribution of Safety Equipment

The safety equipment, personal protective equipment are selected, procured in required quantities based on working condition, hazard potential and nature of activity and distributed to the employees. The usage of PPE by the employee and availability of stocks are monitored.

Appointment of Qualified Safety Officers by Contracting Agencies

Appointment of qualified Safety Officers by all the contracting agencies was made mandatory by incorporating a Special Safety clause in the Letter of Award while awarding the work contracts. The same was ensured during the entire Unit VII construction stage of the plant.

Special Safety Task Force

To exclusively monitor and for effective implementation of safety norms to ensure safe work environment at Unit VII Project under construction, a special safety task force was constituted. A senior executive of Dy. General Manager rank was made the Chairman of the task force with 5 senior executives of Manager and above rank from O&M and from project area as other members.

Disaster Management

To tackle any emergency smoothly and effectively, "Disaster Management Plan" has been prepared with detailed disaster control measures/actions and is being updated time to time. The copies of the plan are distributed to the concerned employees. The synopsis of the plan is prepared in Telugu, Hindi and English and distributed to all cadres of employees to act accordingly in case of emergencies.

Training/Workshops are conducted for all key personnel and members of various teams constituted in the Plan. Appreciation programmes are also conducted for employees of various cadres.

To assess the effectiveness of the Plan, Mock drills involving all teams members including General Manager and Mutual Aid neighbouring industries, are being conducted once in six months and lacunas thus found are corrected for improvement.

Fire Fighting System

NTPC Ramagundam is having a full-fledged fire station operated by CISF(Fire Wing) personnel. The fire station is fully equipped with all the facilities to handle the fire promptly and actively. The Plant is protected against fire hazards and well equipped with all sorts of fire safety devices as per the "Tariff Advisory Committee" norms.

Accident Analysis

All the incidents and accidents are investigated as per the NTPC Corporate Guidelines and recommendations are suggested and maintained for their implementation to prevent recurrence.

List of Safety Awards won by NTPC-Ramagundam

Year	*Award*	*Awarding Agency*
1986	Safety Award	Brithish Safety Council, U.K.
1987	Safety Award	British Safety Council, U.K.
1987	National Safety Award	Min. of Labour, Government of India
1988	Sword of Honour	British Safety Council, U.K.
1989	National Safety Award	Min. of Labour, Government of India
1990	National Safety Award	Min. of Labour, Government of India
1991	NTPC Safety Award	NTPC, Corp. Centre, New Delhi
1995	Safety Excellence Award for Health and Safety Training	Council of Safety Professionals, A.P
1996	Safety Excellence Award for Fire Safety Management	Council of Safety Professionals, A.P
1998	Safety Award	British Safety Council, U.K.
1999	Safety Award	British Safety Council, U.K.
1999-2000	NTPC Safety Award	NTPC, Corp. Centre, New Delhi
2001	Prashamsa Patra	National Safety Council of India
2001-2002	NTPC Safety. Award	NTPC-Corp. Centre, New Delhi
2001-2002	Green Tech Safety Silver Award	Green Tech Foundation, New Delhi
2003-2004	Green Tech Safety Gold Award	Green Tech Foundation, New Delhi

POWER GRID

SAFETY PRACTICES OF POWERGRID

The Organization

POWERGRID, the Central Transmission Utility of the country, undertakes transmission of energy through Inter-State Transmission System. It is the largest transmission utilities in the world, maintaining transmission network of more than 53,000 Ckt km consisting of 765 kV, 400 kV, 220 kV and 132 kV AC systems and 500 kV HVDC systems, along with 91 Sub Stations with over 50,000 MVA of transformation capacity spread across the country. This system is being maintained at over 99.5 per cent system availability, which is comparable to best international standards. About 45 per cent of total power generated in the country, i.e. more than 200 billion units worth Rs. 40,000 crores per year, is being transmitted across the country over POWERGRID's transmission network.

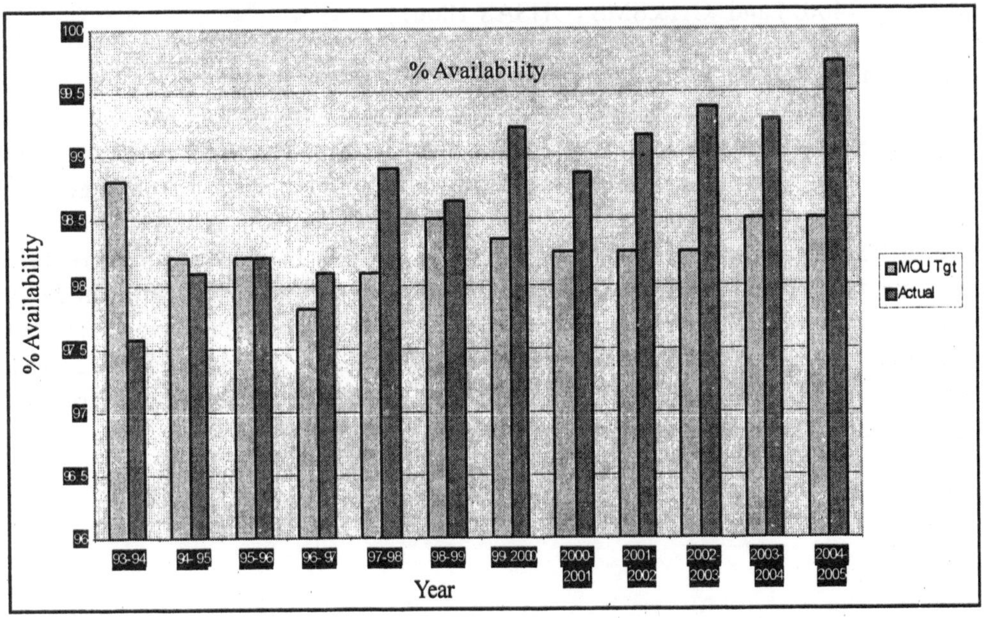

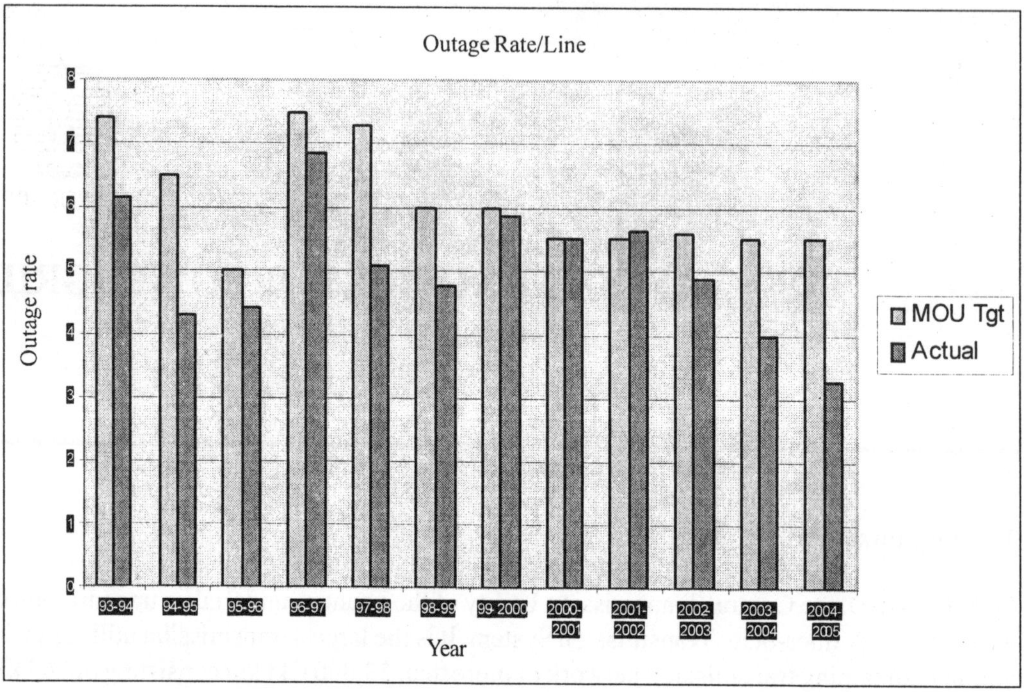

POWERGRID has adopted a safety policy since inception because safety concerns are of paramount importance in power transmission sector due to high risk jobs. POWERGRID has a safety group at Corporate centre, at its regional head quarters and at each group head quarters. Safety has been made as a culture in POWERGRID. Our emphasis is on making safety everybody's responsibility Our safety practices are of International Standard. We have implemented the Integrated Management System comprising of QMS (ISO 9001), EMS (ISO 14001) and OH&SMS (OHSAS 18001).

People: Our Real Asset

POWERGRID's main strength is its professionally competent and well-knit manpower, which is around 7000 strong and is spread all-over the country. Around 2000 executives are Engineers, 500 executives have various other professional degrees, around 1020 Supervisors have professional diplomas and about 1300 workmen have education greater than matriculation. POWERGRID's high productivity is demonstrated by an average annual growth of about 39 per cent in the asset base of the company with a manpower growth of only about 1.7 per cent per annum. Profit per employee has gone up from Rs. 4.2 lacs in FY 1992-93 to Rs. 10.95 lacs in FY 2003-04. POWERGRID recognizes the performance of the employees and rewards the same.

Visionary Leadership

Top Management—Commitment & Involvement: POWERGRID has evolved a safety culture through effective leadership, Senior Management of the organization provide vision and inculcate sustainable values for safety of all stakeholders, which inspire and motivate employees at all levels. This inner drive of the employee under effective leadership sets goal oriented directions for safe working, under which POWERGRID has been moving at fast pace, year after year and improving its own safety record.

Clearly Articulated

The Integrated Management Policy states,

- Establish and maintain an efficient and effective "National Grid" with due regard to time, cost, technology and value additions.
- Sustainable development through conservation of natural resources and adopting environment friendly technology on principles of Avoidance, Minimization and Mitigation.
- Ensure safe, occupational hazard free and healthy work environment to the satisfaction of stakeholders in all areas of its activities and shall endeavor to continually improve its management systems and practices in conformity to legal and regulatory provisions.

These Policy pronouncements helps in formulating a set of short/medium/long term safety objectives so as to formulate specific strategy and action plans to achieve better safety record. These are also aligned with organizations safety objectives and cascade down to various functional goals, which are formulated and updated on a continuous basis through interaction between corporate safety committee, Regional level safety officer and group level safety in charge. These goals and objectives by top management team for further refinement and improvement. Safety mission of POWERGRID is continually re-emphasized by the top management in various other forums such as Open-house sessions with all the employees in POWERGRID's/National Bi-partite Committee, Regional Bi-partite Committee and through similar channels at sub-regional level, touching upon each and every employee.

As a regular feature, top management also takes sessions on safety in Management training programmes, with specific objective of communicating safety mission, objectives, achievements and expectation. This helps in inculcating safety as a culture.

Leading by Example

POWERGRID has the responsibility of transmission system management, for which the Regional Load Despatch Centres (RLDCs) are manned by POWERGRID officials round the clock. Besides manning these RLDCs with senior and expert power engineers, commitment of top management is reflected by manning of the RLDCs by very senior officials (including Functional Directors), particularly in Northern Region, in the night during demanding grid situation for its safe operation. Further, in the event of Natural calamities like floods, cyclones, earthquakes, the top management takes initiatives in providing assistance to State Power Utilities for undertaking reconstruction and safe operation of system.

Transparency in Operation

POWERGRID, has developed System and Procedure Manuals for all functional areas and it is the first utility in Indian power sector to develop Environmental and Social Policy and Procedures (ESPP) with public consultations. Committee of eminent independent experts have been constituted to advise POWERGRID on various strategic issues related to Environmental and Social safeguards, R&D, financial management/procurement/project execution. POWERGRID's safety practices are open to all.

Our endeavor is to develop safety as a culture rather than as a mere compliance to statutory requirements. We carry out safety training programmes not only for our employees but also for Contractor's/Sub Contractor's employees. Normally the responsibility for training of Contractor's/Sub Contractor's employees rest with them but POWERGRID has a system by which they provide training to the employees of Contractor/Sub Contractor.

Safety as Organization Culture

In order to review, maintain and reinvigorate the safety culture, various communication mechanisms are adopted at all levels. This enables a sustainable safety culture and participative leadership style. The mechanism addresses through upward, downward as well as lateral communication within the organization, wherein safety issues, can also be raised. The major systems include—

- Open-House sessions with employees chaired by CMD and Directors,
- Suggestion Scheme for improvement—General suggestions are invited from all employee,
- In-house journals/newsletters at Corporate and Regional levels
- Bulletin Board and news item on POWERGRID Intranet portal "POWERNET". There is also an E-Open house which acts as a tool for upward communication through the internet,
- Special communiqués by CMD to all employees on specific issues,
- National and Regional Bipartite Committee act as important tools of organizational communication also, as they help in cascading the information for safety down to the lowest levels through representatives of supervisory/workman level staff.

Aligning Safety Practices with Values and Beliefs

The safety culture is built in all functional areas of POWERGRID including HR function. This is reflected in the people processes like:

- *Staffing and Selection:* Attitude towards safety and adherence to statutory requirements is selection criterion in selection process.
- *Performance Appraisal System:* Performance Appraisal gives due recognition to safety/accident record. It encourages the culture of setting/achieving safety goals through process of setting evaluating Key Result Areas (KRA). Reward systems encourage team based functioning giving due recognition to safety/ accident record. Individual contributions in identifying near misses are also recognized.
- *Training and Development:* Encourages continuous learning to create a safety culture of learning and inculcate safe performance orientation.

Environment for Continual Improvements

Our senior management provide environment conducive to motivate employees to strive for excellence through:

- An mechanism for communication of company safety policies, achievements and initiatives taken through in-house magazines;
- Any employee can make suggestions for safety improvements on the intranet;
- Well laid out personnel policy, transparent and safe working environment;
- Well established training system with due recognition to safety;
- Open house discussions.

The importance of decision making based on factual data is recognized in the organization and to further strengthen the process of capturing and making the data pertaining to safety available to all concerned while taking any decision.

Powergird a 'Safety Learning Organization'

Learning is vital. We always try to learn from each accident. POWERGRID is a safety learning organization. We encourage our employees to be open to new ideas, communicate frankly with each other, understand thoroughly about our safety practices, form a collective vision for safety, and work together to achieve the safety goal. Real commitment is rare in today's organizations. But in POWERGRID commitment for safety is visible in the actions of all the employees.

For inculcating safety culture, Organization need Analysis and Training need Analysis are conducted, keeping safety issues in view. Further by way of adopting latest technological advances in safety and International Safety Standards, employees get the opportunity to learn more about safety and its implantation organisationwide.

सुरक्षा संदेश

SURAKSHA SANDESH

INHOUSE SAFETY JOURNAL OF POWERGRID, SRTS - II

पावरग्रिड

Volume 2, Issue 2 August - November 200[illegible]

SAFETY DURING EXCAVATION WORK

[illegible]

[illegible]

[illegible]

[illegible]

[illegible]

WATER CONTENT

[illegible]

[illegible]

QUICK SAND

[illegible]

ADJACENT LOOSE FILLS :

[illegible]

Safety Integrated Project Management System

POWERGRID has well established system for planning, scheduling, monitoring and reviewing the construction programmes including annual budgetary targets in order to provide timely feedback to the top management for necessary, corrective and preventive actions and ensure that projects undertaken are completed as per approved schedule and cost. Safety is given utmost importance in these matters. This, *inter alia*, includes:

- Yearly construction targets of various on-going projects are finalized considering safety. Safety issues forms the basis of targets to be set for the Memorandum of Understanding (MoU) to be signed with Ministry of Power after discussions with various Government agencies like Central Electricity Authority (CEA), Ministry of Power (MoP), Department of Public Enterprises (DPE), Memorandum of Understanding (MoU), Task force etc.

The actual performance against set forth targets is reviewed at Senior Management level wherein safety plays an important role. This establishes the involvement and commitment of the senior management in safe and timely implementation of the projects. This is also be demonstrated by the fact that at any point of time POWERGRID handles about 50 projects, which are under various stages of implementation, and most of which are either ahead of schedule or on schedule and accident rate is minimal around six (6) sigma level.

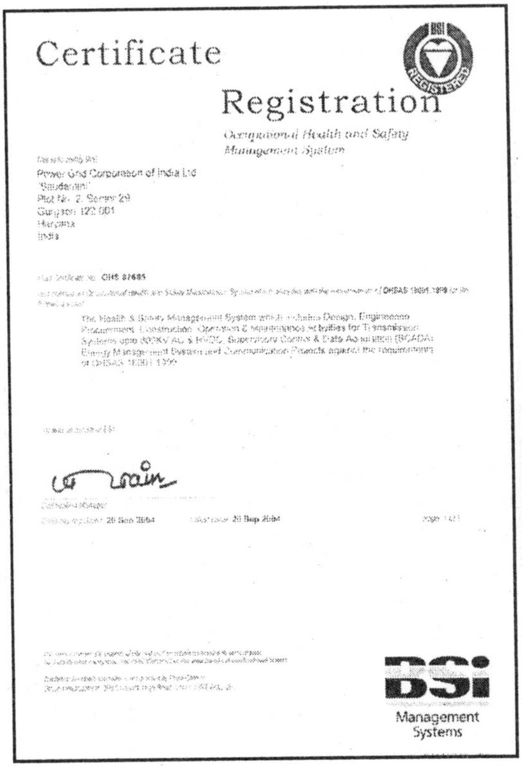
Certificate

Registration

Occupational Health and Safety Management System

Power Grid Corporation of India Ltd

Gurgaon 122 001

Haryana

India

OHS 87685

29 Sep 2004

BSi Management Systems

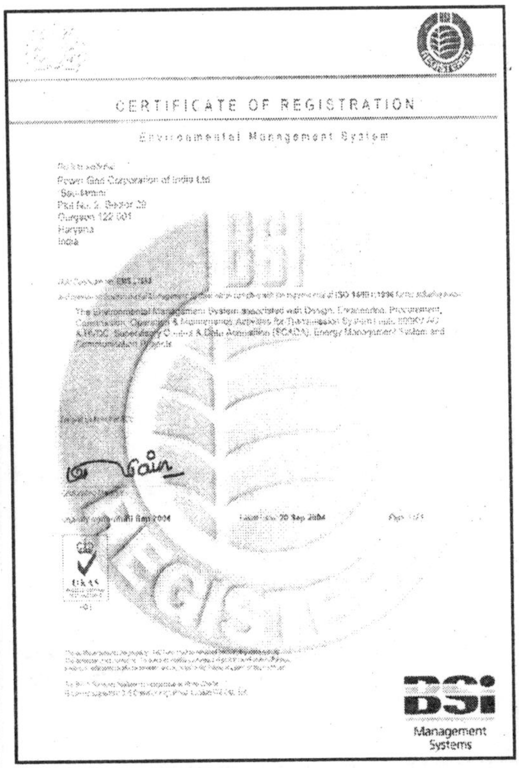
CERTIFICATE OF REGISTRATION

Environmental Management System

Power Grid Corporation of India Ltd

Gurgaon 122 001

Haryana

India

20 Sep 2004

BSi Management Systems

Continual Improvement in Safety Practices

POWERGRID has developed a system of continuous improvement by identifying unsafe practices, undertaking in-depth analysis, and prepare action plans for risk mitigation and minimization. These are monitored by senior management. The step wise activities are:

- In the departmental review of safety practices, trend analysis is done and areas for improvements are identified. Depending on the nature of risk identified, the decision is taken to initiating corrective and preventive actions for risk mitigation either at the departmental level or cross-functional teams are constituted to analyze the problem areas and determine corrective actions.
- Corrective actions undertaken are analyzed, to determine the results of improvements.
- Evaluation of effect of action on other activities is also done.
- Efficacy of actions is also reviewed.
- Records associated with the benefits accrued by the actions taken are maintained for further reference.
- All concerned are informed of the corrective and preventive actions to ensure that safety is not compromised.

In addition to above, an Integrated Management System by integrating the requirements of ISO 9001: 2000 (Quality Management System), ISO 14001: 1996 (Environmental Management System) and OHSAS 18001: 1999 (Occupational Health and Safety Management System) has been implemented. This system ensures that specific, measurable,

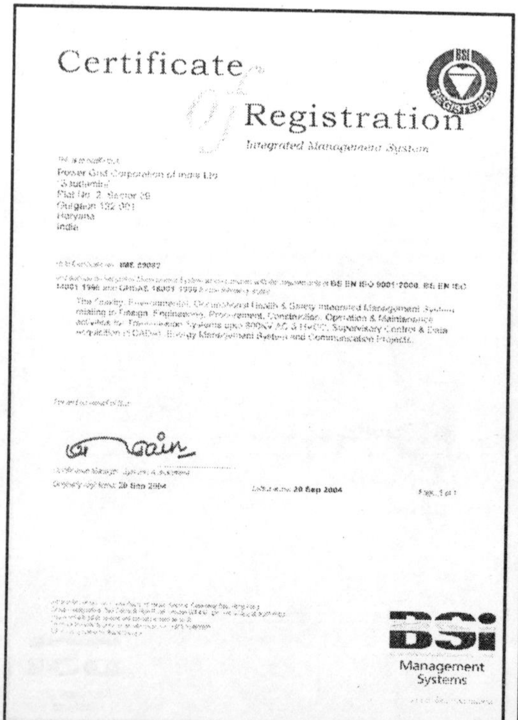

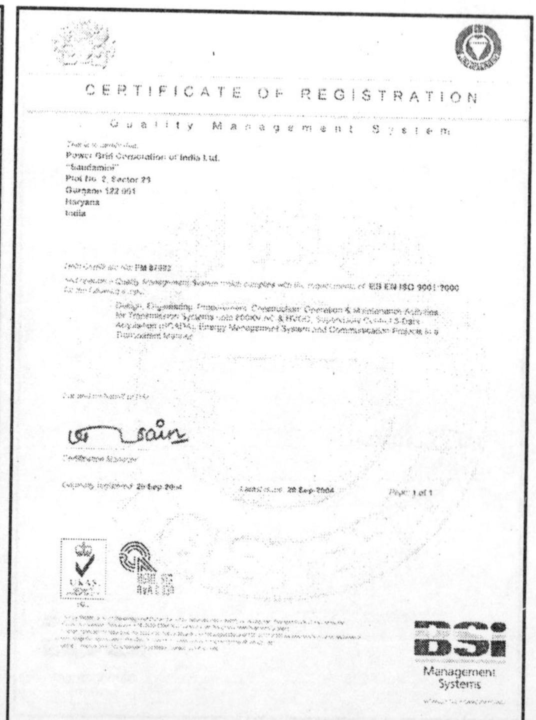

achievable, repeatable and time bound targets are set and achieved. The parameters once achieved are revised further so that continuous improvements are made and sustained. The performance of POWERGRID is reviewed by Ministry of Power (MOP), through MOU. POWERGRID has been able to achieve excellent rating in the MOU signed with MOP since 1992-93.

Total Safety Culture

Extensive use of safety tools and root-cause analysis is carried out for risk mitigation and accident reduction. The respective departmental procedures and acceptance norms are detailed out. In case, any near miss / safety problem/deviation from safety norms is observed from the acceptable norms, an efficient and effective corrective action is taken by the respective departmental head. This involves:

- Depending on the nature of un safe practices / non-conformity (non-conformities include accidents and incidents and impacts caused), decision is taken either to initiate corrective action at the departmental level or cross-functional corrective action team is constituted. The corrective action can also be initiated on the outcome of management review.
- Data is analyzed to determine the root causes for occurrence of accident/non-conformities. While brainstorming for problem solving, use of tools like Fault Mode and Effect Analysis (FMEA), Pareto analysis, Total Employees Involvement, Cause and Effect diagrams are employed.
- Evaluation of the need for action is carried out to ensure that nonconformities do not reoccur.

Mar. 16 2005 12:29PM P2

Mohammad Sharif Niaz

MINISTER OF POWER
J&K STATE

Sgr.: 2476488, 2479475 (O)
2478290 (R)
Jmu.: 2549670, 2565249 (O)
2559166 (R)

No.: PJD-II/AC/30/05

Date 16.....05.....2005

My dear Sayeed Sahib

This is to express my sincere thanks for all the support, we received from your Ministry during the grave crisis faced by the State as a result of unprecedent snow fall in February 2005. Shri R.V.Shahi, Union Secretary Power was in touch with the State administration on a regular basis and extended his full support to the State Power Development Department in carrying out the vital task of restoration of the damaged infrastructure. The manner in which various agencies of central Power Ministry and agencies of the State Power Development Department worked jointly as a team during the crises period to mitigate the sufferings of the people, represents an ideal model for any one to follow.

I would also like to place on record appreciation of the State Government for the support we received from Shri Yogendra Prasad, CMD NHPC and Shri R.P.Singh, CMD PGCIL during the period of crises. Dr. N.S.Saxena, Executive Director, PGCIL Jammu and his team of officers in particular, deserve all appreciation for all the support they offered in restoring the damaged power infrastructure particularly 220 KV Kishanpur-Pampore line which is the main link connecting the Valley with the Northern Grid. The early restoration of this important link has been possible due to their untiring efforts and resourcefulness.

Please convey my appreciation to all the concerned who worked on this difficult assignment in inclement weather and adverse conditions.

With best wishes.

Yours sincerely,

(Mohammad Shafrin Niaz)

Shri P.M. Sayeed,
Hon'ble Union Minister for Power,
Room No.201,
Shram Shakti Bhavan,
New Delhi-110001.

- Determination and implementation of actions is undertaken.
- Review of effectiveness of corrective action over a period of time is carried out.

The concerned departmental head maintains the records associated with the corrective action taken. The effectiveness is reviewed in the regular meetings and / or in the management reviews.

Responsibilities Towards Society

POWERGRID's activities are spread throughout India. Every citizen is a stakeholder. When we lay our transmission lines, we come across areas like snow bound mountains, big rivers, agriculture fields, railway tra[illegible] roads etc. Safety is a paramount importance in such works.

At times, no approach road is available and a large number of people has to move in tandem to carry a single large piece (say of conductor to cover one stretch). There are cases, when two km long, single conductor length weighing about four (4) tons was to be carried for a distance of few km by hundreds of people moving together in synchronism. This situation arose, when two mountain peaks are to be connected by a joint less connection. This requires extreme care for safety. Slightest carelessness may result in a fatal accident. However, in POWERGRID, number of safe practices has been specified to take care of such eventualities.

Recently, due to snow avalanche in J&K, main transmission line in J&K state was completely uprooted. The Power Department of J&K expressed its inability to carry out this task. POWERGRID at the advice of MOP took upon itself the task of re-erecting the line at minus (-) temperature in completely snow clad mountains. POWERGRID completed this task by involving its employees in a record time without any accident. This speaks volumes about POWERGRID's achievement in safety.

It is the first public sector utility to develop a comprehensive "Environmental Social Policy and procedures" (ESPP) for its project through a process of National consultation. The basic principle on which ESPP has been developed are Avoidance, Minimization and Mitigation and it outlines. All projects are implemented according to these plans. These plans are executed at site and monitored at the highest level.

POWERGRID has a well developed disaster management plan for safe restoration of power. At every occasion, POWERGRID had demonstrated its capabilities in restoring the transmission lines with the help of Emergency Restoration System (ERS) in shortest possible time. POWERGRID is notified as Nodal for Disaster Management in Power Sector by Government of India.

Conclusion

Safety awareness is the prime requirement to reduce accidents. Most of the accidents take place by adopting shortcuts/ignoring safety guideline. Employees are being educated to adopt safe practices. Safety checks are prepared for all the activities. Random inspection of the activities is carried out during implementation. All lapses on safe practices are being investigated thoroughly to eliminate the route cause. Implementation of these measures helps reducing accidents. By safety awareness / adoption of safe practices, accident during O&M is almost negligible.

During construction activities, the execution is being done through turn key contracts. To adopt safety measures, strict supervision is being carried out. Safety awareness programmes are being organized to educate the workers of Contractors/Sub-Contractors. Annual testing of load bearing T&P is also being carried out. Inputs were taken from OHSAS 18001 while developing strategy for safety. POWERGRID leaves no stone unturned in complying the requirement of OHSAS 18001 and other statutory authorities and want safety to prevail in all its operations.

Acknowledgement

The authors are grateful to POWERGRID management for granting permission to make this presentation. The views expressed in this are of the authors.

SHE AT NARORA

Narora Atomic Power Station (NAPS), a Unit of Nuclear Power Corporation of India Limited is engaged in generation of electricity. NAPS is a twin Unit module of 220 MWe each of pressurised heavy water reactors. The reactors use natural Uranium as fuel and heavy water as moderator and coolant. NAPS have total 1276 employees including 750 Operation and Maintenance Employees. NAPS achieved Longest Accident Free Period of 1811 Days as on 31st August 2005 with effect from 16th September 2000 and has won various Safety Awards from different renowned organizations like National Safety Council, Ministry of Labour, Atomic Energy Regulatory Board, Greentech Foundation etc. The latest award won by NAPS is the National Safety Award-2004 which is conferred by Ministry of Labour and Employment through DGFASLI, Mumbai. All these awards have been possible by taking innovative steps at NAPS for achieving excellent safety performance. Some of these steps are listed below:

1. All Safety Engineers regularly inspect the field areas and report the unsafe act and unsafe conditions to the concerned officers with a copy to Planning Engineer. The follow up of these activities is done through various plans/job lists which are being discussed in sectional/management meeting. Planning Engineer keeps record of these activities till the job is completed in all respects.
2. Every section is having a Sectional Safety Committee, which comprises the Section Head as a Convener and working engineer and staff as a member with a Industrial Safety representative. During execution of the work any feedback requiring additional platform, toe guard/handrail is being discussed in the Sectional Safety Committee. This committee meets once in a month and gives their recommendations. The follow up of these recommendations is being done by Planning Section.
3. During the annual outages, when more persons are working in round the clock shift and the jobs are to be done at height, in closed vessels, welding jobs or any jobs where unsafe condition is envisaged, a safety permit is being taken from Safety Engineer before starting the job. Safety Engineer provides round the clock coverage for such type of jobs.

4. All the department employees are given an exhaustive Industrial Safety Training, which is mandatory, and all employees are retrained in every three years. A half-day training is organized for contract workers in every six months. During this training, all industrial safety related procedures, work permits, personal protective equipments are discussed in detail along with practical demonstration of various types of portable fire extinguishers and personal protective equipments.
5. It is mandatory to apply Industrial safety work permit for all the jobs carried out at height. Medical fitness of all workers is ensured by the departmental medical officers prior to permitting them for height jobs. Special physical fitness checking facility has been developed for height jobs where a worker climbs and undergoes physical fitness test in presence of Safety Engineers. All workers perform the height jobs after wearing safety belts. Safety nets are also used when high risk is involved. All height jobs are done in day time only.
6. During start up of the Unit (before raising system temperature) after completion of maintenance jobs, a Fire Safety Committee takes round to ensure that no material is left in the area, which may cause fire or Industrial Safety Hazard. All the deficiencies are rectified before raising system temperature.
7. At Chemical addition station, Morpholine and Hydrazine, which are used in Feed Water System, are stored in required capacity bottles instead of keeping them in big containers. These avoid any accident during chemical handling. Further Material Safety Data Sheet is displayed at all appropriate locations.
8. A large number of posters in Hindi and English depicting safe working culture are displayed at various locations in the field. The continuous display of these posters has helped in inculcating the strong feeling in individuals that attitude towards safe working is utmost important for achieving a safe working atmosphere.
9. Occupational Health and Safety Policy and Environment Policy are displayed at various locations. During the promotion interview of an employee various questions related to policy are asked. By this way the employees understand the importance of these policies and role of individual in matters of safety.
10. Inter Sectional Competitions on "Best Industrial Safety Performance" and "Best Housekeeping performance" is organised. The assessment committee visits the section on quarterly basis and reviews the safety performance on various parameters like Accident Statistics, percentage of safety deficiency attended, number of shop floor meetings conducted, use of personal protective equipments (PPE's) and cleanliness.
11. Various safety competitions like Safety lecture, Safety quiz, Safety slogans, Safety posters etc. are organised for motivation of the employees.
12. Home Safety Training programmes are also arranged for Housewives.
13. All Cranes, Hoists, Chain pulley blocks etc. are load tested once per year. The safe working load along with next due date of testing is displayed.
14. A committee headed by Head (F&IS) comprising persons from different sections including HPU representative take regular rounds for checking proper use of personal protective equipments such as safety belts, safety goggles, aprons, hand gloves and identification of unsafe conditions. Head (IS&F) has been empowered to stop

any job being carried out in an unsafe manner. Penalty clauses bearing financial impact have been incorporated in work orders for violation of Industrial Safety. It has been made clear to all persons that safety of worker is the responsibility of the immediate supervisors.

15. A new practice of monitoring and trending of near miss incidents has been instituted. These incidents are discussed in detail during the monthly sectional training where total strength of the section along with representative of Training and Industrial Safety Section are present.

IndianOil

SAFETY, HEALTH & ENVIRONMENT MANAGEMENT IN INDIAN OIL CORPORATION LIMITED

Indian Oil's policy is to conduct its business in a manner that protects the safety of employees, others involved in its operations, customers and the public. The Corporation strives and will continue to strive to prevent all accidents, injuries and occupational illnesses through the active participation of every employee and is committed to continue efforts to identify, eliminate and manage safety risks associated with its activities.

Accordingly, the Corporation's endeavour is to:

- Protect life and property and take every reasonable measure to provide a safe work environment.
- Design and maintain facilities, establish management systems, provide training and conduct operations in a manner that safeguards people and property.
- Identify and manage risks associated with its products and chemicals used in the operations and not to use chemicals when it is not possible through proper design, procedures and practices to provide an appropriate level of safety for people and the environment.
- Take specific precautions required in storage, handling and transportation of its products and take steps to communicate them to employees, customers and other stakeholders.
- Maintain a comprehensive emergency response plan and respond quickly, effectively and with care to emergencies or accidents resulting from its operations, co-operating with industry organisations and authorised government agencies.
- Comply with all applicable laws, regulations and standards, and apply a responsible framework of rules where laws, regulations and standards do not exist.
- Work with government agencies and others to develop laws, regulations and standards based on sound science and consideration of risk.
- Conduct and support research to extend knowledge about the safety aspects of its operations, promptly applying significant findings and, as appropriate, sharing them with employees, contractors, government agencies and others who might be affected.

- Stress to all employees, contractors and others working on its behalf their responsibility and accountability for safe performance on the job and encourage safe behaviour off the job.
- Undertake appropriate reviews and evaluations of its operations to measure progress on safety-related activities.
- Identify and evaluate health risks related to its operations that potentially affect its employees, contract workers or the neighbouring population.
- Implement programmes and appropriate protective measures to control such risks, including appropriate monitoring of its potentially affected employees.
- Communicate, in a reasonable manner, to potentially affected individuals or organisations the scientific knowledge about health risks gained from its health programmes and related studies on Occupational Health.
- Determine at the time of employment and thereafter, as laid down, the medical fitness of employees to do their work without undue risk to themselves, others or machinery.
- Provide or arrange for medical services necessary for the treatment of employee occupational illness or injuries and for handling medical emergencies.
- Comply with all laws and regulations and make necessary medical rules where laws and regulations do not exist.
- Work with government agencies and others to develop responsible laws, regulations and standards based on sound science and consideration of risk.
- Conduct and support research to extend knowledge about the health effects of its operations.
- Undertake appropriate reviews and evaluations of its operations to measure progress on Occupational Health activities and to ensure compliance with the S, H & E Policy.
- Provide voluntary health promotion programmes designed to enhance employees' well-being, productivity and personal safety. These programmes supplement, but do not interfere with, the responsibility of employees for their own health care or their relationship with their personal physicians. Information about employees obtained through the implementation of these programmes is considered confidential.
- Organise structured programmes on occupational Health, and First Aid and disseminate information on occupational health hazards during shop floor safety meetings.

Indian Oil's policy is also to conduct business in a manner that is compatible with the balanced environmental and economic needs of the communities in which it operates. The corporation is committed to continuous efforts to improve environmental performance throughout its operations by the following endeavours:

- Comply with and excel all applicable laws and regulations.
- Encourage concern and respect for the environment, emphasise every employee's responsibility in environmental performance and ensure environment-friendly operations and training.

- Work with government and industry groups to foster timely development of effective laws, review of existing environment laws and regulations based on sound scientific knowledge considering risks, as well as review of cost, benefits and effects on energy and product supply.
- Manage its business with the goal of preventing incidents and controlling emissions and wastes to below permissible levels; design, operate and maintain facilities to this end.
- Conduct and support research to improve understanding of the impact of its business operations.
- Communicate with the public an environmental matters and share its experience with others to facilitate improvements in industry performance.
- Undertake appropriate reviews and evaluations of its operations to measure progress on implementation of environmental management systems.
- Respond quickly and effectively to environmental incidents resulting from its operations, co-operating with industry organisations and authorised government agencies.

Safety, Health and Environment Management Systems

Safety Systems

Petroleum refining has undergone considerable changes due to advancement in technologies involving severe process conditions. Plants have grown in size and complexity and the concentration of stored energy has increased. Therefore, safety in high technology process operations, which involve high pressures and temperatures, is accorded priority and the safety systems and procedures are updated from time to time. Besides taking care of the conventional safety systems, systems approach to safety is adopted that includes:

- A comprehensive corporate Safety, Health and Environment Policy.
- Guidelines on Safety, Heath and Environment Management Systems on 17 identified elements.
- Adoption of proven technology and good engineering practices.
- Process hazard identification, quantification through HAZOP & QRA studies and in-built security system to mitigate the same.
- Institutionalisation of operating, inspection and maintenance procedures.
- Measures to ensure mechanical integrity of equipment throughout the lifecycle of the plant.
- Quality Assurance Plan.
- Management of change.
- Risk Analysis and Risk Communication for Process Safety.
- Application of process safety measures.
- Environmental Impact Assessment.
- Pre-commissioning safety checks. Periodic audits to ensure system effectiveness.
- Investigation of near-misses and incidents and learning from incidents.
- Emergency planning and disaster mock drills, upkeep of lines of defence.

- Experience Sharing through Safety Talks.
- Review by top management.
- Safety Award Scheme to motivate employees.

Safety Training

Human error is the single largest contributor to accidents. This is best controlled by enhancing safety training to both our own employees and contract workmen to use the laid down operating and maintenance procedures. Safety is an integral part of induction course to all the employees.

Field demonstrations, mock drills on simulated emergencies, safety briefings at gate, safety talks and feedback from past incidents supplement classroom training. Audio-visual aids including computer based training aids are used to facilitate learning.

Every incident teaches us some lessons. Each and every incident including a near-miss is analysed to ascertain the underlying causes. Such findings are not only disseminated among the concerned personnel but also shared with others at different locations.

We view all incidents relating to fire or accident seriously. We regret to report that during the year, there were three fatalities, one lost time accident, four fire incidents of which, two were of minor nature, in seven refineries.

Process Safety

In the aftermath of fire/explosion incident at Gujarat Refinery on 29.10.2004, process safety efforts have been intensified in all the refineries through the following measures:

Four training programmes on "HAZOP", two programmes on Process Safety Management, one programme each on "Behavioural Safety" and "Safety Integrity Levels" arranged during the year with the help of national/international experts.

- HAZOP revisit for the existsing units has been initiated.
- Updation/revalidation of operating manuals is going on and shutdown manuals of process units are under preparation.
- Task observation and analysis of the employees has started.
- Technology-based training for officers and staff on regular basis has been initiated.
- A check list for Field and Panel activities, required during start-up/shutdown, specific emergencies like failure of Cooling Water/Power/Instrument Air/Digital Control System (DCS)/Critical Equipment etc., have been prepared and made available with the panel operator/officer.

Our efforts are to streamline Safety Management System at construction sites. Considering the socio-economic background of the construction labour force, various measures including intensification of safety training, preparatory safety talks, use of Personal Protective Equipment (PPE), safety slogans, punitive action on errant contractors etc., have been intensified.

Safety Management

- Every Indian Oil refinery has a well-designed fire-water network, storage and pumping system to tackle two major fires simultaneously.
- Mobile fire fighting facilities like foam/water tender, DCP tender and foam nursers are kept in combat readiness and manned by well-trained crew round the clock at the fire station(s).
- Crude oil and product storage tanks are provided with protective systems like water sprinklers, foam pourers, high volume long-range water-cum-foam monitors, etc. for effective fire control.
- The entire refinery is covered with Sirens and Fire Alarms placed at convenient and strategic locations.
- Our employees and security personnel are trained to fight fires.
- We also train contractors' personnel, truck drivers and khalasis to raise their safety awareness level.
- Modern electronic gadgets like toxic gas detectors, hydrocarbon detectors, smoke detectors, etc. provide early warning for speedy response.
- We carry out regular mock drills and surprise checks to ensure our preparedness and correct response at the time of emergency.

Disaster Control

- All refineries and other installations have well defined on-site disaster management plans based on Risk Analysis. Mock drills are conducted regularly to ensure effectiveness of the disaster management plans.
- All units have off-site disaster management plans duly prepared by the local authorities. Annual off-site mock drills are conducted at all locations.

We have mutual aid arrangements with neighbouring industries who along with district authorities and other agencies render assistance during emergencies.

Occupational Health

We accord high priority to Safety and Occupational Health of people at work and maintain Occupational Health Centres that are fully equipped. We take both preventive and curative measures.

Preventive/Protective Measures

Following in-plant measures are taken to avoid/mitigate health risks arising out of work environment:

- Proper design of facilities with safety norms
- Ensuring use of Personal Protective Equipment
- Provision of safety shower, eye fountain, first aid and fire fighting facilities

- Use of toxic gas monitors for work environment/personal exposure monitoring
- Provision of emergency kits for handling ammonia, chlorine leaks
- Adequate ventilation, illumination and protection from noise in work areas
- Ensuring compliance of safe work practices
- Training and awareness

Environment Management

Prevention of Water Pollution

We are consciously monitoring quality of our water resources. As per the laid down Minimal National Standard (MINAS), refinery effluent has to meet quality and quantum limits. Our refineries are fully complied with the MINAS norms.

Our refineries are equipped with well-designed Effluent Treatment Plants(ETP), the salient features of which are segregated collection of various effluent streams and elaborate treatment involving physical, chemical and biological steps. We have installed and commissioned clean and more efficient treatment systems like Hydrogen Peroxide treatment/ Wet Air Oxidation for treating caustic streams.

Water Conservation

We recycle and reuse the treated effluent for various purposes, i.e. for fire water and cooling tower make-up, coke cutting in Delayed Cokers and also for captive irrigation in our green belts and eco-parks in refineries. This has helped in reducing discharge of treated effluent outside and to that extent, fresh water consumption for our operations. At Panipat Refinery, we have been fully reusing treated effluent and no discharge has been made to water body since the refinery commissioning in 1998. About 50 per cent of the treated effluent has been reused as make-up to process cooling water system.

Besides internal reuse, local farmers near Mathura and Gujarat Refineries have been traditionally using our treated effluent for irrigation of crops. Scientific studies conducted by Aligarh Muslim University for Mathura Refinery, for over 14 years, have affirmed that use of treated effluent has no adverse effect on the quality and yield of crops.

Prevention of Air Pollution

We accord priority to controlling emissions from our operations. The major sources of gas emissions are flue gases from boilers and heaters, FCC regenerators and Sulphur Recovery Units. Hydrocarbon leaks and evaporation during storage, handling and transportation of petroleum products and crude oil are also potential sources of emissions to air. Our refineries have adopted various measures to control emissions and effectively disperse the flue gases.

Some of these measures are:

- Use of low sulphur fuel oil/sweet natural gas/desulphurised refinery gas in our boilers and heaters to minimise SO_2 emission.

- Hydrogen Sulphide generated during desulphurisation of refinery gas is converted to elemental sulphur in Sulphur Recovery Unit.
- Sulphur dioxide emissions from Catalytic Crackers are controlled by effective feed sulphur management.
- Flue gas scrubbing at FCC units at Haldia and Barauni Refineries for arresting sulphur dioxide emission.
- Carbon monoxide from FCCU regenerator is incinerated in CO Boilers for steam generation.
- Use of floating roof tanks for crude and other light product services and mechanical seals in pumps for minimising fugitive emission of hydrocarbons.
- Use of closed blow-down vessels and safety release to flare system for arresting any emission of hydrocarbons during all situations, normal/abnormal as well as emergencies.
- Continuous reduction in fuel consumption by Energy Conservation (ENCON) measures.

Technology Upgradation

Indian Oil has been progressively inducting advanced technology and cleaner processes in its operations. This has largely been done to make our operations environment-friendly both in terms of processes as well as products. Our initiatives in this direction include:

- Installation of Hydrotreater at Digboi, Guwahati and Barauni Refineries for improving quality of Diesel and to phase out hazardous liquid Sulphur dioxide solvent process for extraction of aromatics from Kerosene produced from Assam crude oil.
- Installation of Wax Hydrofinishing unit at Digboi refinery to phase out Acid/Clay treatment and avert hazardous solid waste generation.
- Installation of Solvent Dewaxing unit at Digboi Refinery to phase out conventional labour and energy-intensive Pressing and Sweating process of Wax extraction.
- Installation of Hydrocracking technology for upgradation of heavier ends into higher quality distillates at Gujarat, Mathura and Panipat Refineries and to reduce SO_2 emissions from Catalytic Crackers.
- Installation of high efficiency SRUs (more than 99% efficiency) for reduction of SO_2 emission from refineries.
- Production of Food/Polymer Grade Hexane with Benzene content less than 100 ppm (parts per million) with the technology developed by Indian Oil's R&D Centre.
- Installation of INDMAX unit at Guwahati Refinery, based on a new technology developed by Indian Oil's R&D Centre, for conversion of heavy petroleum stocks to LPG and high-octane gasoline.
- De-odorised Kerosene production.
- Installation of mounded storage for LPG in place of above ground storage for eliminating the risk of BLEVE.
- Installation of double seals in Floating roof tanks for minimising VOC emissions from hydrocarbon storage.

- Installation of Belfield Decantation Valve in HC tanks to arrest escape of hydrocarbon during draining of water.

Auto Fuel Quality Improvement

Indian Oil has progressively upgraded quality of auto fuels in terms of emission-related parameters and to meet the quality requirement of new generation vehicles. Our refineries have implemented various projects for meeting this objective. These include :

- Catalytic Reformers (CRU) at Mathura, Panipat, Barauni and Digboi Refineries for supplying lead-free Motor Spirit. CRU is in use at Gujarat and Haldia refineries since inception.
- An ISOSIV unit installed and commissioned at Guwahati Refinery for production of lead-free Petrol.
- Diesel Hydro-desulphurisation (DHDS) Units at Mathura, Haldia, Gujarat and Panipat Refineries for production of extra low sulphur (0.25% wt)/Ultra low sulphur (0.05% wt) Diesel.
- Diesel Hydrotreatment (DHDT) units at Barauni, Guwahati and Digboi Refineries for production of improved quality Diesel with lower sulphur content and other properties.
- Diesel Hydrotreatment projects are under implementation at Mathura and Panipat Refineries for production of improved quality Diesel with lower sulphur content and other properties.

International System Certification

All Indian Oil refineries have been certified with ISO-14001 Environment Management System, Occupational Health and Safety Management System (OHSMS) and are audited as per the International Safety Rating System (ISRS) of M/s DNV and 5-Star audit of British Safety Council, UK. It is noteworthy that Mathura Refinery was certified with BS-7750 / ISO-14001: 1996 in July 1996 and is the first refinery in Asia and the third in the world to achieve this distinction. Similarly, Mathura Refinery is also the first refinery in the world to receive the OHSMS certification. Gujarat Refinery has achieved Level-9 of ISRS in September 2003 and is the first refinery in the petroleum industry to achieve this distinction in the country. During the year 2004-05, Panipat Refinery has also achieved Level 9 of ISRS. Barauni, Panipat, Gujarat and Mathura refineries have achieved the highest 5-Star rating from British Safety Council (BSC), UK.

Awards and Recognitions

Refineries Division

- Panipat Refinery declared Runners Up for National Safety Award-2003 by Ministry of Labour, Government of India under Scheme-I (Lowest Average Frequency Rate) and Scheme-II (Longest Accident Free Period) in August 2004.

- Panipat Refinery achieved level 9 rating in International Safety Rating System (ISRS) in September, 2004.
- Gujarat Refinery won the prestigious "Shreshtha Suraksha Puraskar", the second highest safety award of the "National Safety Council of India" for the year 2003.
- Barauni Refinery received "Oil Industry Safety Directorate Award" (Second Position) for safety performance for the year 2002-03 in August 2004.
- Barauni Refinery received the TERI Corporate Environment Award on for 2002-03 (Category 3), for a case study on "Installation and Commissioning of flue gas scrubbing unit in RFCCU of Barauni Refinery to remove particulate matter and SO_2.

Pipelines Division

- SMPL : Oil Industry Safety Award for 2002-03
- MJPL : British Safety Council Award for 2003
- HBCPL : National Safety Council of India Award for 2003
- KAPL : Gujarat Safety Council Award for 2003

Marketing Division

National Safety Awards -2003 received in 2004-05 from Directorate General, Factory Advice Service and Labour Institute, Ministry of Labour, Government of India.

1. Cuddapah LPG Plant - Winner - Scheme III
2. Ennore LPG Plant - Runners up - Scheme IV
3. Jaipur Terminal - Winner - Scheme VI
4. Dhanaj LPG Plant - Runners up - Scheme VI
5. Salem LPG Plant - Runners up - Scheme III
6. Cuddapah LPG Plant - Winner - Scheme IV

R&D Centre

The IP Award 2003 of Energy Institute, UK (previously run under the auspices of the Institute of Petroleum) under 'Technology Award' category was won by Indian Oil for the 'Oilivorous' for oil sludge disposal. The IP Awards are given by the Energy Institute is recognition for excellence and innovation.

The 'Oilivorous' Technology has also won National Petroleum Management Programme (NPMP).

'Award for Excellence for Creativity and Innovation—Team Category (R&D)' for the year 2002-03, instituted under the auspices of Ministry of Petroleum and Natural Gas, Government of India.

SAFETY INITIATIVES BY HINDUSTAN LATEX LTD.

Hindustan Latex Limited has integrated safety measures of employees and public along with the production activities of the company. The Safety and Loss Prevention Department has a role in devising accident prevention strategies, maintaining healthy work environment and also devising strategies to meet emergency situations for achieving total loss control.

Inter-Relationship with Other Departements

The objectives of the Safety and Loss Prevention Department are achieved by participation of all departments by implementation of accident prevention strategies, emergency situation strategies etc. Internal communications, safety audit observations, safety inspection reports, accident investigation reports, fire investigation report, Safety-Environment Forum and Safety Committee meetings are the medium used for interacting with other departments Hindustan Latex Ltd. (HLL) is maintaining a unique and high standard of employee participative schemes title 'Safety-Environmental Forum' (SAFE Forum) with the application of KAIZEN. KAIZEN is a Japanese management system aims to ensure total participation of people and development of any organisation. (In Japanese 'KAI' means change and 'ZEN' means for the better).

Objectives of SAFE Forum

1. To meet the requirements of Safety/Health/Environment Policy of the company.
2. To facilitate as a catalysts to the Management Review And Safety Committee meetings.
3. To involve in energy conservation, waste reduction and quality improvement action.
4. To involve the grassroots, level to the formal activities by ensuring recommendations of voluntary participation.
5. To involve in Emergency Preparedness activities, fire fighting, first-aid etc.
6. To regulate the House-keeping activities, hazard identification and selection of remedy.

7. To persuade colleague to wear safety appliances and to follow regulation in connection with environment protection and safety norms.
8. To bring out suggestions from employees.
9. To set right the factors affecting the quality circle concept and re-vitalise, the employees to join hands with management mission.

The SAFE is a three-tier approach with an integration of Management Review Committee, Safety Committee/Safety Circle and Individual Oriented Schemes, which are self-regulatory, statutory and voluntary forums respectively.

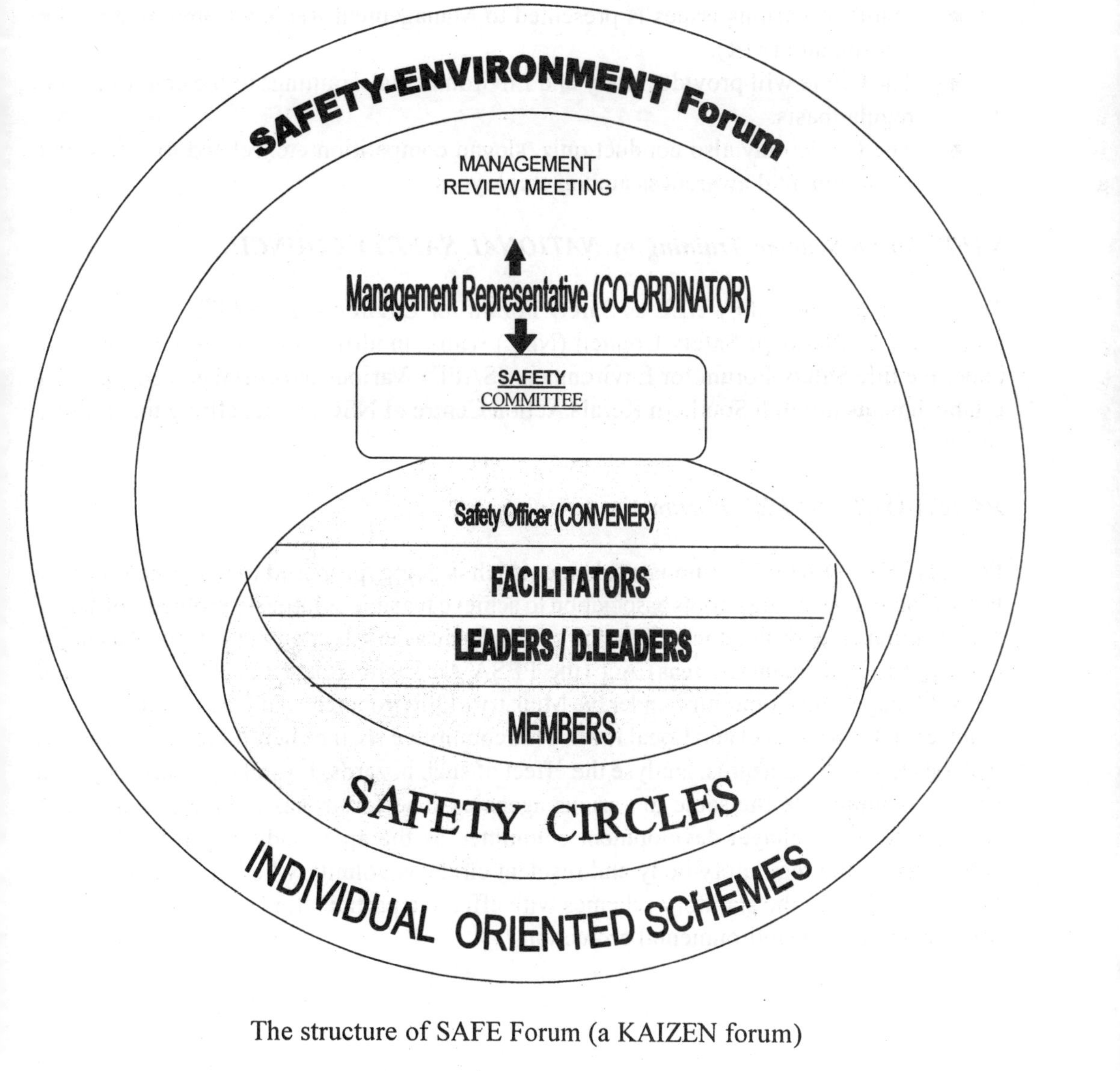

The structure of SAFE Forum (a KAIZEN forum)

Methodology of the 'SAFE' Forum

- General issues at shop floor level are directly dealt with HODs.
- Specific issues are brought to Safety Committee by the Convener, and the implementation is done by HODs.
- Recommendations of Safety Committee on specific issues are presented by the Facilitator of a section to the respective Safety Circle for participation of employees.
- Brainstorming is done in the Safety Circle through Leader/Deputy Leader for arriving at feasible solutions.
- Feasible solutions are implemented at the shop floor level by the Safety Circle directly. Whereas specific solutions are presented by the Leader to the Safety Committee for recommendations/approval.
- Status of various issues is presented to Management Review Committee by the Coordinator (MR).
- The Circle will provide Safety and Environmental Trainings to the employees on regular basis.
- The Circle may also conduct quiz, slogan competition etc. related to safety and environmental awareness among employees.

SAFE Forum Scheme Training by NATIONAL SAFETY COUNCIL

The scheme introduced by HLL as Safety Forum for Environment (SAFE) is now being propagated by National Safety Council (NSC) with a modified form of training scheme under the title Safety Forum for Environment (SAFE). Various industrial and educational establishments through Southern Kerala Action Centre of NSC are benefiting the training scheme.

PANCHAYAT "SAFE" Forum

PANCHAYAT "SAFE" Forum is a scheme which is being proposed to integrate statutory bodies and voluntary grassroots association to achieve feasible solution to problems of safety and environmental protection, mitigation of natural disasters. It interconnects the Panchayat (Extension to Scheduled Areas) Act (the PESA Act) with Local Self-Government and Crisis Management Committee under the Ministry of Environment and Forest: Central level, State level, District level and Local level. This committee shall be helpful to devise various methods to identify hazards, analyse the effect of such hazards, formulate emergency plan and to communicate the same to various agencies. The forum has a three-tier approach with statutory panchayat development committee as the apex body, ward development committee as the regulatory body and resident circle as voluntary body.Panchayat SAFE Forum will fit into the statutory schemes with effective participation of grassroots, public towards safety and environmental protection.

Major Events of Participation with Statutory Authorities and National Safety Council

State Level Safety Drama Competition by HLL

HLL made developed good active relationship with National Safety Council (Kerala Chapter)-NSC (KC) by conducting State Level Safety Drama competition in1993 at Trivandrum.

HLL Opens Office of NSC(KC)

HLL has opened office Southern Kerala Action Centre (SKAC) of NSC (KC) at Trivandrum on 19 March 2001,inaugurated by State Labour Minister Mr. Babu Divakaran, Ex-Officio President of NSC(KC).

HLL Propagates Importance of Safety

HLL has propagated new amendment to Kerala Factory Rules through Department (F&B), National Institute of Personnel Management on 18 December, 2002 by a seminar organised through SKAC.

5 'S' Work Area Improvement System Installed

HLL launched 5's in presence of Kerala State Pollution Control Board on 30 May, 2003 during Environment Week. As a journey towards Total Quality Management we have introduced Five 'S' HOUSE-KEEPING, the Japanese system in following Schedule.

5'S' Element	Date of completion
1 S = Sort	15-02-2003
2 S = Set in Order	15-03-2003
3 S = Shine	31-03-2003
4 S = Standardize	30-04-2003
5 S = Sustain	30-05-2003

5S-5L Work area improvement system

Levels of implementation	*Target Date*
Training =To generate awareness about 5S- 5L among employees	30-11-2004
1L=Floor, passage, corridors, horizontal surfaces etc.	15-12-2004
2L=Walls, columns, ceiling, vertical surfaces	15-01-2005
3L=Inside of rack, shelf, boxes, drawers, keys etc.	15-02-2005
4L=Cleaning of equipments/ materials. Labelling of materials, keys, Marking aisles, pathways etc.	28-02-2005
5L=Ensure strict adherence (Through Checklists)	10-03-2005

Skit on 5 Minute - 5S

HLL has brought out a new method for generating awareness of 5 'S' among people through a skit of very short duration (5 Minutes), which is displayed by a five employees within each department. The skit, which is presented in a humorous and thought provoking manner, is appreciated as a unique system of Human Resource Development activity. The skit is being performed to propagate 5's and safety awareness.

5 Star SAFE forum trophy introduced in presence of Directorate of F&B-March 2005.

"FIVE STAR" Grading of SAFE Forum for Annual Competition

Sl. No.	*Score %*	*Star Grading*
1.	60-69	Bronze
2.	70-79	Silver
3.	80-89	Gold
4.	90-94	Platinum
5.	95-100	Diamond

Best Performed State Level Safety Committee

HLL has suggested a competition among industries in Kerala State and also has sponsored a Rolling Trophy to NSC (KC) for Best Performed State Level Safety Committee in 2003.

HLL Sponsors Electrical Safety Week-2005

HLL has organised an Electrical Safety Week the SKAC from 1st to 7th May, 2005 in association with KSEB, Energy Management Centre, ANERT and Kerala Power Finance Corporation. In connection with the above an Electrical Safety Exhibition was organised on 5th May, 2005 in Thiruvananthapuram. The function was inaugurated by State Power Minister Shri Aryadan Muhammed. During the inaugural function a book titled "World Class Productivity & Safety" was also released by the Minister. He has handed over the copy of the book to P.S. Sukumaran, Chairman (NSC). Mr. Francis Xavior, J D (F&B),—the present Chairman of NSC(KC), accepted the first copy of the released book. Shri P.M.C Nair, Manager (Safety) of Hindustan Latex Limited has authored the book.

SAFE Forum as Work Improvement Team (WIT)

HLL has extended SAFE Forum as Work Improvement Team (WIT) with following structure to deal with continuous improvement activities

Work Excellence Trainers (WET)—Facilitators of SAFE Forum

↓

Work Improvement Trainers (WIT)—Leaders of SAFE Forum

↓

Members | Colleagues

Integrated Management System

Our company believes in Triple Bottom Line approach with Planet, People and Profit as our base. We believe Environmental Protection and Safety activities are dispersed throughout the organization when Line management and Functional management go together and Manufacturing Manager manages HR, Safety, Quality, Environment etc. hence we have Integrated Management System with ISO 9001, ISO 14001 and OHSAS 18001 as organisation system strategies.

March Towards World Class Productivity

Delivering value to our customers is our primary objective. Our customers count on us to deliver quality products on time every time. Our belief is that by utilising our quality processes, we will give our customer value added product they require and deserve. We achieve this goal by focusing on our quality policy and committing to several key areas and building the company with a World Class Productivity awareness generation through Safety and Loss Prevention Dept; Strategy implementation of World Class Productivity is through Balance Score Card Principle for translating Strategy into Action by an acronym WCP-CLIF, which implies SEVEN DIMENSIONS for world class quality key areas.

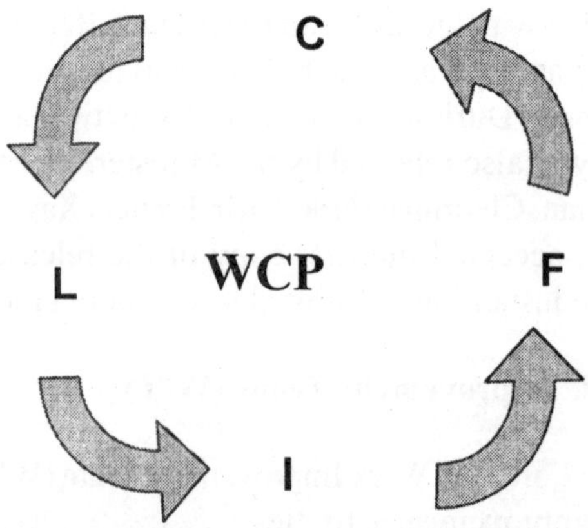

SEVEN DIMENSIONS for World Class Quality

- W = Waste elimination, Work Environment (Six Sigma, 5 'S') to identify a 'black belt' and 'green belt' team leaders for Six Sigma
- C = Classic Quality (TQM, Innovation group, Six Sigma)
- M = Maintenance Excellence (TPM)
- C = Customer Driven approach (Vision, Good Marketing Net work, Partner's meet, Erasing Customer Complaints, Innovation Group, Enhancing VALSALYA as CSR)

- L = Learning and Growth (Training Cell, SAFE Forum facilities). Loss Control- to enrich SAFE Forum and suggestion schemes. To develop 'WE' (Work Excellence) trainers
- I = Internal Perspective, Information flow and Inventory Management. (SPC, JIT techniques, Visit to other units by employee representatives, monthly Union meetings)
- F =Financial Perspective (long term with augmentation, master plan and new product Introduction)

SHE MANAGEMENT AT JINDAL STAINLESS LIMITED

Jindal Stainless Limited, Hisar, since its inception has been very conscientious about maintaining Healthy, Safe and Pollution free work environment. It has always recognized the need to ensure healthy and clean environment to care of the employees and surrounding community, which is reflected, in its Environment, Occupational Health and Safety and Quality Policy. A well-defined World Class system on Environment Management System ISO-14001, Occupational Health and Safety OHSAS-18001 and Quality Management System ISO-9001 are in place.

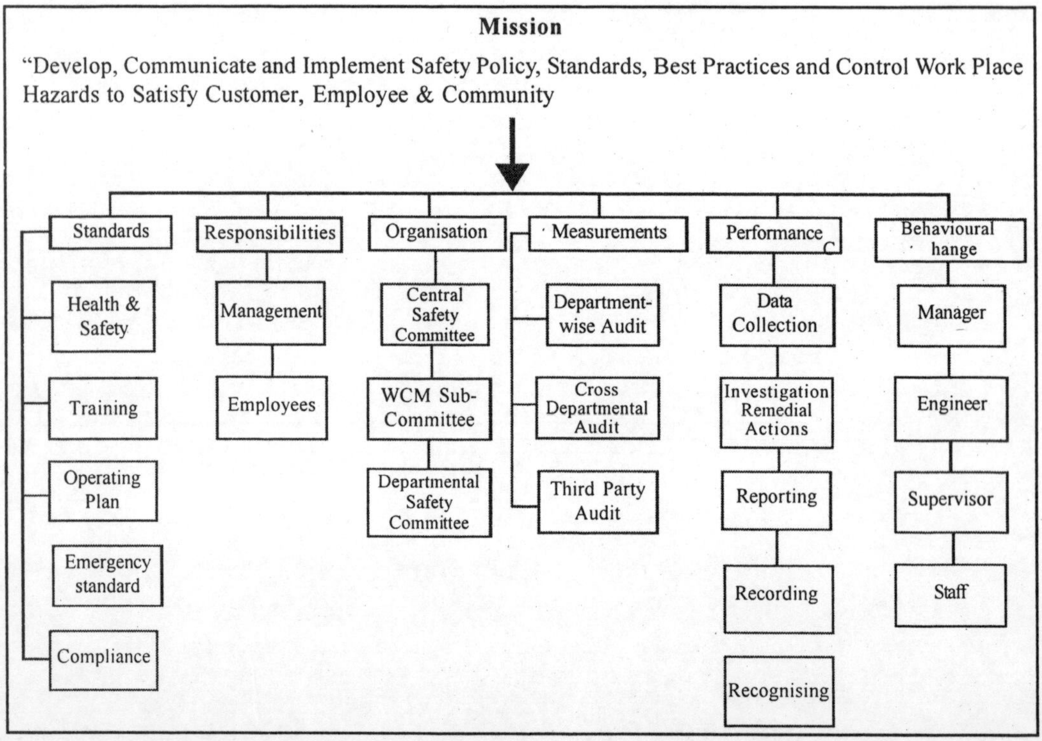

Safety Approach and Deployment of Safety Practices

The organization has adopted a systematic strategic approach in implementation of its key manufacturing areas (KMFAs) to ensure zero accident and a clean and healthy working environment.

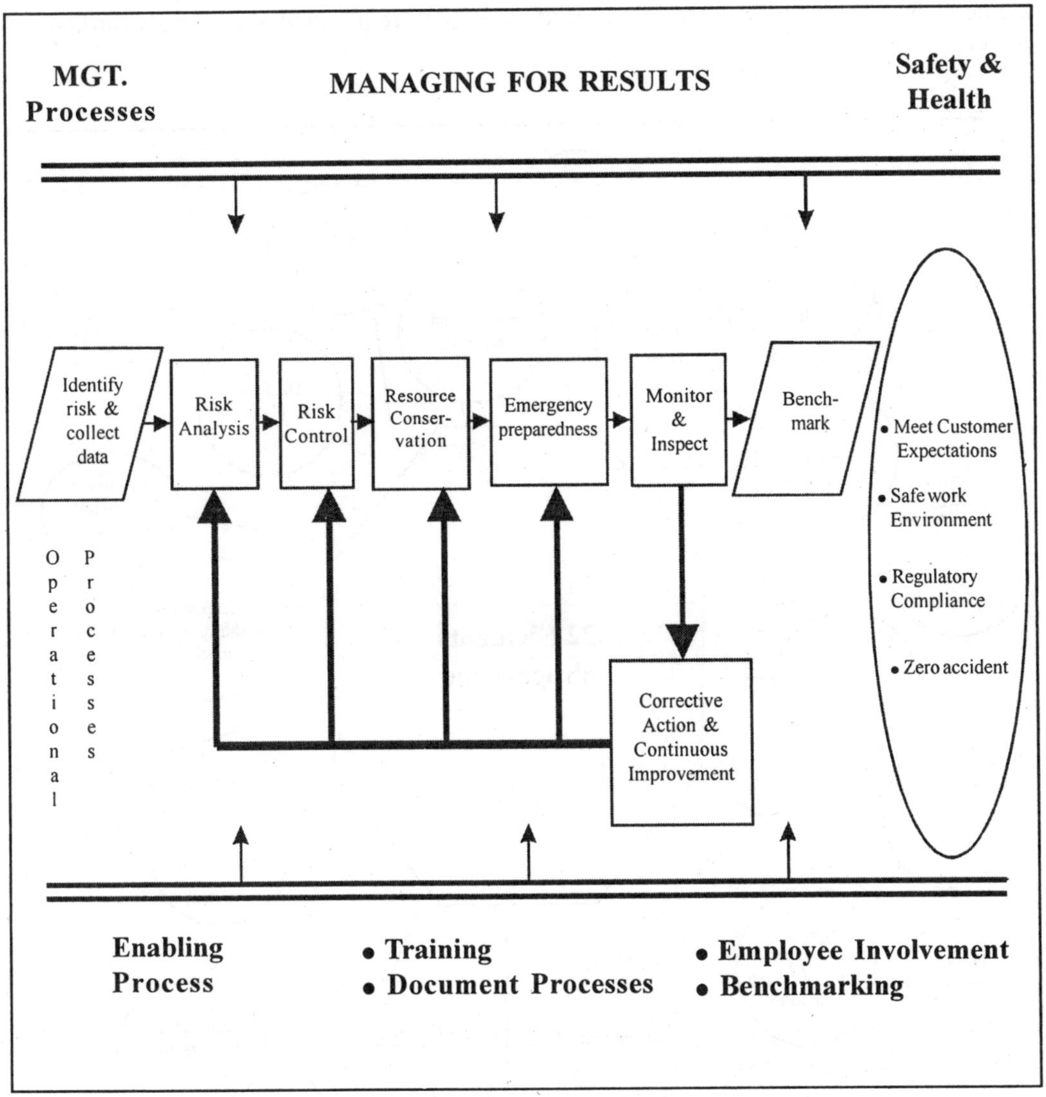

Safety Strategic Plan

Organization has chalked out a detailed SAFETY STRATEGIC PLAN encompassing all physical and chemical risk prone areas and other emerging areas like ergonomics and work organization with high level of safety motivation for implementation and management at its efficacious accomplishment and constant monitoring.

Every year in the beginning of the year an OPERATING PLAN encompassing 22 elements is prepared with targets and safety activities are planned accordingly throughout the year.

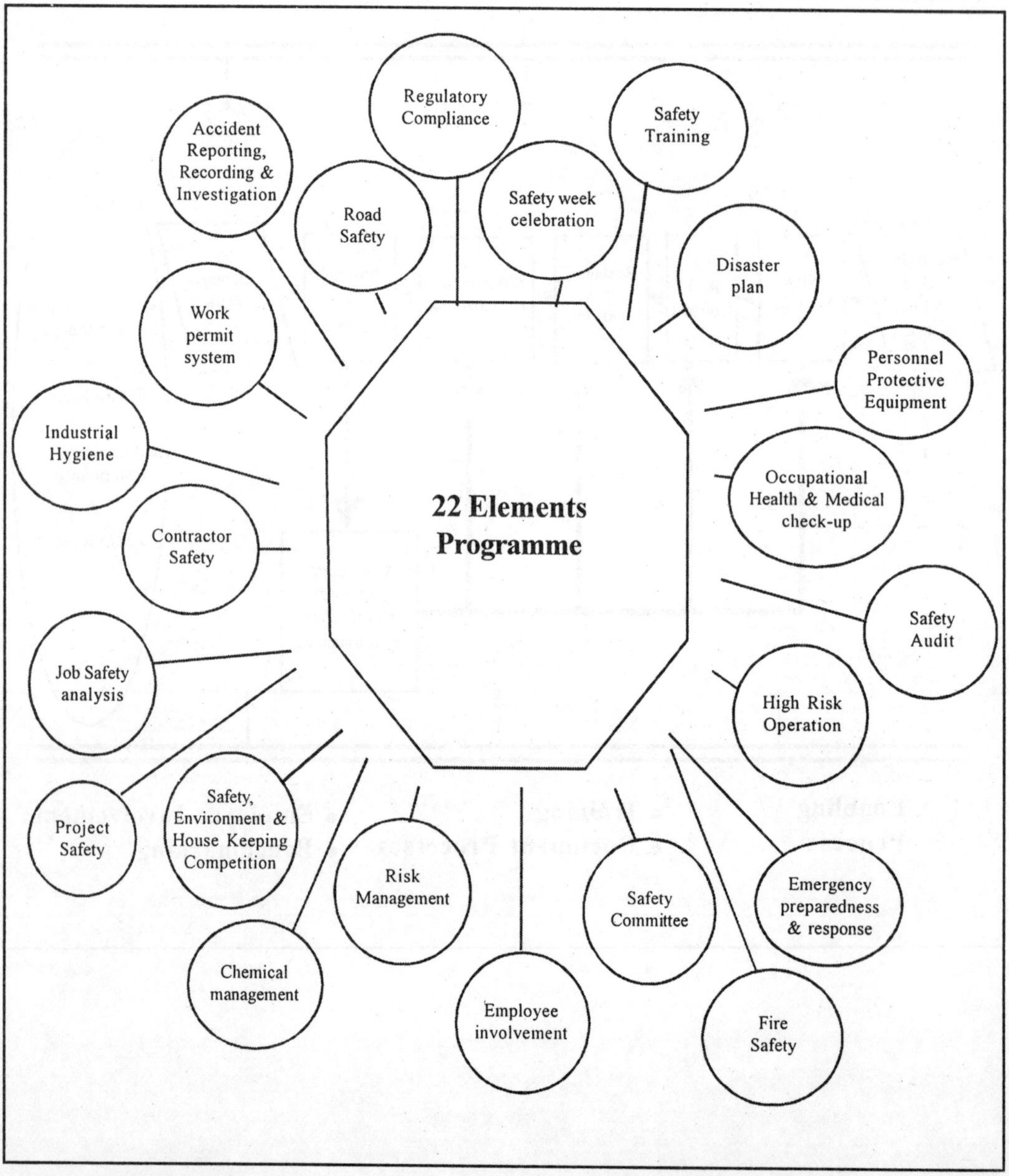

Risk Assessment

Group Risk Assessment :- Group Risk Assessment of all new activities is done before start of the job to ensure safe execution of work.
Risk Analysis Studies :- Risk analysis studies by external consultants are organized regularly.

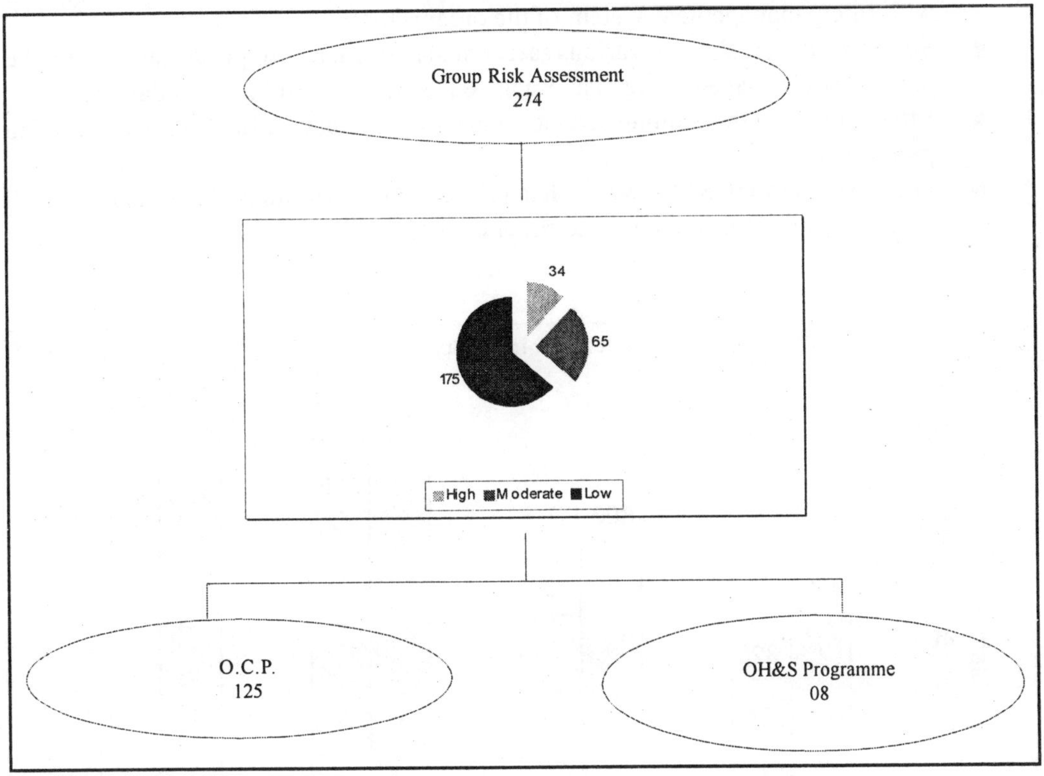

Safety Review Mechanism

A three tier well structured safety review mechanism is in place to review the plant safety performance and safety related issues involving all levels of employees.

S. No.	*Committee*	*Chaired by*	*Members*	*Frequency*
1.	Central Environment, Safety and Health (EHS) Committee.	Unit Head	All HODs.	Monthly
2.	Sub-committee on Safety, Health & Pollution Control/OHSAS System Implementation Committee	G.M. Operations/ Maintenance	HODs/Middle Management Executives	Monthly
3.	Departmental EHS Committee	HOD	Staff & Work.-nen	Monthly

Safety Training

- Safety training is a regular feature and imparted to all levels of employees as per the Yearly Safety Training Calendar.
- All new incumbents including contract workmen are imparted Safety Induction and Orientation Training before inducting them into the job to make them familiar with safety management system of the organization.
- Specialized workshop on various safety management techniques/systems with the help of external agency are also organized as per the training calendar.
- Effectiveness of the training is done conducting Evaluation Test after each training programme.
- Four (4) hours OHSAS-18001 General Awareness Training was imparted to all levels of employees including contract workmen.

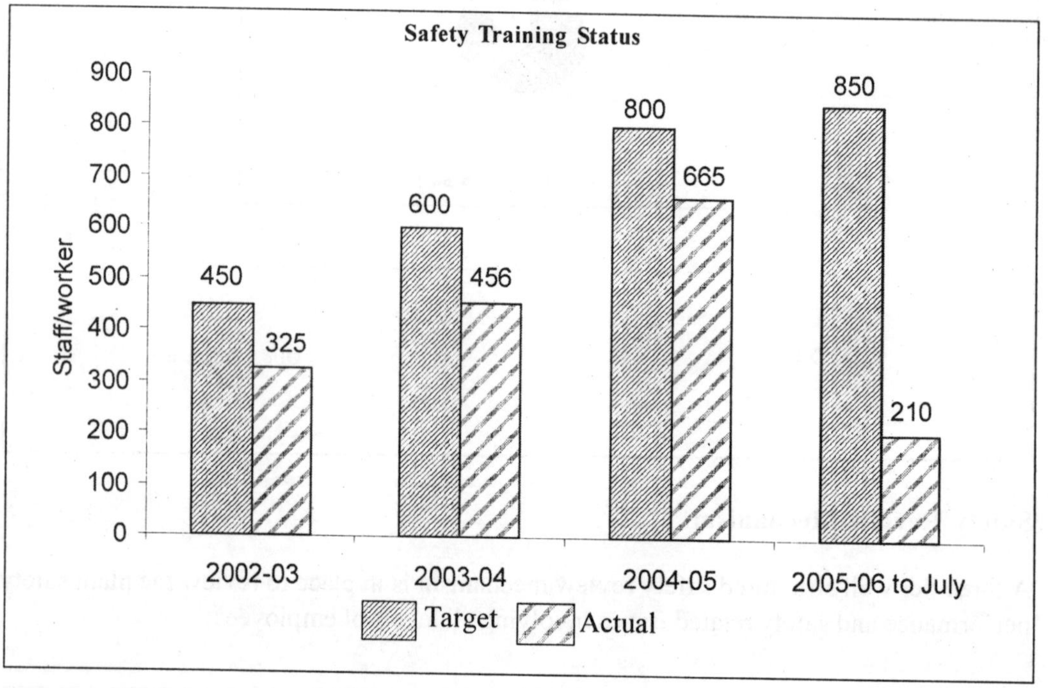

Safety Promotional Activities

- JSL, Cold Rolling Division has pursued optimum degree of safety by evolving a tailor made approach, methodology, procedures and practices to the safety system and strewn around safety exhortation and Safety Promotional Activities in the Factory.
- One Day Safety Exhibition organized on the occasion of National Safety Day on 4th March every year is unique in India.

- Our Safety promotional activities are not only limited with the plant premises but also spread over to all sections of society which includes colony residents, school children and nearby population and this has inspired all the people in and around the colony to a greater extent in building safety culture.

Activities Covered During Safety Weeks

- Safety Model Competition/Safety Shield Competition /House Keeping Competition Safety Case Study Competition/Safety Speech Competition/Safety Poem Competition
- Safety Quiz Competition Safety Poster Competition/First Aid Competition
- Fire and Safety Training Demonstration on Fire Fighting Equipments
- Safety Bulletins on various topics are distributed to employees regularly. Safety Posters have been displayed in all departments and at prominent locations.
- Safety Films are regularly telecasted through Dish Antenna to arouse safety awareness amongst employees.

Employees Health and Hygiene

- A system for Medical Check up of Employees.
- Visit of Plant by a team of medical officers.
- Health check up record of individual employee is maintained.

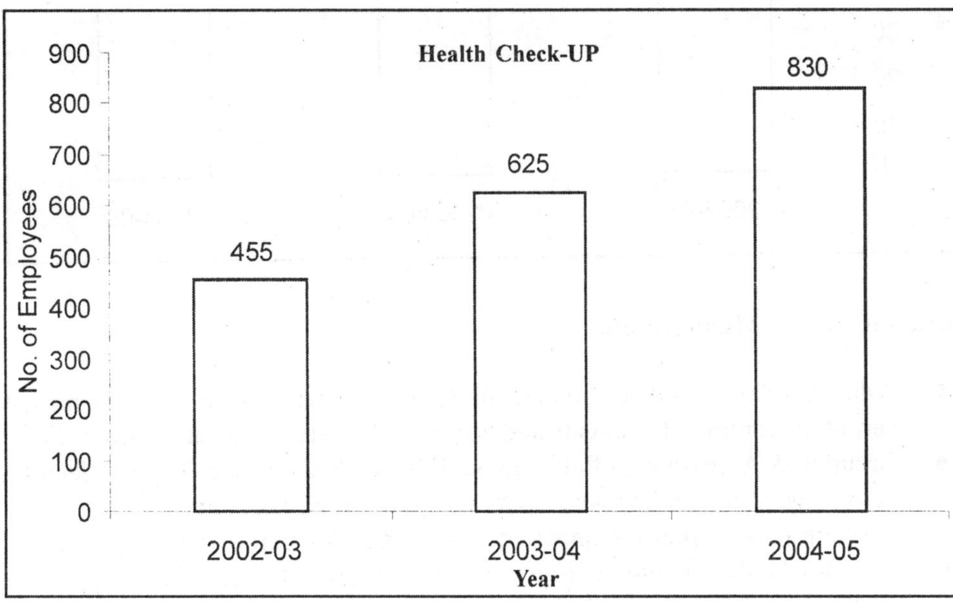

Hygiene Management

Hygiene Teams are functioning to look after hygiene related activities areawise and to ensure good hygiene standards. Hygiene Teams are as follows:

- Plant Area
- Colony Area
- Guest House and Canteen Area
- Health Centre Area

Regular hygiene audits are being carried out by the Hygiene Team to evaluate Hygiene standards and determining Hygiene Index.

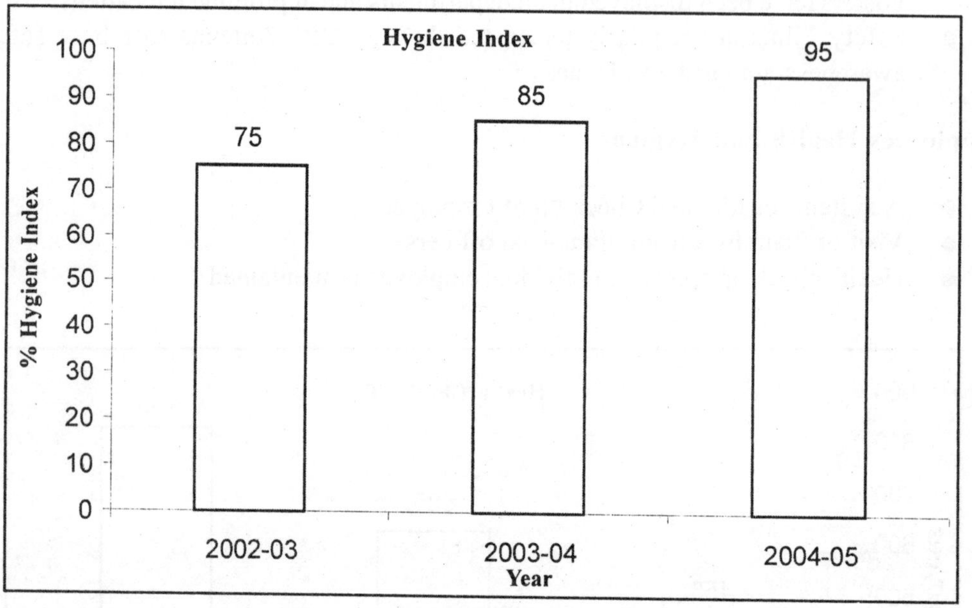

Contractor Safety Management

- A detailed Safety Code of Conduct for Contractors is in place, which is mandatory part of all contract documents and ensures all safety norms at contract sites.
- Group Risk Assessment of all contract jobs is taken on by concerned department before awarding contract and safety precautions are taken accordingly.
- All contract workmen are imparted Safety induction before putting them on job.
- Structured Safety Training Programmes are being organized at Training Centre for all Contract Workmen and Contractors and also at the site itself by the Safety Personnel/departmental engineers.
- Weekly meetings are organized along with Contractors to review the Safety issues of the site.

- Safety code for Visitors and Transporters are in place and they are allowed in side plant only after briefing of plant layout and safety instructions.
- Regular meetings are organized with vendors and transporters on safety issues.

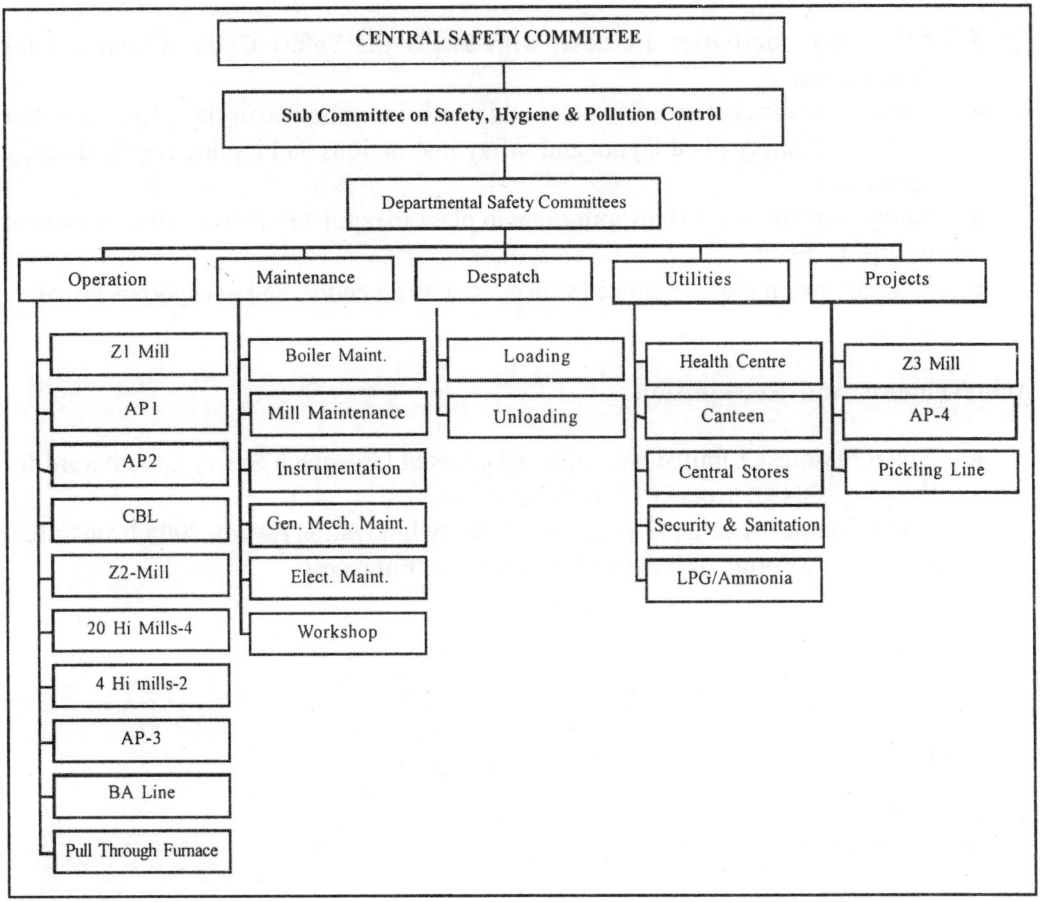

Safety Audit System

A three tier Safety Audit System is in place to identify potential hazards and risk and to develop control measures to eliminate the hazards/risks.

1.	Self audit	Departmental Safety Committee	Monthly
2.	Cross Audit	Senior Officers/ OHSAS Auditors	Quarterly
3.	Third Party Audit	External Agency	Annually

All the audits are conducted as per standard checklist and audit schedule. A time bound action plan is made by departments on audit observations. The action plan is reviewed and monitored by the Apex Safety Committees regularly.

Extension of Safety Practices to Contractors, Vistiors and Vendors

Safety activities are not limited to company employees only but are extended to contract workmen, vendors etc. as follow:

- All contract activities are dealt with as per the Safety Code of Conduct for Contractors.
- Safety code for visitors are in place and visitors are allowed in side plant only after briefing of safety plant layout and safety instructions and issuing required safety appliances.
- Safety code for Road Transporters are in place to regulate safe vehicular movement in side the plant.
- Regular meetings are organized with contractors, vendors and transporters on safety issues.

Achievements and Recognition

- Jindal Stainless Limited has achieved coveted Greentech Safety Gold Award for the year 2004-05 from Greentech Foundation.
- Jindal Stainless Limited has achieved Safety Innovation Award, 2005 from Safety and Quality Forum of India and Institution of Engineers.

MOTHERDAIRY

Initiatives for Safe Working Environment

Mother dairy is a safety conscious organization following Safe Work place practices as part of our daily routine since we have a fleet of more than 100 road milk tankers, one of the biggest ammonia based refrigeration plant in Delhi and many heavy dairy equipments and machines at our plant.

Through our sustained efforts and safety measures taken we could achieve

- Zero Fatal Accident since inception of the Dairy.
- Reduction in total accidents (Major & Minor) by 32 per cent.
- Elimination of all identified near misses.
- Total man-hour loss due to accident has reduced by 30 per cent.

Mother Dairy, Delhi is an active member of National Safety Council and participates in all the activities like observance of National Safety Week, Agni Shaman Week, Road Safety Weak and other such events with complete involvement of employees.

Mother Dairy has been an active participant in EHS Missions organized by CII with presentation of its case studies.

Steps Taken by the Dairy for Environment Protection

Mother Dairy is conscious of its social responsibility towards the environment. Though the effluents from the Dairy are non-toxic, they are nevertheless ecological pollutants. Hence Mother Dairy has installed an Effluent Treatment Plant (ETP) with a capacity to treat 15 lakh litres of effluent per day. Through the process of water audit and optimum utilization of water, effluent generated has reduced to a considerable extent. Now only 7.0 lakh litres of waste water is generated from the dairy and the treated water meets the specifications laid down by pollution control authorities. About 2.5 lakh litres of the treated water is being reused for gardening in the Dairy.

Complete fleet of more than 100 road milk tankers is maintained to comply with "Pollution Under Control" norms as per the standards laid down for control of air pollution through an in-house vehicle pollution testing set up which is authorized by the Delhi State Transport Authority.

Noise within the dairy plant is brought down to within the acceptable noise level as specified in the Noise Pollution Control by providing acoustic lining as noise barrier to isolate the noisy areas. Staff working around heavy machines is provided with earplugs.

HACCP principles are followed to ensure safe and hygienic product to customers. Mother Dairy has been awarded Food Safety Management System Certification by BVQI, which is recognition of our concern for the health of our customers.

Much cleaner fuel, i.e. LDO is used that has low content of sulphur and ash with nil sediment level thus controlling the exhaust air pollution. The exhaust meets the statutory pollution control standards.

To create green environment, steps are taken to promote horticultural activity. The Dairy is situated over an area of 1,56,000 sq.metres. Out of the total covered area, building is about 27,000 sq.metres and area covered under bituminous road and concrete paved road are around 27,000 sq.metres. Balance 1,02,000 sq.metres which is 65 per cent of the total land has been developed into lush green lawns.

Ficus, Moulsori, Neem, Silver Oak etc. trees have been grown all-around the dairy and the lush green lawns are bedded with colourful flowers, carefully grafted to add scenic beauty. The trees planted emit comparatively more oxygen, evergreen in nature and shed less leaves. Mother dairy presents a pleasant panoramic view in contrast to the rest of the neighbourhood. In the last three years more than 500 saplings were planted in the Dairy premises.

There are no environmental accidents or issues pending at any forum for adjudication.

Dairy's Responsibility to Public for Health, Safety and Ethical Business Practices

While providing consumers with milk and milk products, Mother Dairy strives for enriching the quality of life of the people coming in contact with the dairy. These are:

Concern for Society

- Adopting environment friendly process techniques.
- Ozone friendly refrigerant.
- Environment friendly milk distribution system.
- Environment protection—water.
- Environment protection—air
- Environment protection—noise
- Towards consumer awareness, Mother Dairy provides information to the consumers and general public on hygienic aspects, nutritive values and knowledge of adulterants in milk and milk products through mass contact programmes (such as School contact programmes, Programmes at Welfare Societies, Programmes at shops, Exhibitions and Social Welfare programmes, Programmes at Hospitals, meeting Doctors/Health specialists, etc.) by using mobile vans, direct contact programmes and arranging visit to Dairy.

Wide publicity is made by Government of Delhi for making Pulse Polio Programmes successful. Mother Dairy field staff are involved in this publicity campaign and posters are displayed and pamphlets are distributed at Mother Dairy shops located all over Delhi.

Safety Policy

(Organizational Commitment)

We at Mother Dairy ensure that as per as reasonably practicable, the work places, machinery, equipment and processes under our control are safe and without risk to health.

We at Mother Dairy will provide, where necessary, adequate protecting clothings and protective equipments to prevent, so far as reasonably practicable, risks of accidents or of adverse effects on health.

We at Mother Dairy ensure that we shall provide, where necessary, measures to deal with emergencies and accidents, including adequate First Aid arrangements.

We at Mother Dairy make arrangements at all levels of our organizations under which:

(a) Workers, in the course of performing their work, cooperate in the fulfilment of the obligation.
(b) Representatives of the workers cooperate in the field of occupational safety and health.
(c) Representatives of the workers are given adequate information on the measures taken by Mother Dairy to secure occupational safety and health.
(d) Workers and their representatives are given appropriate training in occupational safety and health.

Sd /-
Addl. Managing Director

Mother Dairy adhering to the safety provisions of Factory Act has a Safety Committee consisting of

Chairman	Sr. Manager (Manufacturing)
Designated Safety Officer	Asst. Manager (P&A)
Sr. Officers from Milk & Ice Cream Plant	Asst. Manager (Estate)
Estate Incharge	
Workers Representative from Milk Production, Ice Cream, Transport, Stores etc.	

Mother Dairy Has made safety a part of work culture and the awareness towards this has been promoted through various training programmes, Kaizen movement and Small Group Activity.

Major, minor accidents and the near misses along with man hour loss are being monitored and analyzed for countermeasures on monthly basis. These inputs are captured and feed backs are taken from shop floor.

Major safety plan undertaken by the organization is identification and elimination of Unsafe Conditions, Unsafe Acts and near misses through EHS pillar of TPM. Identification/ capturing of these are done at the shop floor level in every shift. All these are taken up for thorough Analysis as learning Incidents. This has become a major safety tool for elimination of shop floor safety hassles.

Safety Audits are conducted once in six months on the basis of Audit Checklist in all the shop floors for its improvements. Actionable Points are addressed within given time frame.

Safety Awareness has improved in the organization as a whole and the development of safety competent operators with respect to human and machine safety has resulted into high morale and reliability and ownership for work place among all the employees.

Role of Management is to ensure that there should not be any hindrance for financial assistance required to implement the Safety measures. Simultaneously Employee also take interest religiously to identify unsafe conditions, near misses to eliminate possibility of accidents.

No Fatal Accident since inception of Mother Dairy, i.e. December 1974. Longest Accident Free period for Mother Dairy in the year 2004-2005 is 5th October, 2004 to 29th January 2005 although Transport section has achieved Zero Accident since last one year.

Safety Manual in compliance with the Rule 61-G of Delhi Factories Rule 50. Apart from this, safety sheets including MSDS (Material Safety Data Sheets) for Machines and hazardous chemical used are also prepared and displayed for development of safety competent operators at shop floors.

Extensive safety training programmes are given under EHS Pillar of TPM for development of Equipment Safety Competent and Process Safety Competent operators. Regular training programmes are conducted.

Mock Drills on Ammonia safety, Fire Safety and other industrial safety are carried out from time to time at Mother Dairy. Awareness Drills for Personal Protective Equipments (PPE) are being imparted on regular intervals.

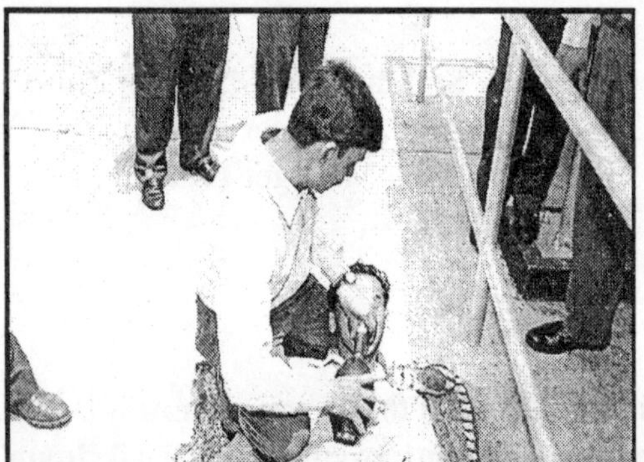

Artificial Respirator

Use of Smoke Exhaust

Mock Drill for Ammonia Safety

Demonstration of Ammonia PPE

Mock Drill of Fire Safety

Mock Drill for Ammonia Safety

Employees are being recognized through presenting their Improvements in Centralized Kaizen Presentation and SGA presentation to the Top Management and getting mementos for the same.

Kaizen on Engine Mounting and Dismounting Trolley (Prepared in-house)

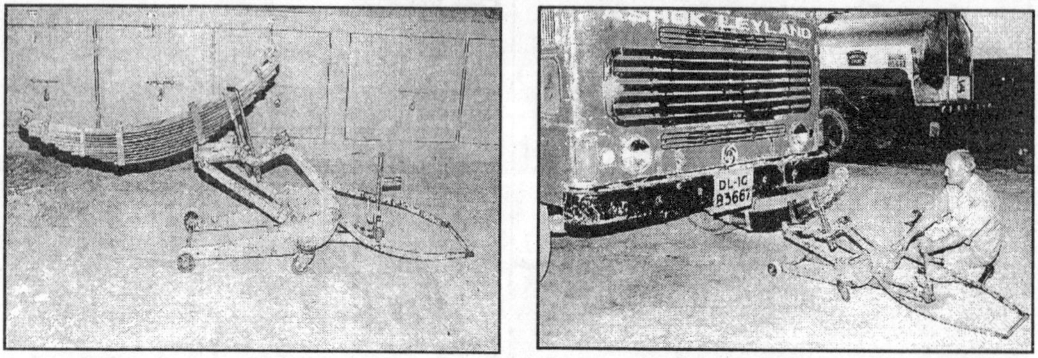

Kaizen on Gear Box Mounting Trolley for Leaf String Mounting with in-house modification

As a Major Safety Initiative following are the steps taken at Mother dairy:

- Extensive visualizations at and around shop floors with relevant details through visual posters, photographs, instructions, safety details and advisory notes.

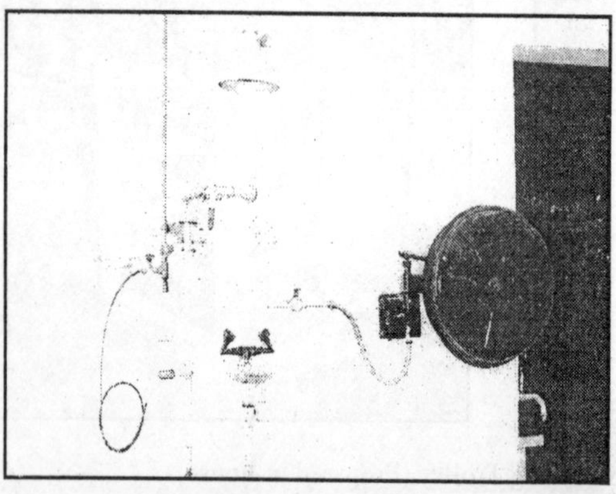

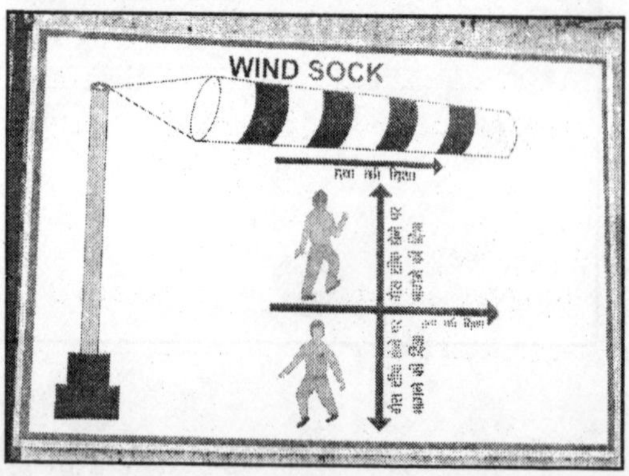

- Displays of right procedures of hazardous chemical handling emergency preparedness plan direction boards, emergency assembly points etc. as shown below:

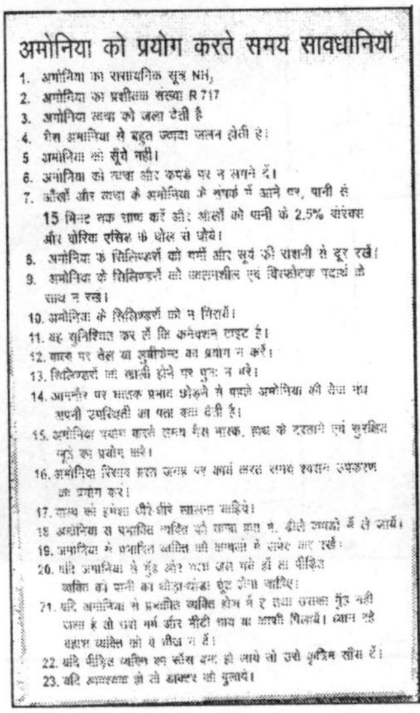

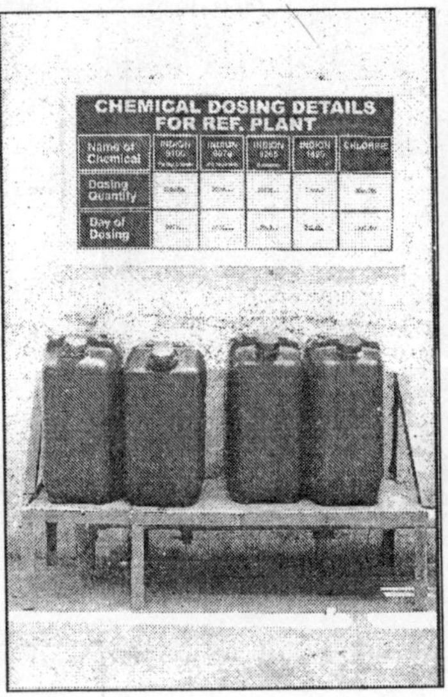

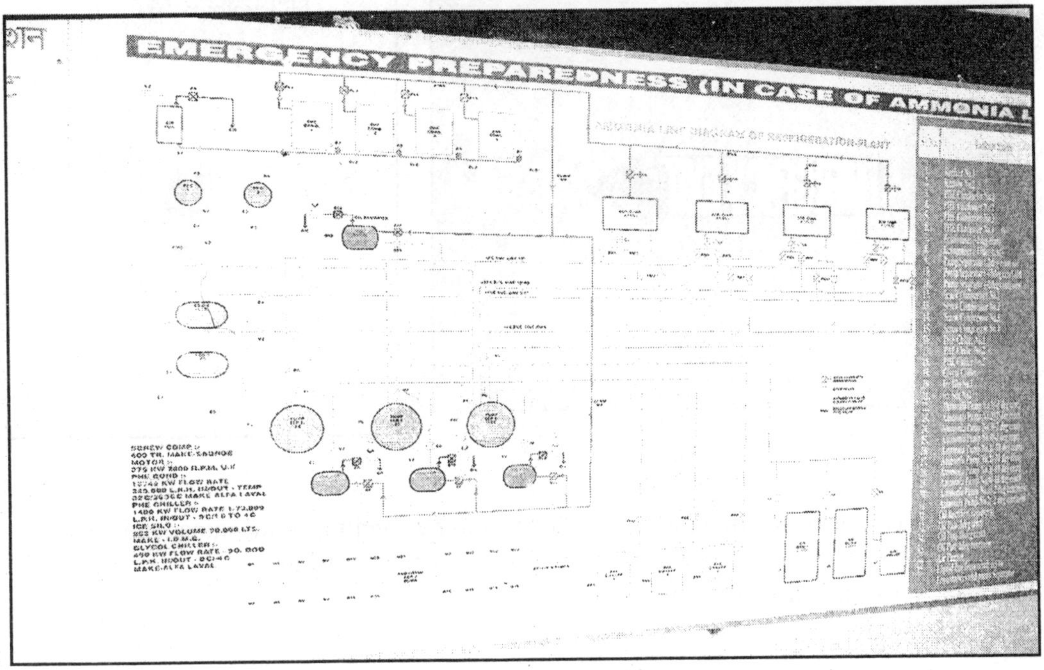

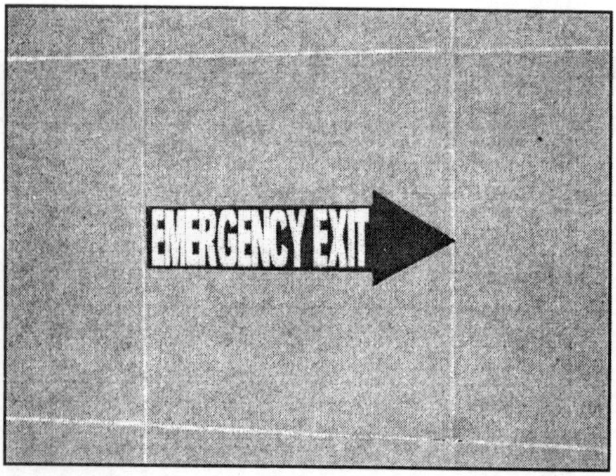

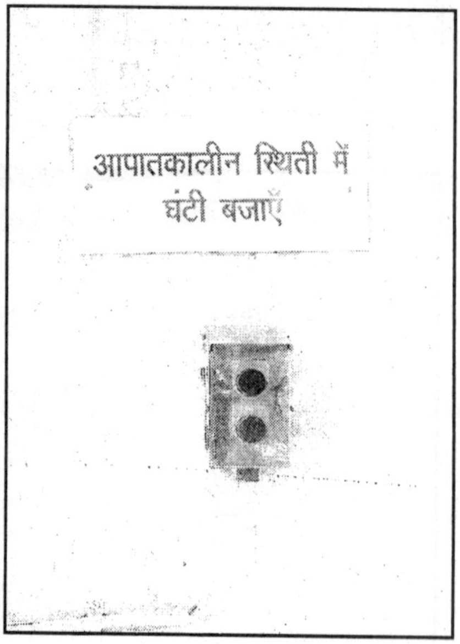

- Creating infrastructure for eliminating unsafe handling of machines and equipments helped us in reduction of minor accidents of day-to-day at shop floor. Some of these are shown below:

- Restricting vehicular traffic speed to 10 kmh and movement only in clockwise direction within the campus.

- Benefits are more of intangible, i.e. improving the moral of the people at shop floor, increased reliability of equipment/workplace.

But there are some tangible benefits achieved as follows:

(i) Fatal Accident is Zero since inception of the Dairy, i.e 1974.
(ii) Total accident (Major & Minor) has reduced by 32 per cent.
(iii) Near misses identified has reduced by 91 per cent.
(iv) Total man-hour loss due to accident has reduced by 30 per cent.

- Mother Dairy organizes regular Need base Safety Programmes.
- Mother Dairy conducts two-fold training programme on safety.
- Training for development of Equipment / Process Safety Competent Operator which is focused and customized to our own equipment and process safety under EHS Pillar of TPM by internal resources.

Awareness on PPE

- On general industrial safety, i.e. handling of Hazardous Chemicals. Ammonia Handling, Electrical safety, uses of different PPEs as per the needs are also being organized through External Agencies periodically.
- We have a monthly meeting of EHS Pillar to review all the EHS concerns.
- Top Management Level person is responsible for safety in Mother Dairy, i.e. Sr. Manager (Manufacturing) and Factory Manager are heading the safety movement.
- Safety at Mother Dairy has been taken as a whole, i.e Human Safety, Environment Safety, Product Safety and Equipment Safety.

Integrated Safety Cycle

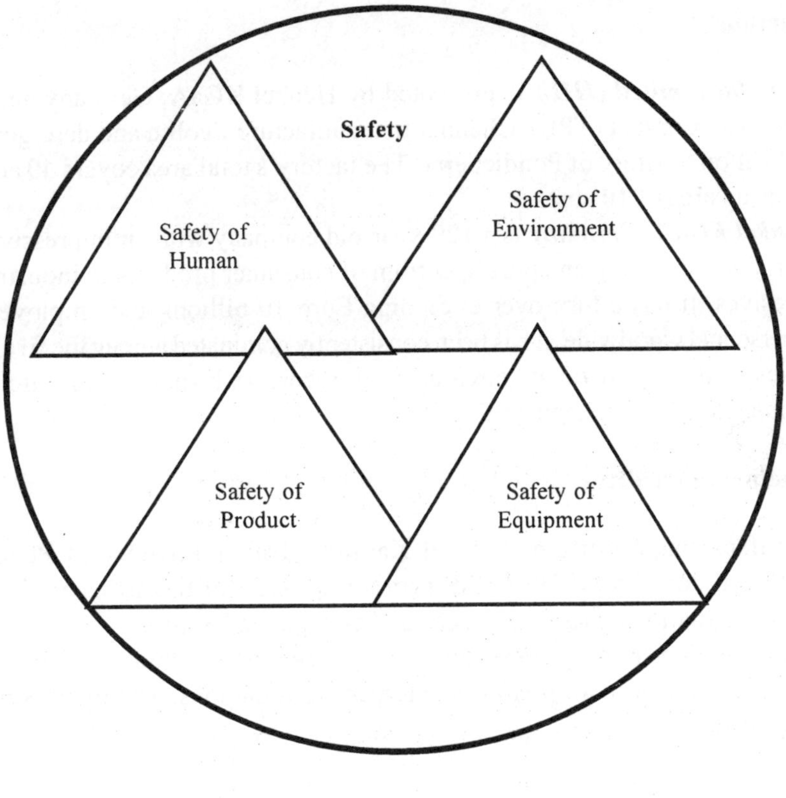

Mother Dairy is certified for Environment Management Systems, i.e. ISO 14001.

Mother Dairy has been called for participating in EHS Mission of CII for presenting its case study.

HENKEL INDIA LIMITED

Introduction

Henkel India Limited (HIL) is promoted by Henkel KGaA, Germany and Tamil Nadu Petro Products Limited (TPL), Chennai to manufacture Zeolite and detergents in India at Karaikal Union territory of Pondicherry. The factory's total area covers 40 acres including plant built-up area of 10 acres.

Henkel KGaA, Germany is a 129-year-old company with an impressive portfolio of 12000 products covering an entire spectrum of consumer products to industrial chemicals and adhesives. It has a turn over exceeding Euro 10 billions and employees more than 50,000 personnel worldwide. It has been consistently nominated among the five most admired companies in the world for its envious track record in business and Safety Health and Environment (SHE) Management.

Manufacturing Facility

The manufacturing facility of HIL, at Karaikal, Union Territory of Pondicherry, was commissioned during year 1993 and incorporates state-of-the-art technology capable of producing International quality detergents of all types and Zeolite.

Zeolite is Eco-friendly substitutes for conventional Sodium Tri Poly Phosphate, an important component in all premium detergent formulations. The plant comprises of the following sections.

Section	*Capacity*
Detergent Powder unit	15,000 MT
Detergent Bar unit (2 streams)	15,000 MT
Zeolite unit	10,000 MT
Liquid detergent	3,600 KL

Henkel India Limited, factory manufactures a range of world class detergents and cleansers making available international brands like 'Henko', 'Pril' and 'Mr. White' to the Indian household. HIL has an excellent export marketing of zeolite to various Middle East Countries and Asia Pacific.

Safety Health and Environment Performance

In line with the global policy of the parent company Henkel KgaA, on 'SHE' performance, production process incorporates Occupational Health, Safety and environmental protection at various stages to fulfil the legal and other requirements.

Compliance to Legal and Other Requirements

HIL comply with all legal and regulatory measures. In addition to the compliance of regional and national legislation, HIL follows global Safety Health and Environmental (SHE) guidelines prescribed by parent Henkel KGaA. Legal updations are reviewed in management committee before implementation.

Safe Work Culture

Safe work culture is part of the job of each and every employee. Right man at right job principle adopted in safe work culture. All the employees are educationally qualified for their respective responsible area.

Safety Health and Environmental Policy has been implemented inside the factory. Suitability of the policy will be reviewed once in two years by the management and if requires suitably modified. To fulfil the policy requirements, objectives are derived with the help of risk assessment.

Risk Assessment

Systematic Risk Assessment was carried out at all the work places inside the factory. Totally 1450 risks were noticed and 41 risks were identified as potential risks. Identified potential risks were taken up for eliminating the risk or minimizing the risk level.

Safety Organization

Managing Director is the occupier of the Henkel India Limited.

Vice President (Technical) is the over all head of technical team, lead the safety management system and is the chair person of Central Safety Committee (where trade union members staff participated the committee).

Chief Manager (Works) is the site in charge responsible for all the 'SHE' related activities.

Deputy Manager SHE Department is the responsible person for implementation and monitoring of 'SHE' related matters.

Communication

Communication plays a vital role to improve the SHE performance. There are two ways of communication—Internal and external communication.

Internal Communication

Every Monday safety pledge is performed among work force in each shift beginning, which contains the previous week's safety issues, safety messages and any new implementation. This forum activities are well received by the workforce. Safety posters are displayed throughout the plant and periodically changed.

To motivate and educate the safety, various awareness programmes are conducted during national safety week celebration and National pollution prevention and conservation of resources day celebration in grand manner.

Worldwide SHE related messages are published by Henkel for internal circulation.

'SHE' Committees

SHE Committees are performed at various levels.

Level I : 'SHE' Steering Committee—Chaired by Managing Director.
Level II : Central Safety Committee—Chaired by Vice President Technical.
Level III : Emergency Coordination Committee—Chaired by Chief Manager Works.
Plant Safety and Hygiene Committee—Chaired by Manager (Production).

External Communication

Our SHE Performance like Safety day celebration and OHSAS 18001 certification are published through state/national level newspapers. HIL participated state level exhibitions conducted by Governmental organizations. Various Social responsible programmes like farmers development are performed to the benefit of nearby community.

Major Equipments, Facilities and Technologies

HIL adopted cleaner technology by in-built bag filters for detergent and zeolite spray towers and world class enzyme handling Laminar down flow booth for enzyme dosing system thereby enabling compliance to stringent occupational health and safety norms of Henkel. Well equipped fire protection and prevention technologies are in place.

HIL has various critical equipments and its fail safe trips and interlocks are checked on regular intervals.

Systematize Safety Practices

Safe Work Procedures

All the departments follow safe work procedures which contains safe work instruction. Systematic preventive maintenance schedules are followed for all the equipments. Systematic maintenance requisition systems are followed which ensures proper handing over and taking over of equipment. Electrical clearance system is in place to ensure electrical safety during maintenance. Comprehensive safety systems are in place to ensure safety. New labours and rescue team members identification and exclusive DG set for fire water pump are some of our initiatives.

Reports

Various reporting systems like Accident report, Near miss occurrence report and Red star report are in line to analyze and to take corrective and preventive measures.

Any modification or change in the process, management of change note report will be raised to capture safety health and environmental matters.

Audits

OHSAS 18001/ISO 14001 Internal audit will be conducted once in six months. Certifying agency will conduct audit once in a year. A part from this Henkel SHE audit team will audit our system once in two years to ensure the compliance.

Outlook

Continual improvements on SHE, minimizing potential risks, formalizing the transparent SHE Management System to comply with the best available standards are some of the continuing efforts towards effective SHE management.
Build corporate image through National and International recognitions on SHE Performance.

46

CULTURE AND LEADERSHIP FOR IMPROVED OHS PERFORMANCE

J.S. Ahluwalia*

Introduction

Since the advent of industrial revolution way back to seventeenth century, the issue of industrial safety has been a matter of great concern for industries and business all over the globe. In concern with efficiency, cost effectiveness and reputation of the company, new safety management theories evolved and are being modified time and again to strike an optimum balance between productivity and safety. However, experiences with the industrial accidents show that there has been a serious gap in whatever is being propagated in safety management theories and the practical realities. The need of the hour is to bridge the gaps between safety aspirations and reality by translating the concept into real goal of accident prevention at every level in the industry.

As a result of some major accidents in various parts of the world, the consequences had a powerful impact on the public mind. Awareness in modern society of the hazards around it has reached a significant level in recent times. Today, 'risk management' and 'risk assessment' are phrases used in diverse industries, companies and activities. In the commercial world, a structured approach to risk management is requried for success in increasingly volatile and difficult external environments.

Industries involved in High Risk and High Tech production activities such as petroleum exploration and production, oil refining, chemical manufacturing with high potential of the risk and hazards associated with their operations, are susceptible to disasters. Such industries need to, adopt safety management practices which can effectively prevent industrial disasters, since disasters can only be prevented through proper planning and sustained efforts of both the management and the employees.

*Lt. General, PVSM (Retd.)

OHS Business Model

Business Model

An "OHS System" is defined as the total set of men, equipment, and procedures specially designed to be superimposed on an industrial system for the purpose of increasing safety. This safety system may eventually make radical changes in the industrial system.

The system safety concept envisions a preplanned, organized effort with a primary goal to conserve resources associated with the given system, product or operation by:

- The pre-accident identification of potential hazards,
- The timely incorporation of effective safety related design and operational specifications, provisions and criteria,
- The early evaluation of design and procedures for compliance with applicable safety requirements and criteria,
- The continued surveillance of overall safety aspects throughout the total life-span, including disposal.

In general, a typical system will pass through the following phases in life-span:

- The conceptual phase, when the basic purpose, mission and preliminary design approaches are considered and formulated. Unfortunately, many existing systems were evolved without passing through any effective system-level conceptual phase.
- The engineering phase, where the objectives requirements and criteria developed during the conceptual phase are translated into actual equipment and procedures. This phase includes testing and analysis for evaluating compliance with stated criteria and goals.
- The operational phase, which starts when the equipment, personnel and procedures function together in order to achieve the intended purposes or mission of the system. The disposal phase which begins when the equipment and personnel are no longer needed and must be effectively disposed of or placed into storage.

Health and Safety issues now affect all organizations to the extent where the potential risks and legal obligations cannot be ignored. To comply with changes in legislation and to protect staff, more organizations are looking to introduce the Occupational Health and Safety Management System based on OHSAS 18001.

One of the such well-known OHS Business Models is the British Safety Council Business Model given at Figure–1 below. Its Five Star Protocol and Methodology quantitatively evaluate 82 elements of management performance. It recognizes what you are doing well and the areas for improvement. The five star audit can be applied to any business and organization.

Rating System

The international quality-assured rating systems evaluate any company or organization's health, safety and environmental performance, through a structured audit system that provides a safety plan for continuous improvement.

The Five Star profile covers all 82 elements in the audit and it can be applied to any business and organization. It also cross checks your systems vertically through the whole of your business, so it evaluates the effectiveness of these systems in the workplace and feeds back to you.

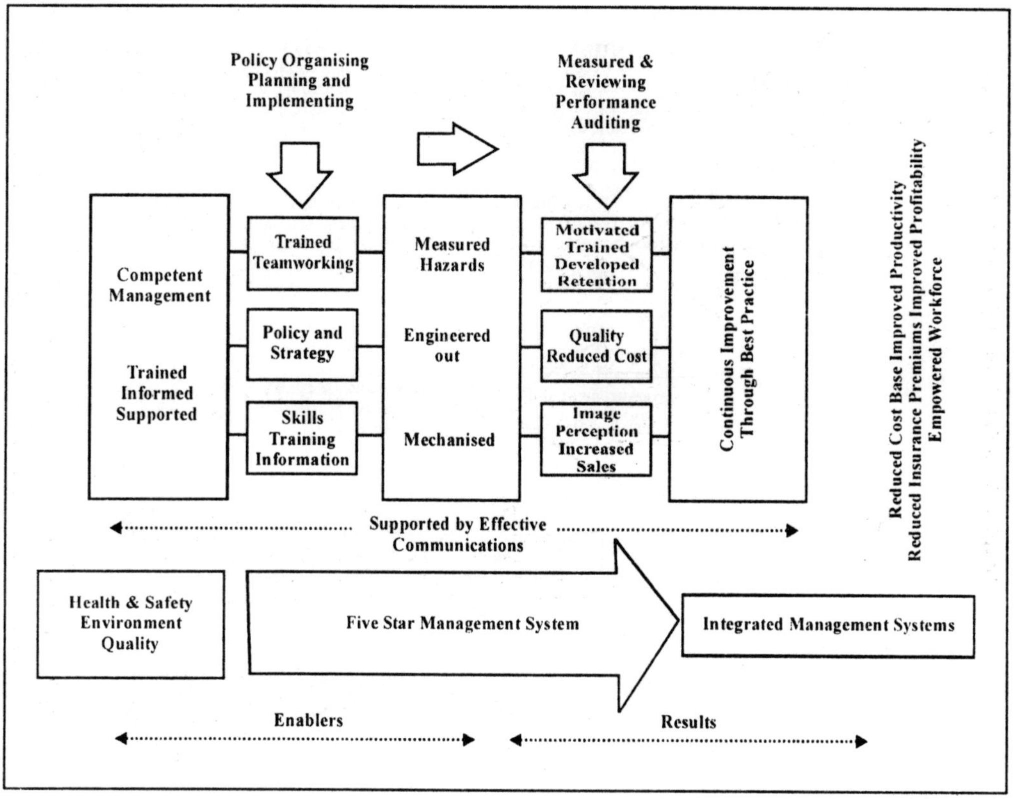

Fig. 1

Areas Covered

The six sections covering the 82 elements provide a detailed report which recognize best practice and areas for further improvement. It provides a clear benchmark and an ability to measure improvement through the report and safety plan.

There are a number of essentially similar off-the-shelf safety management systems available. Each has a number of key element areas against which the corporation is periodically audited and scored. The scoring primarily focuses on documentation for meeting the criteria of the system. Unfortunately insufficient weight is given to actual safety performance results. Too often workplace practice bears little resemblance to the defined system.

Safety needs to be one of a company's core values, and given at least equal importance as other core values like production, quality, customer service and profitability. Safety must be integral to the way business is conducted—not an afterthought or bolt-on extra.

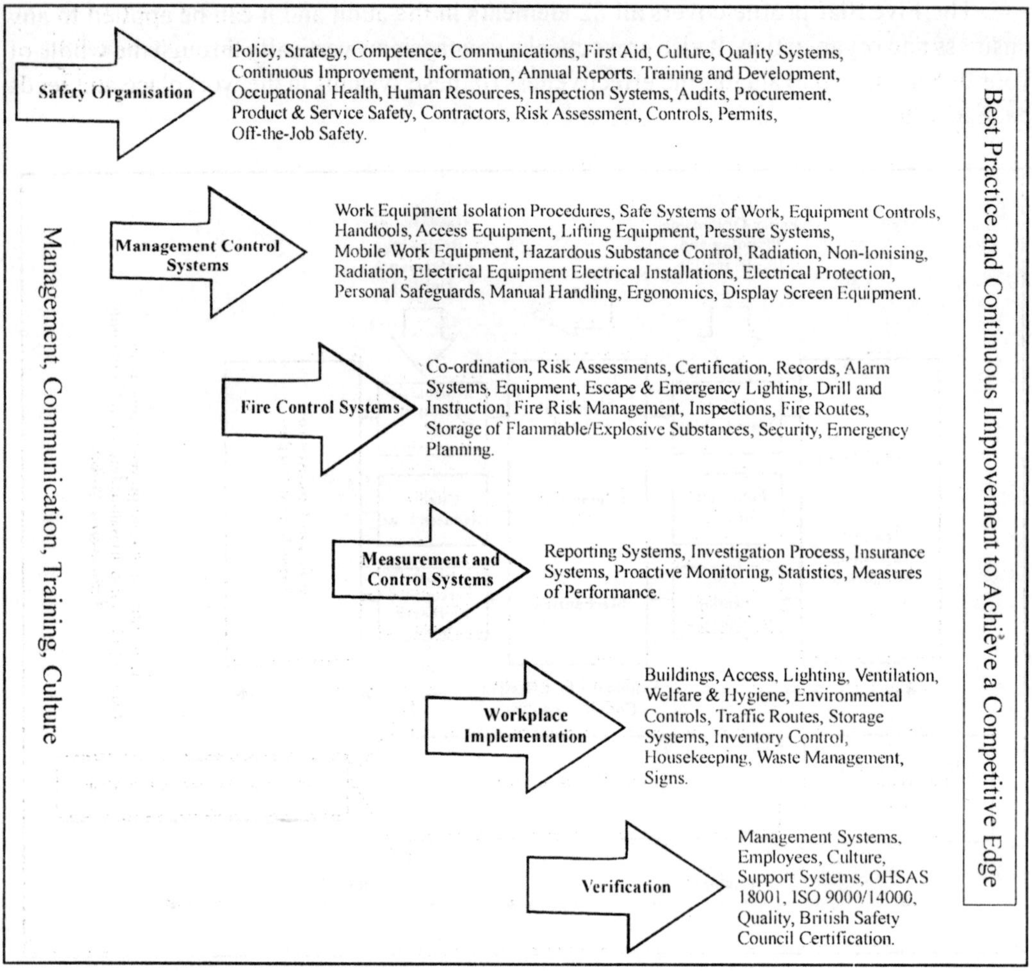

Fig. 2

Operating discipline is a reflection of the safety behaviour of everyone in the company. It's a cliché, but it really does start at the top or it doesn't start at all—and the stronger it is the better the response and performance of the rest of the organisation. Its their leadership and personal behaviour that's the all-important ingredient. Safety management is not something that can be delegated. Senior management must provide the lead through their own personal involvement in the safety management process.

To achieve safety excellence and drive towards "best practice", it is necessary to have a two-pronged strategy. One being the development and implementation of a safety management system that's the company's own and the other being safety behaviour modification to "put legs on the system" and develop a strong safety culture. The focus on behaviour sterns from the fact that about 96 per cent of all injuries occur because of behaviour—not because of flawed procedures or inadequate guarding of equipment.

Integrated Systems

Only a few companies have achieved a fully integrated management system covering quality, health, safety and the environment. By applying the principles of Total Quality to the management of health, safety and the environment within a business, managers cannot only reduce costs but explore opportunities to improve their performance.

Our vision of the future anticipates where the ball is going to land so that you can be there in front, rather than going where it has already landed. Integration will help expose areas of waste, duplication and nonvalue-added activities and provide an opportunity for rationalization.

The techniques for integration involve a comprehensive review of the business in terms of the way it manages quality, health and safety and environmental aspects. This review needs to be established at executive level and it should identify initial integration opportunities for business improvement.

- *Strategy and objectives* take into account all aspects of business performance including quality, health and safety and environmental requirements.
- *Reviews* when integrated allow all business risk levels to be evaluated.
- *Risk management* addresses those matters which affect the company image, product, public and employee liability, taking into account not only the regulatory requirements but also those of insurers and the associated costs.
- *Processes* are the essence of business. Benefits of integration are the identification of all process inputs and outputs, the interfaces between them, stakeholder requirements and customer expectations.
- *Documentation* is simplified, concise and focused.
- *People* are involved in the continuous development and implementation of the management system. We believe the person doing the job knows best. This leads them to a greater understanding of their individual responsibilities and the effect of their actions on business performance and the reduction or elimination of risk.
- *Performance indicators* are focused on key tasks and objectives and have measurable outcomes for matters which affect company image, product, public and employee liability, physical and consequential loss, taking into account not only regulatory requirements but those of insurers and the associated costs.
- *Tools and techniques* which already apply to quality, health, safety and the environment, traditionally used for single facets of the business, are applied to all aspects of management.
- *Culture* in any business directly relates to how employees are managed. Employees are the single most important asset in any company and, through involvement, training and effective teamwork, can ensure continuous improvement at all levels of the business.

Safety Management System

The safety management systems not only improve our knowledge of hazards and risks but also are the most effective way to capture, retain, and transfer valuable knowledge by embedding that process into the workflow minimizing downtime. The cornerstone of these safety, environment and occupational health management systems is sharing best practices and lessons. The adoption of safety management systems reflects a growing strategic importance of health safety and environment issues.

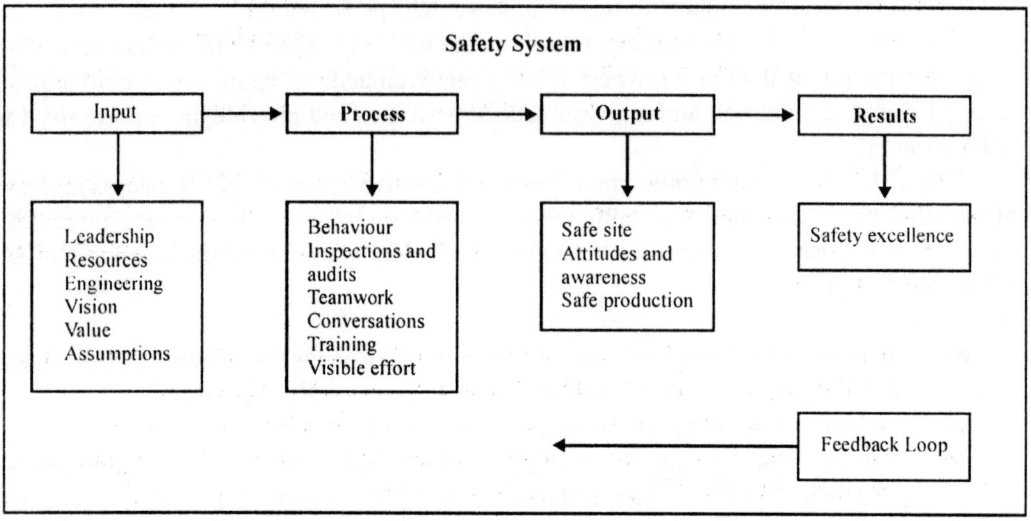

Fig. 3

The single most important characteristic element of the system safety concept is a pre-accident hazard evaluation. In order for this to be accomplished in practice, two conditions must co-exist.

- There must be a management structure and attitude which promotes the asking of a question. "What are the potential hazards, and what are risks to the conservation of resources associated with this contemplated endeavour?"
- There must co-exist the technical capability within the management structure to effectively analyze the problem to provide an acceptable, accurate answer to this management question, so that the necessary decisions as to how to proceed can be resolved.
- It is important for safety professionals to become adequate trained and equipped to answer such management questions, as it is to work and create a management environment that will promote the asking of the proper questions.
- System safety analysis comprises of repertoire of technique helping to organize an effective safety programme around the "hazard potential" rather than around a panic reaction to some major accident or catastrophe.

The risk faced by workers are assessed through the quantitative investigation of workplace exposures; by the development of a hazard surveillance system, by the routine use of the occupational health history, and by the implementation of occupational disease surveillance programmes. The other methods include:

Stop (Safety Training and Observation Process)

It is a methodological way of observing people at work, having a systematic conversation about the observations and then reporting the conversation in a very brief and non-personal way on a STOP Card electronically, into any networked computer with which a computer based database is generated to guide us on corrective actions to be taken. It can also be used in the same way to report unsafe conditions.

ASA (Advanced Safety Audits)

Advanced Safety Audits is based upon open, unthreatening and respectful intervention by a manager. Unless you audit, you will never know when complacency has set-in, until the accident happens. It is specifically designed for Leaders (CEO, Managers, Directors, Supervisors) since it believes that managers are essential to the safety effort, unless they do something different nothing will change in the workplace.

TRA (Task Risk Assessment)

Task risk assessment is a method of systematically examining individual work assignments (tasks) to identify the hazards, evaluate the risks and specify appropriate controls. This is very powerful Tool.

World-Class Safety Requires Continuous Attention in three Domains

These factors are dynamic and interactive. Changes in one domain eventually impact on the other two. For example, behaviours that reduce the probability of injury often involve workplace change, and lead to attitudes consistent with safe behaviours. The behaviour and personal factors represent the human dynamics of safety. The basic principle is that both behaviour-based and person-based factors should be addressed in order to achieve a world class safety culture.

PERSON

(Knowledge, Skills, Abilities
Intelligence, Motives,
Attitude, Personality)

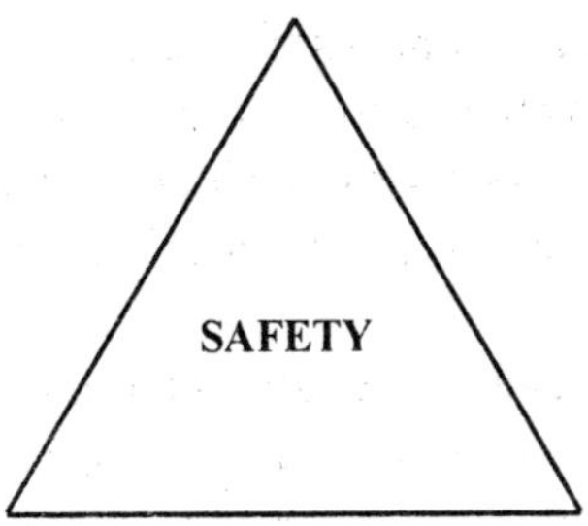

WORKPLACE

(Systems, Procedures)

BEHAVIOUR

(Following procedures, complying with standards using PPE, locking out power, not working under an unsupported roof, coaching others).

OHS Culture

Seven Key Points

The seven key points with the potential to provide professionals with a clear picture of their role as safety leaders are covered in this part of the paper.

Key Point 1: The Concept of Safety Culture must be Practically Defined to be of Value

A useful definition of safety culture as a product is "that observable degree of effort by which all organizational members direct their attention and actions towards improving safety on a daily basis".

People are neither deterministically controlled nor entirely self-determining. Instead they exist in a state of reciprocal determinism... whereby they and their environments influence one another. Reciprocity does not mean the different sources of influence are of equal strength, neither do the reciprocal influences occur simultaneously... This bi-directionality means people are both products and producers of their environment.

Similarly, safety culture is understood as a system or a process involving four elements: (1) inputs; (2) process activities; (3) outputs; and (4) results. Inputs involve basic assumptions; the process includes values and activities; output is the product; and the results are the performance.

This model can clarify misconceptions about the "causes" of workplace injuries. For example, it is often assumed that workers "bad attitudes" (input) lead to unsafe behaviours (process) causing injuries (output) resulting in poor safety performance. But a proper understanding of the model reveals that attitudes are an output—not an input—of the safety culture process. Poor leadership and management lead to unsafe processes and are one cause of poor attitudes.

Thus safety culture is a process rather than a programme; it is a simple way of viewing a rather complex reality; and it lends itself to the ability to improve the system.

Key Point 2: Safety Culture is an Integral Part of the Larger Organization

Organizational culture can help or harm safety efforts. An organization's culture is more powerful than any individual. As the culture dictates what's important, who's important, and what it takes to be successful, it has enormous impact on the organization's future direction. In short, it is impossible to substantially change the direction of the enterprise without changing the culture.

Recognition of the influence and strength of the existing site culture is important for the SH&E professional trying to initiate change. An understanding of basic underlying assumptions can help prevent needless frustration and wasted efforts. The safety function is not an island isolated from the influence of the broader organization and its culture. One must understand the existing culture before trying to improve the safety culture.

Key Point 3: Safety Culture can be Quantitatively Measured

Safety culture elements can be measured with existing assessment tools and measures, as under:

- *Person:* perception surveys, employee interviews.
- *Situation:* safety audits, measurement of the safety management system.
- *Behaviour:* behavioural sampling.

These elements can be measured directly or in combination, which enables one to quantify safety culture in a meaningful way. Two relatively simple requirements are needed to effectively quantify safety culture across the three reciprocal elements:

(i) Measurement of "matched" factors within each element. For example, one might measure management commitment in a safety audit, then compare it to a question about the safety budget in the perception survey, and to management behaviours in the behavioural area.

(ii) Use a common metric, such as percentages, across each element. Most scores can be easily converted to percentages. Deming emphasized that total quality can only be achieved through a systems approach. The same is true for safety excellence, where the systems approach involves focusing continual attention on all three sides of the triad: person factors, organization factors and behavioral factors. Leadership and management are both art and science. Quantitative measures contribute to a more scientific approach to safety culture development.

Key Point 4: Leadership must Share a Vision for Establishing Safety Excellence

Cultures are created to a great extent by leaders. Leadership is the most powerful component of culture. Leaders affect change, which in turn creates and sustains an organization's culture. A poor safety culture implicates poor leadership.

Leadership is generally viewed as the key determinant of organizational success in all its various endeavours. Leaders create the vision and strategies. They "establish the vision for an organization; managers are the key to making the vision a reality".

Leadership is situation-specific—the characteristics needed are dependent on the situation. The search for a universally correct leadership style is doomed to failure because of cultural variation by country, by industry, by occupation, and by the particular history of a given organization.

Management deals mostly with maintaining *status quo*; leadership deals mostly with change. Positive culture change is up to leadership, but most leaders are not aware of how much the assumptions they take for granted are reflected in day-to-day behaviour by the way they manage the decision-making process.

To create the greatest impact on safety culture, the SH&E professional should be more of a leader and less of a manager.

Key Point 5: Leaders Must Focus on Specific Behaviours to Strengthen Safety Culture

When people talk about culture change, what they really want is behaviour change. When someone talks about changing the culture, what most people hear is 'changing core beliefs', which is nearly impossible. The key is to speak about changing behaviour, not the basic beliefs that underlie those behaviours. Leaders must pay attention to their own behaviour rather than that of their employees. Let's assess what general behaviours impact safety excellence for leaders.

Key Point 6: Leaders Need both "What To" and "Know How" to Established Excellent Cultures

Leaders must possess both the desire to act and a clear understanding of the specific behaviours that lead to excellent,safety cultures. Top management must be committed to excellence and drive the agenda by establishing a vision, value and goals, and by seeing that all line managers have safety improvement objectives; by auditing performance; and by visual personal involvement.

Key Point 7: The SH&E Professional's Role-Influence the Right People to Take the Right Actions

It is not the SH&E professional's responsibility to establish an effective safety culture; in fact, in most situations, that would not even be possible. Instead, the SH&E professional must influence site and corporate leaders and line management to establish a safety culture. The practitioner should be knowledgeable and technically competent to provide relevant guidance, but more importantly, illustrate the way forward and functions effectively as an organizational "leader".

It is not the SH&E professional's responsibility to establish an effective safety culture; in fact, in most situations, that would not even be possible. Instead, the SH&E professional must influence site and corporate leaders and line management to establish a safety culture. The practitioner should be knowledgeable and technically competent to provide relevant guidance, but more importantly, illustrate the way forward and function effectively as an organizational "leader".

The SH&E professionals must

- Recognize and inform management when safety concerns result from common cause (rather than special cause) and advise on solutions to the cause that exist within the management system. Common cause and special cause derive from the concept of variability. Special cause include unusual events and activities that are not part of the everyday system. Most injuries are a result of common cause (behaviours that have occurred previously) rather than special cause.
- Continue learning and applying new knowledge.
- Effectively make the business case for safety and supporting the business function.

Leadership Behaviour

The leadership behaviour for establishing safety excellence in an organization require covering the following seven Es.

Seven Es

1. *Establish Expectations—Leaders Establish Expectations and Accountabilities for Safety Through Clear Vision and Communication.*

Leaders need to translate their vision into clear expectations and accountabilities. Examples of establishing expectations and accountabilities include:

- Intervene as a safety coach. Most organization are over managed and under led, and coaching is almost totally missing.
- Create and sustain a viable future for safety improvement through their conversations.
- Establish accountability for safety performance at all levels of the organization. The success of... safety and health management team being held accountable for performing or not performing their tasks.
- You cannot achieve safety excellence without proper behaviours. You cannot get the proper behaviours without the right culture. Also you cannot get the right culture without accountability. It is not one or the other—you must build all three—and in the proper order to achieve excellence in safety.
- Management accountability requires a system of role, definition, correct measures of performance and adequate rewards contingent on performance. Proactive actions such as these build a culture which says safety is so important that all managers and supervisors have to do something about it everyday.

2. *Engineering Support—Ensure that Engineering Concerns are Handled as Priorities Designated in the Hierarchy of Controls.*

- Leaders should ensure that safety-related engineering concerns are handled as priorities. The hierarchy of safety controls designates the most effective methods of safety intervention: (1) hazard elimination; (2) safeguards and enclosures; (3) warning and instructions; and (4) protective clothing and administrative controls. "The hierarchy recognizes that design, elimination and engineering controls are most effective in reducing risk than lower level controls such as warnings training, procedures and personal protective equipment."
- When possible, organizations should attempt to eliminate, engineer and guard against hazards. Behavioural approaches must not be used to abdicate management responsibility; they must be used to enhance—not replace—the traditional hierarchy of interventions. Experience shows that employees are less likely to meet expectations if management takes no responsibility for correcting potential safety hazards through feasible engineering methods.

3. *Exemplary Behaviour—Leaders set a Personal Example of the Behaviours Required for the Desired Safety Culture.*

- Leaders set a personal example of behaviours required for the desired safety culture. They must follow established safety rules and regulations, and must participate in safety meetings and regularly discuss safety in their conversations.
- Effective leadership of safety requires senior managers to develop and implement a strategic plan for safety that captures the hearts and minds of employees, personally demonstrate excitement and enthusiasm for the changes, and model the behaviours others are expected to follow.

4. *Educate Employees—Leaders Provide Education, Training and Resources to Ensure that Employees are Fully Developed and Prepared to Contribute to Safety Excellence.*

- Leader provide education, training and resources to ensure that employees are fully developed and prepared to contribute to safety excellence.
- Learning culture is the willingness and competence to draw the right conclusions from its safety information system, and the will to implement major reforms when their need is indicated. Learning culture along with a reporting culture, a just culture and a flexible culture result in an informed culture.

5. *Enable Employees—Leaders Grant Employees Authority, Flexibility and Partnership Once they are Educated and Properly Prepared to fulfil those Characteristics.*

- Often, management is unwilling to share power and authority with workers.
- Leaders must establish a trusting climate so that all injuries and potentially even near-hit cases will be reported. Reporting must be held as a higher value than retribution or punishment. High participation in reporting is important because it is impossible to correct problems if no one knows they exist.

6. *Encourage Employees—Leaders Encourage Everyone at all Levels by Positively Reinforcing Desired Behaviours. They also Encourage Self-efficacy. Self-efficacy ('the Capacity to Exercise Control over one's own Thought Processes, Motivation and Action....') is Fundamental to Proactive Safety Culture.*

- Some 350 years ago, Musashi wrote, "The foreman should take into account the abilities and limitations of his men, circulating among them and asking nothing unreasonable. He should know their morale and spirit, and encourage them when necessary". Leaders encourage self-efficiency, which is "the capacity to exercise control over one's own thought processes, motivation and action" and is fundamental to proactive safety cultures.
- Effective leaders encourage everyone at all levels by positively reinforcing desired behaviours.

7. *Evaluate Effectiveness—Leaders Measure and Evaluate the Effectiveness of Their Organizational Strategies and Make Needed Adjustments.*

- Continuous improvements in safety requires that a company measure, evaluate and make appropriate adjustments as necessary. These behaviours should be part of an established system—not simply anything leaders may occasionally decide to practice.
- Leaders' conversations, behaviours and visible support play a major role in establishing and changing organizational culture. The best way to encourage such behaviour is to make it part of the management system.

Behaviour Based Safety Process

Long-term employee participation requires understanding and belief in the principles behind the process. Employees must also perceive that they own the procedures that make the process work. For this to happen it is necessary to teach the principles and rationale first, and then work with the process participants to develop specific procedures. When team members are asked to identify their own critical behaviours, it creates ownership and leads to long-term involvement.

The behaviour-based safety process is essentially a culture change process. To achieve a world-class safety culture and to affect culture change in an organization, the principles underlying behaviour-based and person-based psychology need to be integrated. This is done by focusing on the behaviour of people and, by changing it, bringing about changes in their perceptions and attitudes. Specific techniques can be viewed as possible routes to reach a destination, in our case a world-class safety culture.

Primacy of Behaviour

Whether treating clinical problems or preventing workplace injuries, the focus is on behaviour. The process should focus on three basic questions:

- What behaviours should be increased or decreased to treat or prevent the problem?
- What workplace conditions, including interpersonal relationships, are currently supporting the at-risk behaviours or inhibiting desirable behaviours?
- What workplace conditions can be changed to decrease at-risk behaviours and increase desirable behaviours?

Thus, behaviour change is both the outcome and the means. It is the desired outcome of the process, and the means to solving the identified problem. Behaviour refers to an action by an individual that can be observed by others. Feelings, attitudes, perceptions, motives, thoughts or intentions should not be confused with behaviour.

Behaviours are actions we can see and measure. Whether behaviours are repeated or not depends on their consequences. Actions with positive results tend to be repeated. For behaviour based safety approach to be successful, actions that are likely to generate negative results should have tendency to be avoided. The cycle of Action-Habit-Behaviour is inculcated over a long period of time. A choice of short-cut may be tempting. A reflux response generated out of some choice, may have its bearing on this cycle.

Prompting safe behaviour at work is a critical part of the management of Health and Safety, because behaviour turns systems and procedures into reality. On their own, good systems do not ensure successful health and safety management, as the level of success is determined by how organizations 'live' their systems.

Safe behaviour can be shown to yield benefits. These benefits will in turn reinforce the actions that produced them. In this way, safety becomes a habit. On job sites, positive results can include recognition from management, respect from co-workers, safety awards, and in some cases, incentive programmes. These are the responses that encourage, reward, and reinforce safe behaviour.

The behaviour modelling approach is to target a limited number of critical behaviours, and one-on-one observation. A team defines a critical behaviour or behavioural outcome to observe. After defining their target so that two or more observers can reliably observe and record a particular behaviour, the team members do the following:

- Define and prioritise their critical behaviours.
- Make a pact among themselves that they will conform to their critical behaviours.
- Agree on the frequency of behaviour modelling.
- Agree on the frequency of observation and feedback.

Team members must be observed willingly and anonymously for the target behaviours, using the checklist to maintain daily records of the safe and at-risk behaviours defined by the team.

"Behaviour modelling, observation and feedback is key to achieving a world-class safety culture."

The five basic steps that each team has to take in order to implement a behaviour-based safety process are :

- Define their critical behaviours.
- Make a formal pact to conform to those behaviours.
- Agree on the frequency of their behaviour modelling sessions, and stick to it.
- Agree on the frequency of individual observations and feedback, and stick to it.
- Find solutions to the problems that they have identified.

The dynamics of these steps are as follows:

- Defining the behaviours themselves gives the team a sense of ownership;
- The pact triggers the dynamics necessary to form new team values;
- The behaviour modeling sessions entrenches the behaviours at the both the cognitive and subconscious levels; and
- The problem solving process not only removes barriers to safe working, but also empowers team members to take decisions and have them executed.

Safety Analysis

Primary Goal

The system safety concept envisions a preplanned, organised effort which has as a primary goal the conservation of resources associated with a given system, product or operation by:

(i) The pre-accident identification of potential hazards;
(ii) The timely incorporation of effective safety related design and operational specifications, provisions and criteria;
(iii) The early evaluation of design and procedures for compliance with applicable safety requirements and criteria;
(iv) The continued surveillance of overall safety aspects throughout the total life-span, including disposal.

Basically a system is the sum total of all elements working together within a given environment to achieve a given purpose or mission. System safety analysis comprises a repertoire of techniques helping to organize an effective safety programme around the "hazard potential" rather than around a panic reaction to some major accident or catastrophe.

The single most important characteristic element of the system safety concept is a pre-accident hazard evaluation. In order this to be accomplished in practice, two conditions must co-exist:

(i) There must be a management structure and attitude which promotes the asking of a question,"what are the potential hazards, and what are risks to the conservation of resources associated which this contemplated endeavour?"
(ii) There must co-exist the technical capability within the management structure to effectively analyse the problem to provide an acceptable, accurate answer to this management question, so that the necessary decisions as to how to proceed can be resolved.

It is important for safety professionals to become adequately trained and equipped to answer such management questions as it is to work to create a management environment that will promote the asking of the proper questions. The products, plant and machinery need to be designed for optimum inherent freedom from hazard. Where it is not possible to reduce the magnitude of a hazard through design, safety and warning devices, special procedures need to be established.

The Process of Risk Management

Risk management is a process through which a company or individual attempts to minimize the risks they face when they attempt a task, project or simply manage their business. Risk Management is attempted in many different ways by companies and individuals, but it is necessary to follow a sound process, or the illusion could exist that risks have been identified and minimized.

The management of risk is essential to optimize the efficiency of any industrial process and to minimize losses. These risks may be related to:

(i) employee safety;
(ii) public safety;
(iii) environmental damage;
(iv) damage to equipment; and
(v) business performance and robustness.

Eventually all these risks have an impact on the profitability and success of an industrial concern. In other words failures, which affect safety and the environment also have a significant effect on a company's bottom line and are therefore part of the total business risk. For that reason, it is sometimes useful to quantify all risks in terms of Rupee value (impact on 'bottom line') so that they can be put together in a risk profile relating to each other. These risks can be ranked and then portrayed in a graph. Since a thorough risk analysis can be very time consuming and expensive when applied on all issues, it is necessary to identify those issues that when analysed in detail will contribute most to minimizing the total risk.

In risk assessment, different risk analysis techniques can be applied in order to quantify the risk as well as to determine those issues or components that should be changed or managed to reduce the risk to an acceptable level.

Risk Assessment and Risk Management

Risk Assessment is a vital element of risk management. Risk Assessment incorporates Risk Analysis and the first part of the Decision Making process. Risk Analysis, in turn, incorporates Hazard Analysis, Frequency Analysis and Risk Presentation. Risk analysis proceeds from hazard identification, a qualitative process, through consequence and frequency analysis (quantitative analysis) to presentation of the risks in numerical terms. The decision making process, therefore, starts with the assessment of risks against suitable criteria: this enables the acceptability of the risks and the degree of risk reduction required to be determined. Alternative strategies or means of reducing risk, or alternative designs, can be analysed and assessed in the same way; once this has been done, the effects of different risk reduction measures can be compared with the criteria. Once this is set and the risks are known, a Safety Management System (SMS) can be developed on the basis of the known remaining risks. Although this is shown as the 'end point' of Risk Management, it is really one more stepping stone, since it is the SMS that will form the cornerstone of ongoing safe operation of the facility when it has been constructed and commissioned.

Risk assessment is one of several ways of formal Safety Assessment. A number of techniques are commonly used to identify and quantify hazards and risks, and to optimize the costs and benefits of engineering solutions, e.g.:

(i) Hazard and Operability (HAZOP) Studies
(ii) Failure Modes and Effects and Criticality Analysis (FMECA)
(iii) Fault Tree Analysis (FTA)

(iv) Event Tree Analysis (ETA)
(v) Cause-Consequence Analysis (CCA)
(vi) Reliability, Availability and Maintainability (RAM) Studies

Advantages of Quantitative Risk Assessment (QRA)

QRA may be used to provide input on safety issues to the design, operation or regulation of hazardous activities. They can be extremely useful in several ways, including:

(i) Estimating risk levels and assessing their significance. This helps decide whether or not the risks need to be reduced.
(ii) Identifying the main contributors to the risk. This helps understanding of the nature of the hazards and suggests possible targets for risk reduction measures.
(iii) Defining design accident scenarios. These can be used as a design basis for fire protection and emergency evacuation equipment, or for emergency planning and training.
(iv) Comparing design options. This gives input on risk issues for the selection of a concept design.
(v) Evaluating risk reduction measures. QRA can be linked to a cost-benefit analysis, to help choose the most effective ways of reducing the risk.
(vi) Demonstrating acceptability to regulators and the workforce. QRA can show whether the risks have been made "as low as reasonably practicable".
(vii) Identifying safety-critical procedures and equipments. These are critical for minimizing risks, and need close attention during operation.

Taken together, these possible uses of QRA provide a rational structure for monitoring risks and providing guidance for decision-making about safety issues.

Human Error

Safety can be managed more effectively through an insight into the causes of human error, and practical steps to reduce the incidence of errors, incidents and accidents. You can't ignore low probability risks. Lessons learnt from infamous failures and little known close calls can help design jobs for Error reduction and driving organizational and Human errors to zero. Analyzing performance problems highlights the influence of culture on human error. We need to strive for a Safety Perfect culture. Understanding human error could be the Safety's next most profitable frontier.

Occupational Health

Occupational Stress

Occupational stress can no longer be considered an occasional, personal problem to be remedied with palliatives. It is becoming an increasingly global phenomenon, affecting all categories of workers, all workplaces and all countries. This trend-coupled with its rising cost to the individual, to industry and to society as a whole—has greatly heightened awareness of the need for effective and innovative ways of tackling stress. Stress prevention at the workplace has proved particularly effective in combating stress, by attacking its roots and causes, rather than merely treating its effects.

Stress audit improves our understanding about individual vulnerability or risk, so that effective manpower development programmes can be implemented. The need to understand the nature of stress at work can eliminate or moderate the very root of the problem. Certain stressors would respond to change at the macro level of an organization (e.g. a change in culture), whereas others might relate to the micro-environment by dealing with issues of workload and shift patterns.

Exposure Monitoring

Monitoring of worker exposure and monitoring of health effects, these twin methods are to be coupled in a co-ordinated fashion. Routine environmental monitoring of employee exposure is an important step in assuring that workers are not placed at risks. Routine environmental monitoring of employee exposure is an important step in assuring that workers are not placed at risk. The epidemiologist requires industrial hygiene data to quantitatively assess risk factors in occupational disease. Routine monitoring, and record collection and maintenance as aid for future studies is essential.

Hazard Surveillance System

The lack of national occupational health surveillance system which could provide data on the incidence, demography, secular trends, and mortality related to occupational exposures, has been recognized as a problem. Such a hazard surveillance system would identify chemicals in use, the industries and occupations where they are used, and worker exposure to them. To assess accurately the risk of a population of a disease associated with occupational exposure, the following parameters need to be determined:

(i) Identification of chemicals in use by industry.
(ii) Description of the industrial process.
(iii) Number of workers exposed to a particular substance.
(iv) Assessment of the current exposure levels.
(v) Identification of workplace settings in which there is increased risk.
(vi) Assessment of the toxicity of specific agents (animals, human, or short term test data).
(vii) Assessment of the effectiveness of controls

The accurate diagnosis of an occupational disease is important for two reasons:

(i) Determination that an illness or disease has been caused by exposure to a toxic agent will facilitate removal of the patient from exposure, implementation of appropriate exposure—reducing techniques, and treatment of the patient; and

(ii) The ascertainment that a disease is of occupational origin will enable preventive steps to be taken to ensure that other workers are not similarly affected.

The occupational health history is an essential element in the diagnosis of workplace disease. Given the general nature of physical examination findings and laboratory tests, the occupational history may be the only means to identify a work-related medical problem.

Disaster Management

In India, the subject of disaster management is still in its infancy. The concern for Disaster Management came into picture only after Bhopal Disaster on 3rd December, 1984, due to leakage of poisonous gas (its effect on social life is still evident). Only after this tragedy, classification of the disasters as natural or man-made gained prominence. A disaster is an event triggered by natural and man-made causes leading to serious disruption of normalcy/ functioning within the society, causing widespread human, material or environmental losses that exceeds the ability of the affected society to cope using its own resources. It is an event that endangers the life and property of innocent citizens, needing immediate assistance and outside intervention.

Prevention of Disasters

For the prevention of serious incident, prevention process models are available that merge proven quality management techniques with sound risk management practices. According to one of such model by Thomas Burns, eight elements have been identified as essential for inclusion in an effective process for maintaining the workplace condition necessary to sustain serious incident-free operation. These are as follows:

(i) Establishing the serious incident prevention as an organizational policy
(ii) Involvement of employees
(iii) Identifying and understanding the organizations specific risks
(iv) Identifying critical work for controlling the risks
(v) Performance standard to be established
(vi) Measurement and feedback
(vii) Reinforcement and corrective actions
(viii) Improving and updating the process

Disaster Preparedness

Disaster cannot be prevented totally. However, timely warning and planning can minimize the effect of disasters. An accurate disaster management plan needs to be prepared. It may be observed that advancement in Information Technology in the form of Internet, GIS, Remote Sensing (RS), satellite communication, etc. can help a great deal in planning and implementation of hazard reduction measures. GIS can improve the quality and power of analysis of natural hazard assessments, guide development activities and assist planners in the selection of mitigation measures and in the implementation of emergency preparedness and response action. The use of modern technology like GIS and RS can be of vital importance in the preparation of plans by keeping records of vulnerable areas, monitoring of rescue and relief operations, deciding response, managing the database etc. Mutual aid schemes shall be of great help in mobilization of resources, and regular mock drills shall ensure the efficiency and affectivity of response. Training of the various functionaries and an effective public awareness and education campaign involving the communities will ensure that the plans are disseminated to the lowest levels.

Even the mitigation strategy plans of government of India recognize the fact that effective community involvement and public awareness can largely minimize the impact of disasters and, therefore, community based mitigation strategy would go a long way in strengthening and stabilizing the efforts of the administration. The focus should be on capacity building including formation of Community Emergency Response Teams (CERT).

Generally, it is observed that in order to quickly conduct disaster response activities during various disasters, Disaster Management Plan is created with special emphasis on Vulnerability analysis, Hazard analysis, Risk analysis and Resource analysis. Training strategy of the key role players, functionaries and general public both at community and social levels is an essential element for the successful execution of a Disaster Management Action Plan (DMAP). The DMAP cannot be fully operationalised without a training strategy. Manuals for Warning and Evacuation, Emergency Operations Centre (EOC) and District Control Room (DCR) have to be prepared keeping this in mind.

Present Status

The Government of India has created a crisis management group which coordinates and ensures effective relief assistance to be given from Central Government and international assistance. With this in view, the Government of India has made certain ministries responsible for certain types of disasters and these ministries are known as nodal ministries for coordinating such Disaster Management activities. The other ministries of the Government of India support the said nodal ministries as members of the crisis management group, in whatever way they are required to help. The latest list of nodal ministries for different types of disasters is, as presented below:

Disaster	*Nodal Ministries*
Air Accidents	Ministry of Civil Aviation
Civil Strife	Ministry of Home Affairs
Chemical Disasters	Ministry of Environment & Forests
Railway Disasters	Ministry of Railways
Biological Disasters	Ministry of Health
Nuclear Disasters	Department of Atomic Energy

Training

Even to think of giving training to general public and NGOs, spread across India, looks impossible on its face. For the last two years, National Fire Service College could hardly train general public, leave aside NGOs. Training has to be mass movement. Government must also select other NGOs having technical capabilities for imparting training to general public as nodal training agencies.

These NGOs could give training to 20 to 25 per cent of able-bodied villagers in case of remote areas, so that they become organized members of self-help parties possessing skills of basic rescue, basic fire-fighting and first aid. Apart from the above, in the major centres selected to assist in remote area disasters we could thrust the responsibility on skilled NGOs having knowledge and equipments for carrying out skilled jobs of rescue, fire-fighting and first aid. This force could be augmented by the urban teams from nearby locations coming and carrying out needful task by assisting the above NGOs and the local selfhelp parties working on the ground. The needful training should be carried out in all seriousness, as we do in the case of any of the other Rojgar Yojana schemes.

When such a scheme is floated it is but natural that the financial implications will loom large in implementing the programme. However, taking into consideration the hundreds of crores of rupees government spends in relief and rehabilitation schemes a sum of even one per cent of such total amount, if diverted for this useful activity, will go a long way in combating problems of disasters.

Emergency Management System

In the era of industrialization, the level of risk to society and the environment has been rising continuously. This level has to be kept at the lowest. There are various tools available from the fields of engineering, technology and management for the identification of the inherent hazards involved and assessment of the consequential fallout in case a hazardous incident occurs. The Emergency Management System (EMS) is one such tool that can be effectively used to prevent, if possible, and otherwise drastically reduce the risks.

Disasters like the Piper Alpha, major blow outs like Flixborough, Seveso and Bhopal have all brought in sharp focus the hazards that are extant wherever operational activity or chemicals are being handled, and the damage they can cause to people, equipment material and the environment (PEME). There is no such thing as 'Zero risk'. So the need for emergency preparedness and forward/advance planning to contain and restrict damage to the lowest minimum possible is evident and urgent. Advance planning is the key.

There can be no alternative to preparedness and planning with perpetuity in controlling and restricting the damaging effects of any disaster. The ultimate and broad goal of EMS is to improve the quality of life, focusing on optimizing investments by reduction of risks; it has an important role in preventing emergencies, providing combating tools and tackling emergency incidents with the least possible damage.

Most experts today emphasize a comprehensive "all hazards" approach to emergency preparedness. A comprehensive emergency response plan that takes into account potential natural, technological, and man-made threats, and involves key personnel in the planning process can assist an organization to systematically manage emergencies in an effective and efficient manner. The planning process is a key element that forces managers and their staff to explore viable options that can be employed in the event of an emergency or disaster.

Emergency preparedness and planning need a detailed and systematic approach at both the macro as well as micro levels. Usually the macro level aspects get proper attention while the micro level ones are assumed to sort themselves out automatically. This is flawed thinking leading to disasters. To achieve the desired result, planning at the lower levels must be given equal and as much attention.

It is imperative that the plan be fully integrated into the organization's standard operating procedures (SOPs). According to FEMA, SOPs and checklists provide the detailed instructions that an organization or individual needs to fulfil responsibilities and perform tasks assigned in the Contingencies plan. Clearly, a comprehensive checklist that includes major planning, implementation, training/testing, response, and recovery components would be an invaluable asset to any organization's emergency response plan.

Sectoral Needs

Bulk of the safety requirements in different industries are common, however, each will have special specific safety problems to manage, e.g :

- *Fire and Explosives:* There are many hazards that need controlling in a process operation, but avoiding fires and explosions is one of the prime safety objectives. In general, they share common initiating conditions which are the simultaneous presence of a flammable material, ignition source and an oxidant (normally oxygen in air). Operations handling flammable or explosible materials will take precautions that are based on excluding one or more of these initiating conditions.
- *Food Safety:* Managing Food Safety for any modern food handling business is vital. The benefits of application of HACCP (Hazard Analysis and Critical Control Point) systems should be understood by all concerned.
- *Construction Safety:* Construction Safety through Design and Management of site safety and legal issues is vital.
- *Petroleum Sector:* Well logging operations and recommended practices for oilfield explosive cover safety and health risk and proper precautions with respect to particular materials and conditions. Safety precautions in oil production installation (oil mines regulation 1984) prescribe the minimum needs of safety. Many products are hazardous in handling, transportation and storage, because they are explosive, flammable, poisnous or corrosive. The sector constantly requires continued effort to develop technology, engineering analysis and judgment to fit specific applications for well-sites and deep sea effort.

Statutory Requirements

In most part of the world, it is accidents that have driven health and safety legislation as well as improvements in safety technology. The concept of a 'duty of care' makes employer responsible for safety and welfare at work for his employees.

Government legislation and the moral demands of contemporary society have led to the need for an ever increasing focus on workplace safety. Today the acceptance that some workplace injuries are inevitable is constantly being challenged by progressive managers. With the threat of large fines, both for the corporation and personally for directors, managers and supervisors plus the threat of jail sentences, many companies have taken steps to try and manage the risk by developing in-house, or acquiring from another organisation, safety management system.

Our legal system by its nature demands active participation by all those involved if it is to work successfully. The regulatory regime recognises that industry, as the creator of the hazards which expose people to risks, is in the best position to control those risks. Duty holders' absolute responsibility for health and safety is of course fundamental.

Modern legislation is based on goal setting, with the state setting the standards to be achieved. This leaves considerable—though not total—freedom to the regulated to devise solutions. The non-prescriptive risk-based approach of our regulatory regime is intended to enable industry to devise solutions appropriate to the problem at hand and to adapt these solutions as time goes on to reflect technological change.

Industry is required to assess the risks it creates and take action proportionate to those risks to reduce them to a level which is as low as is reasonably practicable. The overall aim must be to keep accidents and ill health to a minimum. To achieve this ideal requires all parts of industry actively to consider, appreciate and understand the importance of risk assessment and risk management as essential requirements of good management. In seeking to fulfil their responsibilities, duty holders will rely heavily on the level of awareness of risk concepts amongst their engineers, both in management and in the engineering of products and processes —from design, through use and maintenance, to decommissioning. Increasingly in more complex, high hazard industries there are requirements in law for duty holders to prepare safety reports. They have to set out how the risks have been identified, what has been done to reduce them as low as reasonably practicable and what systems are in place to ensure that the risks are in fact being properly controlled.

OHS Certification Scheme

An effective health and safety management system must minimize the risks associated with the organization's activities. Central to minimizing risk is the process of hazard identification, risk assessment and risk control. It is, therefore, necessary to examine the risk assessment activities in detail to establish they are comprehensive, have been completed by a competent employee, and that (as reasonably practical) the control measures put in place are adequate.

For a number of years there have been calls for a certification scheme for occupational health and safety (OH&S) to give organisations a vehicle to publicly demonstrate their commitment to safe working practices and act as a benchmarking tool by defining accepted best practices in safety management.

A significant feature of OHSAS 18001 is that it defines the key elements to be addressed by a management system. Its successful implementation depends upon a high level of management commitment. These days, directors are more aware of the merits of taking an active role in the management of OH&S. The ultimate responsibility for any breaches of health and safety regulations lies with them and it is clearly in their interests to, at the very least, ensure that activities under their control comply with relevant work-related legislation. If proof were needed, the dozen or so MDs currently serving prison sentences in the UK because of OH&S negligence, provide a sobering reminder.

The standard also states that all organizations should have an effective means for employees to be consulted and represented in OH&S policy development and implementation.

In addition to routine monitoring of health and safety performance, the standard recommends that the company addresses the following questions:

- Is the organization's overall health and safety management system capable of achieving the required standards of health and safety performance?
- Is the organization (or part of it) fulfilling all its obligations with regard to health and safety?
- What are the strengths and weaknesses of the health and safety management system?
- Is the organization (or part of it) actually carrying out and achieving what it claims to do?

A further key element in OHSAS 18001 is the commitment which the company has to show in order to comply with the appropriate legislation. Health and safety management systems are intended to deliver legal compliance and the policies of the organization with respect to protection of its workforce and when appropriate customers and the public.

In common with our approach to safety management, OHSAS 18001 describes a risk based methodology. The crux of the development of the system centres on the analysis of tasks, the identification of hazards and their associated risks.

Health and safety issues now affect all organizations to the extent where the potential risks and legal obligations cannot be ignored. To comply with changes in legislation and to protect staff, more organizations are looking to reduce the risk of accidents and litigation by implementing an Occupational Health and Safety Management System based on OHS 18001.

As global trading increases, the introduction of an internationally-accepted protocol for OH&S will enable organisations to demonstrate to all stakeholders their integrity and commitment to operate within safe limits. The level of interest in certification around the world has been impressive. As a vehicle for demonstrating due diligence, OHSAS 18001 should be seen as an essential part of an organisation's overall risk-management strategy for the future. Efforts should be made by each organisation to integrate OHS&S 18001 into their Quality and Environmental Systems.

Integrated Systems

Throughout the rapid advance of technology a concerted focus on health, safety and environment issues by adopting various systems like Environment Management system based on ISO-14001, Occupational Health and Safety Management systems based on OHSAS-18001, have played a major part in making operations more efficient and effective:

- These Audits drive learning and improvement in the company.
- The management systems developed over years of research are the framework for innovation to succeed in the risky work environment.

Integration will help expose areas of waste, duplication and non-value-added activities and provide an opportunity for rationalization. The techniques for integration involve a comprehensive review of the business in terms of the way it manages quality, health and safety and environmental aspects. This review needs to be established at executive level and it should identify initial integration opportunities for business improvement.

- *Strategy and objectives* take into account all aspects of business performance including quality, health and safety and environmental requirements.
- *Reviews* when integrated allow all business risk levels to be evaluated.
- *Risk management* addresses those matters which affect the company image, product, public and employee liability, taking into account not only the regulatory requirements but also those of insurers and the associated costs.
- *Processes* are the essence of business. Benefits of integration are the identification of all process inputs and outputs, the interfaces between them, stakeholder requirements and customer expectations.
- *Documentation* is simplified, concise and focused.
- *People* are involved in the continuous development and implementation of the management system. We believe the person doing the job knows best. This leads them to a greater understanding of their individual responsibilities and the effect of their actions on business performance and the reduction or elimination of risk.
- *Performance indicators* are focused on key tasks and objectives and have measurable outcomes for matters which affect company image, product, public and employee liability, physical and consequential loss, taking into account not only regulatory requirements but those of insurers and the associated costs.
- *Tools and techniques* which already apply to quality, health, safety and the environment, traditionally used for single facets of the business, are applied to all aspects of management.
- *Culture* in any business directly relates to how employees are managed. Employees are the single most important asset in any company and, through involvement, training and effective teamwork, will ensure continuous improvement at all levels of the business.

Role of SHE Professionals

I begin with a truism. Economic prosperity and quality of life for all depend to a great extent on the innovative capacity of our engineers and on the acceptability to society of their innovations. Any innovation naturally brings with it risks to a greater or lesser extent. The risks may affect the workplace or the public. HSE regulation of industrial risks has to strike a balance between the need for protection against risks and the needs of society for beneficial economic activity. This balancing act poses challenges for both the regulator and the regulated, i.e duty holders with responsibilities under the law. The regulator should desirably operate within a framework of law that allows that balancing act. Engineering is very much about achieving a balance between differing requirements. It is about searching for the best practicable option and this must include the minimisation of the attendant risks. So the mindset of the regulator of industrial risk and of engineering professionals should coincide. The fundamental belief that underpins our recent approach to explore what should be done to equip engineers of the future to operate in harmony with the health and safety requirements is to ensure that engineers understand the wider context within which they practice and appreciate that business success and management of health and safety risks go hand in hand. Hence engineers need to:

- Recognise and inform management when safety concerns result from common cause (rather than special cause) and advise on solutions to the cause that exist within the management system. Common cause and special cause derive from the concept of variability. Special cause include unusual events and activities that are not part of the everyday system. Most injuries are a result of common cause (behaviours that have occurred previously) rather than special cause;
- To have a sufficient knowledge of the obligations imposed on them and their employers by law, directives/regulations and international agreements;
- To appreciate the benefits of bringing health and safety considerations into account, when first considering a design and during development of systems and hardware;
- To have an understanding of risk assessment, in order to use it as an aid to describing and ranking decision options under uncertainty;
- To have an understanding of cost benefit analysis, in relation to health and safety as a way of ranking options;
- To know how important it is to take into account human factors when seeking sound engineering solutions; and
- Continue learning and applying new knowledge.

Recommendations

This paper has highlighted the importance of Occupational Health and Safety as vital arms of any integrated successful Business Model and rating system. It has stressed on the need for safety culture, behaviour and leadership in every organization. It placed importance on the process of Safety Analysis, Risk Assessment and Management. It projected that, understanding of the causes of human error will be the Safety's next frontier. Disaster management and the importance of emergency management plans and the practical training of all concerned by simulating such scenarios has been highlighted. The role of engineers and OHS certification audits has been brought in. The need for statutory requirements and an Umbrella Act for OHS to ensure compliance has emerged.

Create HSE Awareness

Instead of relying solely on poster campaigns and safety weeks, we need to create an environment of HSE awareness through very effective programme like STOP, which is a system developed by DuPont, Advanced Safety Audits (ASA), which is a system developed by Safety Management Consultants Ltd., U.K. and Task Risks Assessments (TRA) which is a standard industrial practice. The STOP, ASA and TRA programmes are just few of many such systems employed within the highly visible "Get HSE Right" global campaigns.

All these systems are interlinked and they build on each other to generate "an effective HSE awareness".

(a) Case Studies on best safety experiences must be documented and shared with other organizations so that the indigenous developments, and know-how could be made available to all stakeholders and concerned organizations. Such "best practices" should also define the techniques and approaches through and approaches through which success was attained.
(b) "Case Studies" of failures due to negligence or near misses must also be recorded and shared to provide a good resource information for development of preventive measures and establishing a 'Safety Net' in any organization.
(c) Set up an Institute of Safety relating to natural and man-made hazards. The Institute should develop linkage with all promoting activities leading to safety.
(d) Safety awareness is to be inbuilt in the general public by including this as a topic at primary school level. This will help in developing desired safety culture. Also there is urgent need to introduce 'Industrial Safety' as full course for engineering degree, diploma and technician level course.

OHS System—Action Plan

A. National Level

National Campaigns for

(i) Formulation of National Policy on 'Safety, Health and Environment' (SHE) by Government.
(ii) Constitution of National (Tripartite) Commission on 'SHE'.
(iii) Modification of 'SHE' legislations into one Umbrella Act.
(iv) Activisation of National 'Trade Union Forum on Safety.'
(v) Periodic reporting/review of Safety Issues at the national level.
(vi) 'SHE' protection issues affecting the community at large.
(vii) Workers safety in the unorganised sector/unorganised workers, legally or otherwise.

Training

(i) Exclusive National level 'Training the Trainers Programme', 'SHE' and independent programmes covering Safety, Health and Environment separately in detail.
(ii) Health and Safety training should be an examinable part of any engineering undergraduate course, if it is to be accredited.
(iii) High quality teaching aids should be developed to help in imparting an understanding of risk concepts.
(iv) Training of NGOs and general public on emergency handling be organized by the Government throughout the country, and specifically in the remote areas.
(v) Centrally producing posters, guidelines, checklists and training material on Safety. Industry sectoral guidelines be developed in conjunction with respective professional and industrial federations and associations.

(vi) IoE to consider publication of a Quarterly issue of Journal on 'Quality & Safety', through 'Safety & Quality Forum', to widely share knowledge and experience on Quality & Safety concerns.

(vii) Organise cultural programme for educating public on 'SHE'—specific issues and create awareness, with the assistance of concerned NGOs.

Safety Audits

(i) Formalise 'Safety Audit Standard' and encourage its wide implementation.

(ii) Government needs to encourage adoption of Occupational Health and Safety practices by the industry and service sectors and compliance of statutory requirements by encouraging certification to OHSAS 18001.

(iii) Quality Council of India should develop an integrated management system standard to cover Quality, Environment and Occupational Health and Safety System in one standard.

(v) IOD to Institute Golden Peacock National Awards on 'Safety'.

B. Regional Levels

— Disaster Management
  - The existing disaster management plans in the region should be reviewed, at least every 5 years for continual improvement.
  - Emergency Management Plans, both on-site and off-site, should cover all industrial towns and major industries.

— Safety and Quality Forum of Institution of Engineers to act as the informal coordinator and watch-dog over the functioning of factory inspector, industrial estates, fire brigades, and other central municipal bodies engaged in safety.

— Take up the 'Safety Protection' issues, affecting the community at large remedial measures.

— Take up the safety issues of the unorganised sector in the region/state, legally or otherwise.

— Organize the training/awareness programmes with the assistance of engineering bodies and NGOs, relating to safety issues.

C. Local/Unit Level

— Encourage adoption of SHE policy by the organized sector.

— Government to financially support SSIs to go in for certification for OHS 18001 standard (as is being done for certification for ISO 9001 and ISO 14001), to ensure compliance and recognition.

— Observe 'Safety Day / Week' to initiate major local Safety level programmes and rewarding better '5S', Accident prevention suggestions schemes, and Zero accident departments.

— Educate the workers :
  - On specific SHE problems.
  - On workers rights provided under Sec.111a of the Factories Act.
  - Integrate SHE issues in the union workers education programme.
  - On the role of safety committees.

— Conduct Periodic Safety and Health Surveys of hazardous industries/ areas.

— Develop standard formats for recording work-related injuries and illness in industry.

— Measures to ensure effective cooperation and communication such as:
  - *Safety briefings* — Monthly briefing document describing key accidents or incidents in the month the lessons learnt and preventive actions.
  - *Safety updates* — a regular update produced centrally providing information that focuses attention on topical areas of safety, health and environmental issues.
  - *Planning and setting targets* — targets are set annually as part of performance contract. These are developed in consultation with the business units and as a result of analysis of incident, accident and other data measuring its safety performance.

RECOMMENDATIONS

1. Safety in design and operation has equal significance. However, parameters that have a direct bearing on health and safety of the common man should be given priority.

2. Occupational safety and health should be paid immediate attention in activities relating to economic and social development. Intensive awareness programmes need to be launched, at all levels, to ensure stakeholder participation and safer living and working conditions. Involvement of school and college students should be ensured. National Council as Science and Technology Communication, Department of Science and Technology should be approached to guide and support such a programme.

3. Intensive training programmes for different level of workers should be launched to ensure improved effectiveness in operations. Whereas organisation specific training programmes are being conducted by most of the employers, there is an urgent need for establishing general safety training programmes which may be developed sector-wise and conducted by training providers. Such programmes should be launched and conducted periodically jointly by SQF and QCI.

4. It is necessary that programmes be immediately launched for interaction with professionals to create awareness of safety provisions relating to safety in the National Building Code-2005 to ensure safer construction practices and safer habitats. An understanding between SQF and BIS can ensure well coordinated and sponsored programme. SQF should conduct this programme jointly with QCI.

5. Safety in Mega Cities is a major concern. SQF should identify studies to be launched for these cities. Attention should be focused first on Delhi. A Plan of Action on "Managing Delhi Safety" be developed with the involvement of experts, officials, NGOs and RWAs.

6. SQF should identify areas needing focus on safety in the energy sector and undertake programmes leading to safer, secure, reliable and sustainable energy supply for the safety of common man.

7. Safety dimension of disaster management is as important as safety aspects in other areas. Safe approaches in preparatory phase; rescue and relief; rehabilitation and reconstruction stage should be developed and implemented in select areas.

8. The domestic food quality control system needs to be strengthened so that foods for domestic consumption are also of international standard. This would require a revision of standards as well as the conformity assessment structure. Specific focus should be placed on catering institutions including departmental canteens.

9. SQF should undertake studies as identified and sponsored by QCI, BIS and other such agencies.